Seismic Ground Response Analysis

GEOTECHNICAL, GEOLOGICAL AND EARTHQUAKE ENGINEERING

Volume 36

For further volumes:
http://www.springer.com/series/6011

Nozomu Yoshida

Seismic Ground Response Analysis

Springer

Nozomu Yoshida
Department of Civil and
Environment Engineering
Tohoku Gakuin University
Miyagi, Japan

ISSN 1573-6059 ISSN 1872-4671 (electronic)
ISBN 978-94-017-7840-4 ISBN 978-94-017-9460-2 (eBook)
DOI 10.1007/978-94-017-9460-2
Springer Dordrecht Heidelberg New York London

Printed on acid-free paper

Springer is part of Springer Science+Business Media (www.springer.com)

Preface

Earthquake resistant design is an important consideration for designing structures in seismic prone areas including Japan. Recently, following the developments in seismology and earthquake engineering, design earthquake motion is defined at the engineering seismic base layer (engineering bedrock) or the seismic bedrock regardless of the structural type and the ground type. Therefore, in order to get the design earthquake motion at the ground surface or at the foundation of a structure, engineers need to make a seismic ground response analysis.

I graduated from the Department of Architecture and Architectural Engineering, Kyoto University (Japan), and my thesis for doctor of engineering was cyclic behavior of steel braces subjected to earthquake loading. Several years after my post-doctoral life in the university, I joined Sato Kogyo, a general contractor firm, and I engaged in the design of the RCWS (reactor cooling water system) of a nuclear power plant. There, I faced a problem on how to consider the liquefaction of the crushed rock that was used to fill the excavated area during construction just neighboring the grit chamber. I visited Prof. Ishihara, University of Tokyo, to get an idea. He introduced a computer program YUSAYUSA and I engaged to improve it with him, which is now open for public from my website (Yoshida and Towhata 1991), and it was my first technical paper in this field. Soon after, Prof. Ishihara introduced me to Prof. Finn, University of British Columbia, Canada, and I got a chance to stay with him for 1 year. I engaged to develop the computer code TARA-3 (Finn et al. 1986). After coming back to Japan, I decided to change my research topic from structure to geotechnics, especially earthquake geotechnical engineering.

There were many differences between the structure and the soil as engineering materials and I was sometimes overwhelmed by these differences. Why, for example, stress-strain curve is expressed by, so-called, G-γ and h-γ relationships, and why just an approximated method is called an equivalent linear method, etc. Modeling the stress-strain behaviour was not a big issue in structural engineering, but it was a big issue in the new field. The error of the analysis was very large compared with the one in my former field. Things that are believed to be common sense looked curious to me. So, it was a new world for me and I believed that I could contribute something in this field, which is the reason why I decided to move to the earthquake geotechnical engineering.

I started as an academician in the university several years ago quitting my second practical engineering job in Oyo Corporation, a consulting company. I was surprised that there is no time to teach the seismic ground response analysis in the syllabus. There are many topics that should be taught in the university, but time is limited. Seismic ground response analysis is a difficult issue because all other issues taught are necessary in order to understand the behavior during an earthquake.

On the other hand, seismic ground response analysis is an essential tool for the practical engineers. Then I had a question how they study this subject. Just that time, I was asked a lecture titled "seismic ground response analysis for practical use" from Prof. Wakamatsu, Kanto Gakuin University, who was a board member of the Japan Association for Earthquake Engineering (JAEE). Fortunately, the lecture was successful such that the JAEE turned away many people as the room became full. Then I understood that practical knowledge is desired from the practical engineer.

I published a book in Japanese in 2010 (Yoshida 2010) on the seismic ground response analysis based on this lecture. There are already many books on this subject. I feel, however, that there is no book that gives the knowledge or techniques required by the engineering practice or that the engineer can refer in their daily job. Many books deal with only the theoretical field, but there are many issues that cannot be discussed only by theory. So I especially focused on these topics. Soon after the publication of the Japanese book, Prof. Ansal and Dr. Tönük, Bogazici University, Turkey, suggested me to write an English version of this book. I omitted many theoretical descriptions in the Japanese version, but I think it would be better to add some theoretical parts for the foreign readers, which was also a suggestion by Dr. Tönük. So this is quite a new book.

This book deals with the total stress analysis, and the effective stress analysis or the liquefaction analysis is not considered partly because the effective stress analysis is more difficult compared with the total stress analysis and partly because another volume may be required to write the introductory part of the liquefaction analysis.

The author wishes sincere thanks to Prof. Wakamatsu who gave a chance to publish the Japanese book, and Prof. Ansal and Dr. Tönük who suggested the English version. Thanks are extended to Drs. Ohya, Port and Airport Research Institute, and Dr. Miura, Oyo cooperation, for their pre-reading of the Japanese version. Thanks are also extended to Dr. Tönük and Prof. Bhattacharya, University of Surry, UK, who checked my English.

Notes for Reading This Book

This book follows SI unit system (Promotion committee of SI unit 1999). In other words, kN, m, and s are used for force, length, and time, respectively. Unit for pressure is expressed by kPa instead of kN/m^2. In the field of seismology, Gal

(gal) has been used for acceleration, but it is not used in this book although it is possible as exceptional treatment. In the same manner, kine has been frequently used for velocity, but is not used in the book, either, because it is a non-SI unit. When converting the old unit system into the SI units, acceleration of gravity g is usually taken as 9.8 m/s^2, but 10 m/s^2 can also be used. Error of 2 % appears by this interpolation, but it is an acceptable error in many fields of engineering.

All the figures and equations are given using the SI unit. In the field of the geotechnical engineering, many empirical equations have been developed, and they are usually valid only in the specified unit. They are also rewritten in the SI unit system in this book.

Recently, information and data are published not only in the technical paper but also in the website. Here, unlike the technical paper, contents of the web are sometimes revised or erased, in which case the reader cannot find the data. In this book, the valid date of the existing website is shown in brackets [ ].

Abbreviations are used for several Japanese organizations which appear frequently in this book. They are as follows

AIJ:	Architectural Institute of Japan
JAEE:	Japan Association for Earthquake Engineering
JGS:	Japanese Geotechnical Society
JSCE:	Japan Society of Civil Engineering
JSSMFE:	Japan Society of Soil Mechanics and Foundation Engineering, renamed JGS in 1995
PARI:	Port and Airport Research Institute
PWRI:	Public Work Research Institute

Sendai, Miyagi, Japan Nozomu Yoshida

References

Finn WDL, Yogendrakumar M, Yoshida N, Yoshida H (1986) TARA-3, a program for nonlinear static and dynamic effective stress analysis, Soil Dynamic Group, University of British Columbia, Vancouver

Promotion committee of SI unit (1999) JIS 8203 SI unit and its usage -from gravitation system unit to International system unit (SI unit)-, Ministry of International Trade and Industry of Japan (in Japanese); International Bureau of Weights and Measures (2006): The International System of Units (SI), 8th edn

Yoshida N (2010) Nonlinear analysis of ground, Kajima Institute Publishing, 256 pp (in Japanese)

Yoshida N, Towhata I (1991) YUSAYUSA-2 and SIMMDL-2, theory and practice, revised in 2003 (version 2.1), Tohoku Gakuin University and University of Tokyo; http://www.civil.tohoku-gakuin.ac.jp/yoshida/computercodes/eqcode.html

Contents

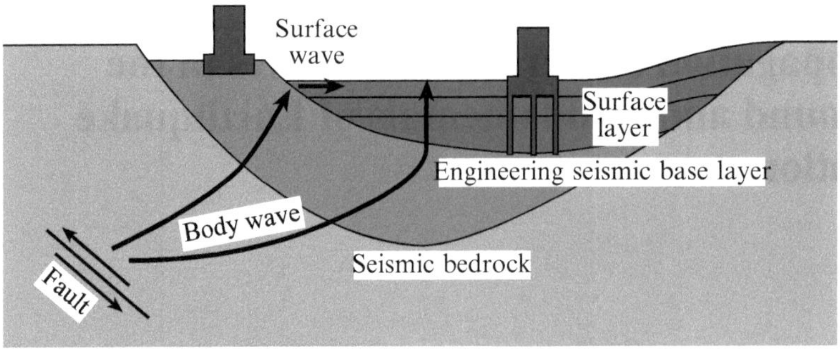

Fig. 1.1 Propagation of earthquake wave

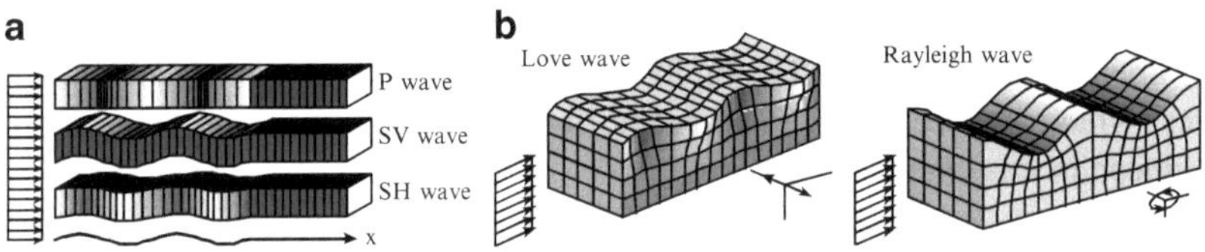

Fig. 1.2 Types of earthquake waves. (**a**) Body waves. (**b**) Surface wave

propagation, the S-wave is called as SV wave, and it is called SH wave when the soil particle vibrates out of the plane. Distinction between two waves is not necessary in the one-dimensional analyses although it is frequently called as SH wave. The SV waves are treated in the ordinary two-dimensional analysis.

As described in the next section (Sect. 1.1.2), earthquake wave propagates vertically when it is close to the ground surface. Consequently, the P-wave creates up–down or vertical vibration, and the S-wave causes horizontal vibration. Among these two body waves, the S-wave is the important wave in the earthquake resistant design. Therefore, almost all the content of this book deals with the S-wave; when a term "wave" appears, it refers to the S-wave. If the boundaries between any two layers are not perfectly perpendicular to the direction of the wave propagation, both P- and S-waves are refracted even though the incident wave is either P- or S-wave. These waves are called as PS or SP converted waves. Additionally, perfectly vertical wave propagation is just an approximation to observed wave phenomena. Therefore, vertical motion cannot be considered to be caused only by the P-wave and in exact sense horizontal motion only by the S-wave. However, it is also true that predominant horizontal motion is caused by the S-wave and vice versa. Therefore, for simplicity, the earthquake wave is assumed to propagate in the vertical direction, and the horizontal motion is caused only by the S-wave through this book.

A surface wave is generated by the interference of body waves radiated from the fault with a free surface. As schematically shown in Fig. 1.1, it is usually generated at an edge of a basin and propagates in the horizontal direction. The name "surface wave" is because it propagates along the ground surface. Similar to the difference between SH and SV waves, there are two kinds of a surface wave. The one vibrates perpendicular to the plane of wave propagation and is called as Love wave, and the other one that vibrates in the plane of wave propagation is called as Rayleigh wave. Wave amplitude of a surface wave is largest at the ground surface and attenuates quickly with depth. Surface waves have another characteristic called dispersion, i.e., wave velocity depends on frequency. Amplitude of the wave is large, but at the same time, wavelength is long. Therefore, acceleration due to surface waves is not significantly considered in the design of ordinary structures. However, in the design of the underground lineal structures, it should be considered because of its large displacement (Committee of gas facility standard 2000). In addition, surface waves travel long distances because of their long wavelengths and may affect ultra-tall structures that have long natural periods. For example, ultra-tall buildings in Tokyo vibrated and were damaged during the 1983 Nihonkai-chubu earthquake (Kinoshita and Ohtake 2000) (epicentral distance is about 450 km) and during the 1984 Nagano-ken-seibu earthquake (Editorial committee of records of Nagano-ken-seibu earthquake 1986) (epicentral distance of about 200 km). Similar phenomena were also reported in Tokyo during the 2000 Tottoriken-seibu earthquake (epicentral distance is about 580 km) and 2003 off Miyagi earthquake (epicentral distance is about 350 km). Likewise, the cause of the oil tank fire at Tomakomai, which is about 200 km far from the epicenter of the 2003 Tokachi-oki earthquake, is resonance due to the long period wave (JSCE Earthquake Committee 2003). Therefore, they require consideration in the earthquake resistant design, but this is not a subject in this book partly because it is not considered in the current design specifications and partly because the analysis is very difficult for practicing engineers.

1.1.1 Path of Wave Propagation and Analysis Region

It is a complicated problem to analyze the whole region shown in Fig. 1.1, i.e., from fault to site, in a single analysis. The problem is not easy partly because computer power is still not sufficient enough and partly because soil data in whole region is not sufficient. This whole region is divided into three subregions by setting two base layers which are called the seismic bedrock and the engineering seismic base layer. The latter is the generally used concept in Japan, but may not be common outside Japan although similar concept is used as explained in Sect. 3.1.

Seismic bedrock is widely known as the base layer. Two following concepts are introduced regarding the definition of seismic bedrock (Toki 1981):

1. Layer that behaves individually regardless of local site structure
   This definition indicates that the local site conditions have to be excluded in the definition of bedrock as the local site condition affects the earthquake motion at

the ground surface significantly. The seismic bedrock in this definition should satisfy the following two conditions:

(a) The seismic bedrock spreads in certain extent, and mechanical properties in this layer are homogeneous.
(b) Variation of mechanical properties and structure of the layers below the seismic bedrock is minor than those above it.

2. The shallowest layer that can reflect the earthquake motion characteristics at the fault in the earthquake resistant design of structures

This definition comes from the structural design point of view and indicates that the base depth is selected providing that the thickness of the superficial layers is sufficient if the natural period above the base is a little longer than the natural period of structures.

The earthquake motion observed at the interested site $R(t)$, as a function of time t, can be evaluated from the occurrence $Q(t)$ at the source (fault), behavior $P(t)$ from source to the bedrock, and amplification characteristics $G(t)$ from bedrock to the ground surface (AIJ 1987) as

$$R(t) = Q(t) \otimes P(t) \otimes G(t) \tag{1.1}$$

where $\otimes$ indicates convolution. Here, $G(t)$ is separated into two parts as shown in Fig. 1.1, i.e., a path from the seismic bedrock to the engineering seismic base layer and a path above it. It is noted that Eq. (1.1) holds when there is no interaction between each part. In other words, earthquake motion is assumed as propagating in one way from seismic bedrock to the engineering seismic base layers, and reflected wave from the surface layer does not alter the incident wave to the surface layer. This definition is compatible with the first definition of the seismic bedrock described above. It is, however, noted that this assumption does not always hold (Yoshida et al. 2005), which will be discussed in Sect. 15.8.

It is necessary to understand the general feature of the S-wave velocity structure in order to understand the earthquake wave propagation. Earthquakes occur at a fault in the earth crust above the upper mantle. The representative value of the S-wave velocity is about 3.5 km/s in the upper earth crust (Kinoshita and Ohtake 2000). The seismic bedrock is mainly composed of granite for which the S-wave velocity is about 3 km/s (Irikura 1978). The S-wave velocity of the engineering seismic base layer is defined to be between 300 and 700 m/s in Japan and will be explained in detail in Sect. 3.1.

The S-wave velocity decreases in the superficial layers. It is about 100 m/s in the soft soil sites in the urban area. Since man activities densify the subsurface layers, 100 m/s may be the minimum value in the urban area, but it can have much smaller values in the country or undeveloped areas. In general, S-wave velocity becomes smaller as the depth becomes shallower partly because the soil is consolidated and solidified more at greater depths and partly because the elastic modulus as well as the wave velocity depends on the confining stress.

Although both density and wave velocity are necessary in discussing the wave propagation characteristics, bedrock is defined only by the wave velocity since the densities of the rock and soil are similar to each other, of the order of 2 t/m^3.

1.1.2 *Path of Body Wave Propagation*

One of the important features shown in Fig. 1.1 is that the path from the fault to the site is not straight but curved. This behavior is strongly related to the aforementioned wave velocity structure.

A boundary between two layers with different S-wave velocities V_1 and V_2 is schematically shown in Fig. 1.3a. Snell's law indicates the relationship between the incident angle θ_1 and the refraction angle θ_2 as follows:

$$\frac{\sin\theta_2}{\sin\theta_1} = \frac{V_2}{V_1} \tag{1.2}$$

This law can be applied not only to the earthquake waves but also to other waves such as sea waves and lights. As a comprehensive example, let us consider the situation shown in Fig. 1.3b. You are sitting at point A in the shore and see a drowning person at B in the sea. The problem here is that which path is the fastest to arrive at. A straight line AB is not an optimal path because distance in the sea (very low speed) is longest. Although the distance in the sea is shortest in the rectangular path ACB, distance in the shore is the longest. Therefore this path is again not an optimal path. The optimal path ADB lies between these two ultimate paths, and it can be found from Eq. (1.2).

Change of the refracted angle evaluated from Eq. (1.2) is shown in Fig. 1.4 by two different expressions. For example, setting $V_1 = 3{,}000$ m/s (seismic bedrock), $V_2 = 150$ m/s (soft ground), and $\theta_1 = 45°$, refracted angle is computed as $\theta_2 = 2°$, which indicates that the wave propagates nearly in the vertical direction. This can be demonstrated again by adapting to sea waves.

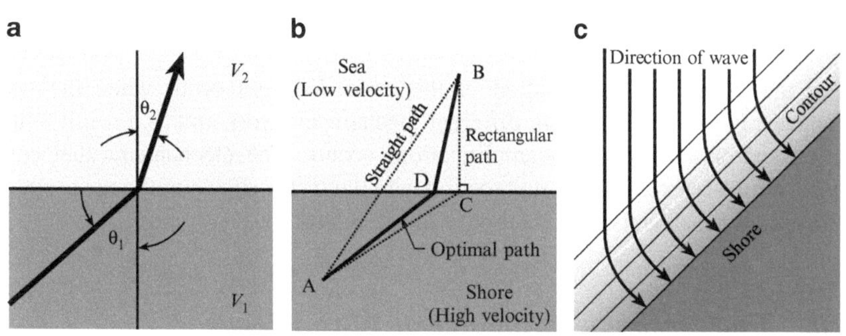

Fig. 1.3 Refraction of waves. (**a**) Refraction of wave. (**b**) Action near shore. (**c**) Sea waves at shore

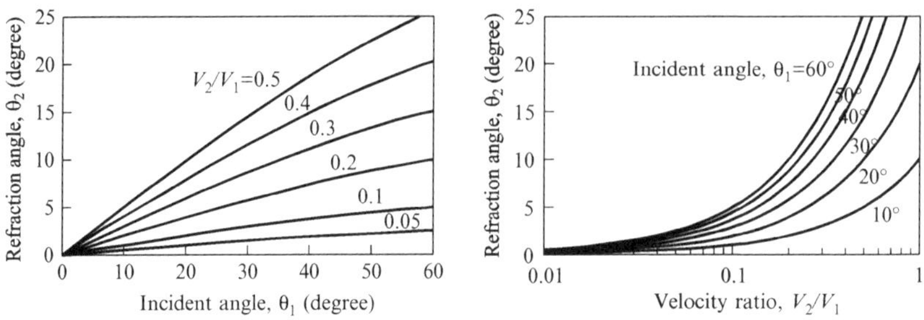

Fig. 1.4 Incident angle vs. refraction angle

When you see sea waves coming toward you on the shore, you feel that the wave front is parallel to the shore line. According to the wave theory, the wave velocity is obtained by $\sqrt{gh}$ where g is the acceleration of gravity and h is the depth of the seabed; wave velocity becomes smaller as closer to the shore. Therefore, as shown in Fig. 1.3c, the path curves are perpendicular to the shore when the waves come close to the shore since equi-depth contours are parallel to the shore. As a conclusion, phenomena that earthquake waves propagate in vertical direction as they approach the surface and that sea waves travel perpendicular to the shore are the same mechanism.

1.2 Amplification of Earthquake Wave

There are significantly different wave propagation characteristics between the path from the fault to the seismic bedrock and the path from the engineering seismic base layer to the ground surface. The wave radiates in all directions in the former case, whereas it propagates in one direction in the latter case.

When the wave radiates in all directions, the amplitude becomes smaller as the wave front expands, while the distance from the fault increases. This phenomenon is known as attenuation of the earthquake waves, and an example is shown in Fig. 1.5. Here PGA and PGV in the ordinate denote the peak ground acceleration and velocity, respectively.

On the other hand, attenuation of this kind does not occur when the wave propagates in one direction, but different mechanisms work and as a result amplification and/or attenuation (deamplification) occurs. The mechanisms that cause amplification are explained in this section, and the rest is shown in the next section. There are four mechanisms that amplify the earthquake wave.

1.2.1 First Mechanism: Change of Wave Velocity

Let us consider the sea wave case again to understand this mechanism. When the sea wave propagates toward the shore, wave velocity becomes smaller as explained

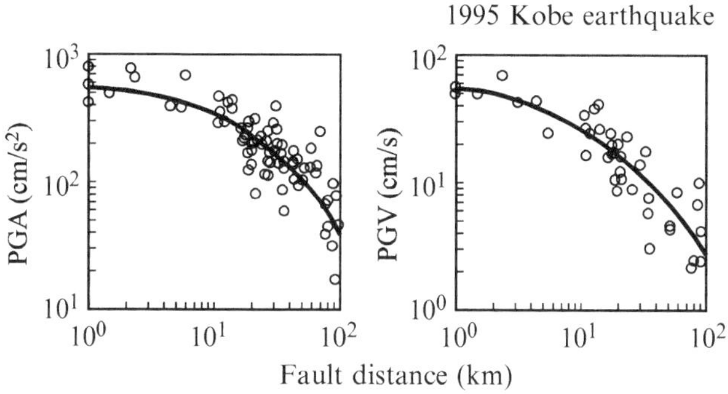

Fig. 1.5 Example of attenuation of wave by distance (Si and Midorikawa 1999)

in the previous section. As a result, the wave front reaches to the forward waves, which results in larger energy density than before. This accumulated energy must be dissipated somehow. It is dissipated by gaining potential energy as a result of increase in the height of the wave. Thus, the amplitude of the wave increases as the wave comes close to the shore.

The same mechanism occurs for the earthquake waves; energy density is larger near the ground surface since the wave velocity is smaller. Unlike the sea waves, however, the energy cannot be dissipated by the potential energy under the S-wave propagation because only horizontal movement occurs. Instead of it, it is dissipated by the strain energy, by which again amplitude becomes larger.

As a summary, decrease of the wave velocity causes amplification of the amplitude both for the sea waves and for the earthquake waves. There are, however, some differences between them. Amplitude of sea wave cannot become infinitely large; wave collapses at certain height. Another difference is that the sea wave disappears by running up to shore. Some of the waves may reflect, but they radiate toward the ocean and never come back again. On the other hand, all the earthquake waves are reflected at the ground surface, and this causes the basis for the next amplification mechanism.

1.2.2 Second Mechanism: Reflection at the Ground Surface

The earthquake waves propagated from engineering base layer are reflected when they reach the ground surface. Figure 1.6a schematically shows one period of a wave whose front just passes the ground surface. If there were no boundary, the incident wave (chained line) would propagate without any change as shown by dotted line. Since the ground surface works as a free end, the wave is reflected as mirror symmetry against the boundary as shown by the dashed line. Therefore composite wave (solid line), sum of the incident and reflected waves, becomes larger than the incident wave. The earthquake wave is amplified twice when it is reflected at the ground surface. Phases of the incident and reflected waves are identical

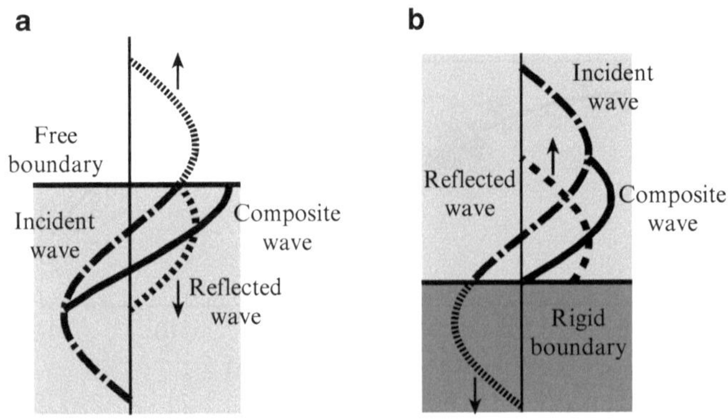

Fig. 1.6 Reflection of wave at ground surface and at base. (**a**) Reflection at free boundary. (**b**) Reflection at fixed boundary

at the ground surface, but phase lag occurs in the underlying layers. Therefore, amplification becomes smaller as it gets deeper.

Let us consider another example: if there is a structure on/in the ground, reflection that results in double amplitude does not occur. Therefore, the design that uses the earthquake wave at the free surface as the input motion for a structural analysis is not rational, because the earthquake wave at the ground surface is larger than the wave that hits to the structure. This is one of the reasons why soil–structure interaction is required to be considered in the structural design.

1.2.3 *Third Mechanism: Reflections from Underlying Layers*

The wave reflected at the ground surface will be reflected again at the interfaces between underground layers. Unlike the total reflection at the ground surface, the wave is partially reflected; some is transmitted and the rest is reflected at a layer interphase. The ratios of the reflected and transmitted waves are controlled by the impedance α ($=\rho V$, where ρ is density and V is wave velocity) of the two interfacing layers.

When a wave propagates from the layer with impedance α_i to the layer with impedance α_o, reflectivity R and transmissibility T are calculated as

$$R = \left(\frac{\alpha_i - \alpha_o}{\alpha_i + \alpha_o}\right)^2, T = \frac{4\alpha_i\alpha_o}{(\alpha_i + \alpha_o)^2} \tag{1.3}$$

In the homogeneous media ($\alpha_i = \alpha_o$), $R = 0$ and $T = 1$, which indicates that there is no reflection and the entire wave is transmitted. As α_o becomes larger, more waves are reflected and all the waves are reflected when $\alpha_o = \infty$ as $R = 1$ and $T = 0$. This type of boundary is called a fixed end, and it is termed as a rigid base when considering the earthquake wave propagation.

The behavior of the wave at the rigid base is schematically shown in Fig. 1.6b. The wave is reflected with a phase difference of 180° and with point symmetry. Then, the displacement at the boundary becomes zero, which is one of the reasons why we use relative displacement with respect to the base in the theory of vibration.

The wave reflected from underlying layers return to the ground surface. Energy of the earthquake motion is accumulated and trapped by the multiple reflections between surface and underlying layers. Therefore, both amplitude and duration of earthquake motion increase.

1.2.4 Fourth Mechanism: Resonance

In order to understand this mechanism, knowledge on the vibration explained in Chap. 9 is required. Only conclusion is explained in this section.

Amplification characteristics of a single-degree-of-freedom system subjected to the sinusoidal excitation is shown in Fig. 1.7a, where ω_0 is a constant only determined from the mass and the spring constant of this system and is called natural circular frequency. The largest amplification occurs when the circular frequency ω of the sinusoidal input coincides with ω_0, which is called as resonance.

The same phenomenon occurs in the ground as well. The difference is that soil is a continuum or an infinite-degree-of-freedom system. If certain conditions are satisfied, a stationary wave is produced during the multiple reflections between the ground surface and underlying layers, and the earthquake wave amplifies. In the case of a rigid base, displacement of the stationary wave is zero at the base and becomes largest at the surface. A quarter of the sinusoidal wave satisfies this condition for the homogeneous ground. Then, the period T and the circular frequency ω_0 of this stationary wave are

$$T = \frac{4H}{V_s}, \omega_0 = \frac{2\pi}{T} = \frac{\pi V_s}{2H} \tag{1.4}$$

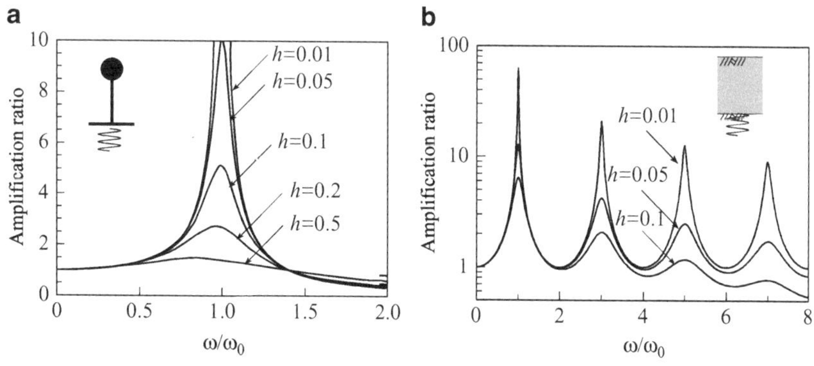

Fig. 1.7 Amplification of earthquake motion. (**a**) Single-degree-of-freedom system. (**b**) Continuum or infinite-degree-of-freedom system

where H denotes thickness of the surface layer, V_s denotes S-wave velocity, and T and ω_0 are called natural period and natural circular frequency. The earthquake wave is amplified around this period.

A quarter of the wavelength equals to the thickness of the surface layer in this case. The same phenomena occurs when H equals to 3, 5, 7, ... quarters of the wavelength. Natural periods of the higher modes become smaller, and the largest amplification occurs at the first mode or when H is a quarter of the wavelength, which is seen in Fig. 1.7b. If the ground is not homogeneous, waveform of the stationary wave is not sinusoidal, and natural periods are not at constant multiples of wavelength, which is seen later in Fig. 14.6 as an example.

An eigenvalue problem must be solved in order to obtain the natural period. In engineering practice, however, the predominant natural period (the first natural period), which is the most important period for almost all structures, is frequently evaluated by the approximate equations. The following two equations are frequently used:

$$T_v = \sum_{i=1}^{N} \frac{4H_i}{V_{si}}, \quad T_w = 4H / \sum_{i=1}^{N} \frac{V_{si} H_i}{H} \tag{1.5a, b}$$

where N and H denote number of layers above the base and the thickness of the soil profile, respectively, and V_{si} and H_i are the S-wave velocity and the thickness of each layer, respectively. Equation (1.5a) can be derived by equalizing the wave arrival time to the surface with the homogeneous profile arrival time, whereas the second equation uses average S-wave velocity weighted by the thickness. These equations show significant errors in some cases. A more accurate approximation can be, for example, derived by considering the wave reflection.

If only single reflection at the ground surface and at the base is considered, the following equation is obtained (Sawada and Kishimoto 2001):

$$T_r = \frac{3\sum_{i=1}^{N} S_i t_i^3 + \sqrt{9\left(\sum_{i=1}^{N} S_i t_i^3\right)^2 - 8\left(\sum_{i=1}^{N} S_i t_i^2\right)\left(\sum_{i=1}^{N} S_i t_i^4\right)}}{4\sum_{i=1}^{N} S_i t_i^2} \tag{1.6}$$

where

$$t_i = \sum_{k=1}^{i} \frac{4H_k}{V_{sk}} = \sum_{k=1}^{i} \left(4H_k \sqrt{\frac{\rho_k}{G_k}}\right), \quad S_i = \frac{\sqrt{\rho_i G_i} - \sqrt{\rho_{i+1} G_{i+1}}}{\sqrt{\rho_i G_i} + \sqrt{\rho_{i+1} G_{i+1}}} \tag{1.7}$$

and ρ and G are density and shear modulus in each layer, respectively, and subscripts indicate layer numbers counted from the ground surface. If the value in the square

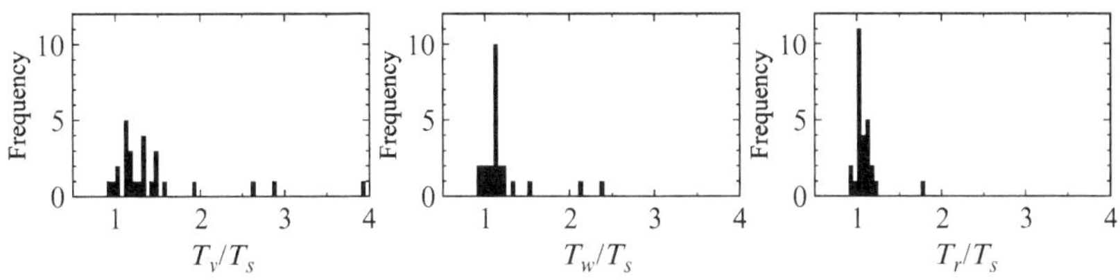

Fig. 1.8 Comparison of natural periods

root of Eq. (1.6) becomes negative, number of layer, N, is reduced one by one so that it becomes positive. It occurs when a layer boundary with significantly different impedance exists, and the reduction of N works to neglect the deepest layer. Figure 1.8 compares three natural periods T_V and T_W, in Eq. (1.5a, b), and T_r in Eq. (1.6), as a ratio over the predominant period T_s obtained from the theoretical transfer functions. Generally, natural periods are overestimated, and methods by Eq. (1.5a, b) sometimes overestimate natural period several times larger than the actual periods.

1.2.5 Example of Earthquake Motion Amplification

Earthquake damage is known to be larger in the soft soil sites (or soft ground). This indicates that earthquake motion is larger in the soft ground, and it is a good example of the amplification of the earthquake wave. Clear evidences can be seen in many past earthquakes, among which some of them are introduced in this section.

Figure 1.9 shows contours of the peak acceleration observed during the 2000 Tottoriken-seibu earthquake, Japan. Since seismographs were installed not only on the ground surface but also in the deep depth in the KiK-net system (Digital Strong-Motion Seismograph Network KiK-net 2010), contours of the maximum acceleration can be drawn both at the ground surface and in the deep depth. Areas with high earthquake motions expand on the ground surface compared with that in the underground, which indicate that the earthquake motion was amplified on the surface. Another important feature of the ground shaking is also seen in Fig. 1.9. Areas subjected to higher acceleration levels extend south of the epicenter because rupture of the fault is toward south, which is a good example of directivity.

More clear evidence of amplification due to soft site conditions observed during this earthquake is shown in Fig. 1.10 (Nozu 2003) on four acceleration records. The four stations are located at almost identical longitude, and distance between the stations is 20 km at maximum. However, acceleration time histories are quite different from each other, especially for Sakaiminato and Sakaiminato observatory records, although the distance between two stations is only 1.2 km. There is about

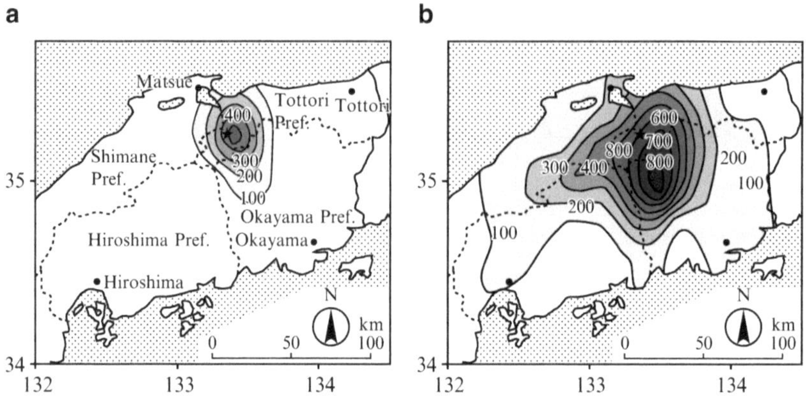

Fig. 1.9 Amplification of ground motion during the 2000 Tottoriken-seibu earthquake, Japan, where numbers indicate acceleration in cm/s^2. (**a**) Deep depth. (**b**) Surface (Modified from JGS reconnaissance team 2000)

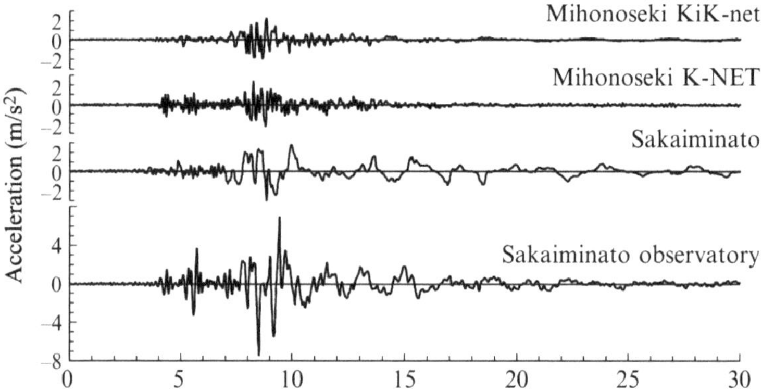

Fig. 1.10 Earthquake records near Sakaiminato during the 2000 Tottoriken-seibu earthquake

900 m thick surface deposit at Sakaiminato and Sakaiminato observatory stations, whereas the other two stations are located in the mountain area.

Peak ground acceleration (PGA) in San Francisco during the 1989 Loma Prieta earthquake, shown in Fig. 1.11a, is another example. The PGA recorded on the Holocene is about double of the PGA recorded on rock outcrop (Tokyo Prefecture 1990). Figure 1.11b shows aftershock observation by US Geology Survey (Kameda 1990). Amplification of the earthquake motion is small at the marble rock site where impedance is the largest. On the other hand, amplifications are very large in the Holocene soft deposit (alluvium). It is also noted that the duration of the ground

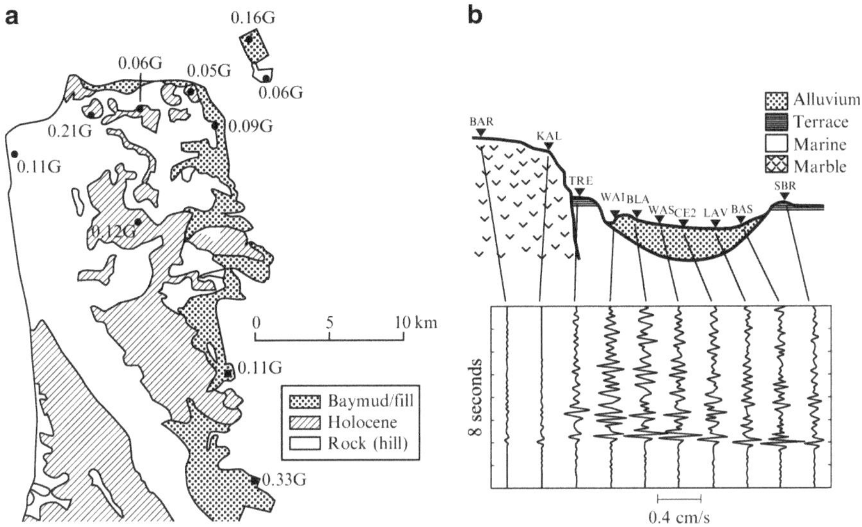

Fig. 1.11 Earthquake observation in the 1989 Loma Prieta earthquake. (**a**) PGA at main shock (Tokyo Prefecture 1990). (**b**) Aftershock (Modified from Kameda 1990)

shaking in the soft ground is longer than that in the marble rock site because of the multiple reflections at the base. Earthquake motion at the terrace has medium values.

1.2.6 Amplification of P-Wave

The P-wave has been believed not to amplify. The P-wave velocity in the saturated soil is strongly controlled by the porewater because the bulk modulus of the water is much higher than that of the soil skeleton. The P-wave velocity of the water is about 1,500 m/s, but observed P-wave velocities of soils are 1,300–1,400 m/s in many cases as the soil is not fully saturated. Since the wave velocities do not vary with depth, the first and the third mechanisms of the amplification previously mentioned do not work for P-wave propagation. This is the reason why P-wave does not generally result in amplification.

However, P-wave can amplify when the amplification mechanisms become active. A typical example is seen in the vertical array record at Port Island during the 1995 Hyogoken-nambu (Kobe) earthquake. The Port Island is a man-made island. The vertical array observation system had four seismometers, and the location of the observation station is shown in Fig. 1.12a.

Soil profile at the site is shown in Fig. 1.12b. The profile up to about 20 m from the ground surface is filled by decomposed granite, called “Masado” in Japan, and

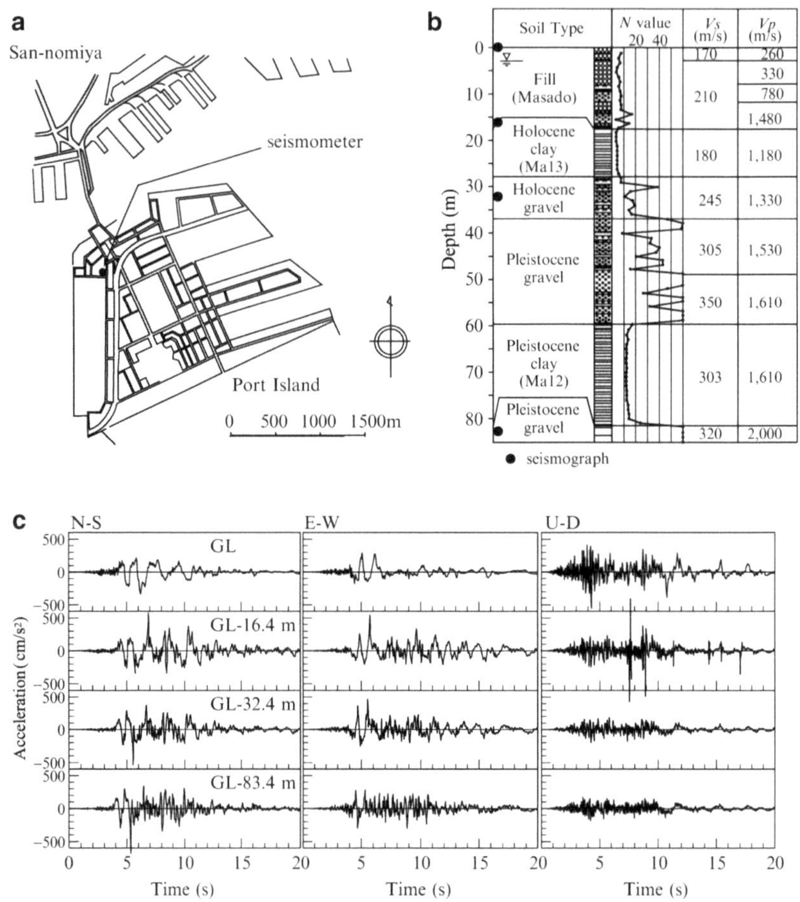

Fig. 1.12 Earthquake record at Port Island during the 1995 Kobe earthquake. (**a**) Location of seismometer. (**b**) Soil profiles. (**c**) Acceleration time histories record on the vertical array (Modified from Kobe City Developing Department 1995 and Yoshida 1995)

soft Holocene clay layer, called Ma13, lies beneath it. Beneath, there are Holocene and Pleistocene gravels. Seismographs are set at GL, GL-16.4 m, GL-32.4 m, and GL-83.4 m. The vertical motion amplification near the ground surface can be seen in the right side column in Fig. 1.12c. The reason of this amplification becomes clear by looking at the P-wave structure in Fig. 1.12b. P-wave velocity is slightly higher than 1,300 m/s and nearly constant in deep layers. Therefore amplification does not occur as seen in the record at GL-32.4 m and 83.4 m. On the other hand, the P-wave velocity in the Holocene clay is 1,180 m/s, and it drops smaller value of 780–330 m/s in the fill. This clearly indicates that fill layers are still unsaturated although about 20 years have passed since the reclamation, and it is natural for the P-wave to amplify because the previously mentioned amplification mechanism holds for this abrupt change of P-wave velocity.

1.3 Attenuation of Earthquake Wave and Upper Bound Earthquake Motion

Mechanisms causing amplification are explained in the previous section, whereas opposite behavior is explained in this section.

Figure 1.13 shows relationships between the accelerations on the soft soil sites and the accelerations on the associated rock sites, originally drawn by Idriss (1990). As explained in details at Sect. 3.1, the rock site in the abscissa has almost the same meaning with outcropping engineering seismic base layer. At the time when this paper was written, earthquake records from two earthquakes, 1989 Loma Prieta, USA, and 1985 Michoacán, Mexico, were available. Accelerations were amplified on the ground surface with respect to bedrock in both earthquakes. Since strong ground motion records were limited, data were supplemented by the numerical analysis, and it was found that acceleration at the soft ground becomes smaller than that at the rock site beyond intense shaking levels. The authors added data recorded during the 1995 Kobe earthquake (Suetomi and Yoshida 1998), which are shown by ▽ and ▼ marks in the figure. These new data also show the same feature of attenuation on soft ground during intense shaking.

Figure 1.14a shows maximum responses at Port Island evaluated by the numerical analysis (Kobe City Developing Department 1995). Although liquefaction was a big issue at this site as liquefaction was observed almost whole islands, behavior of the soft clay layer beneath fill layer is discussed here. The maximum acceleration attenuates from deeper layers to the ground surface, which is an opposite feature of the amplification explained in the previous section. The maximum acceleration decreases significantly at GL-28 m, boundary between the Holocene clay layer and the gravel layer beneath it. The stress–strain curve at the bottom of the Holocene clay layer is shown in Fig. 1.14b. The maximum strain reaches about 2.5 %, and the stress–strain curve has a plateau region at this strain level. This indicates that the stress reaches close to the shear strength.

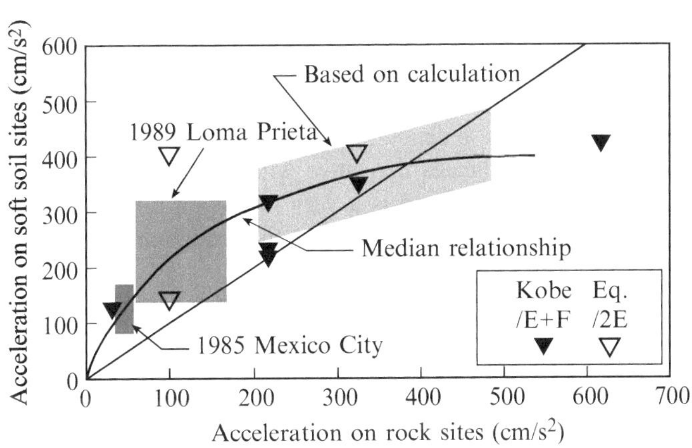

Fig. 1.13 Change of amplification characteristics associated from nonlinear behavior

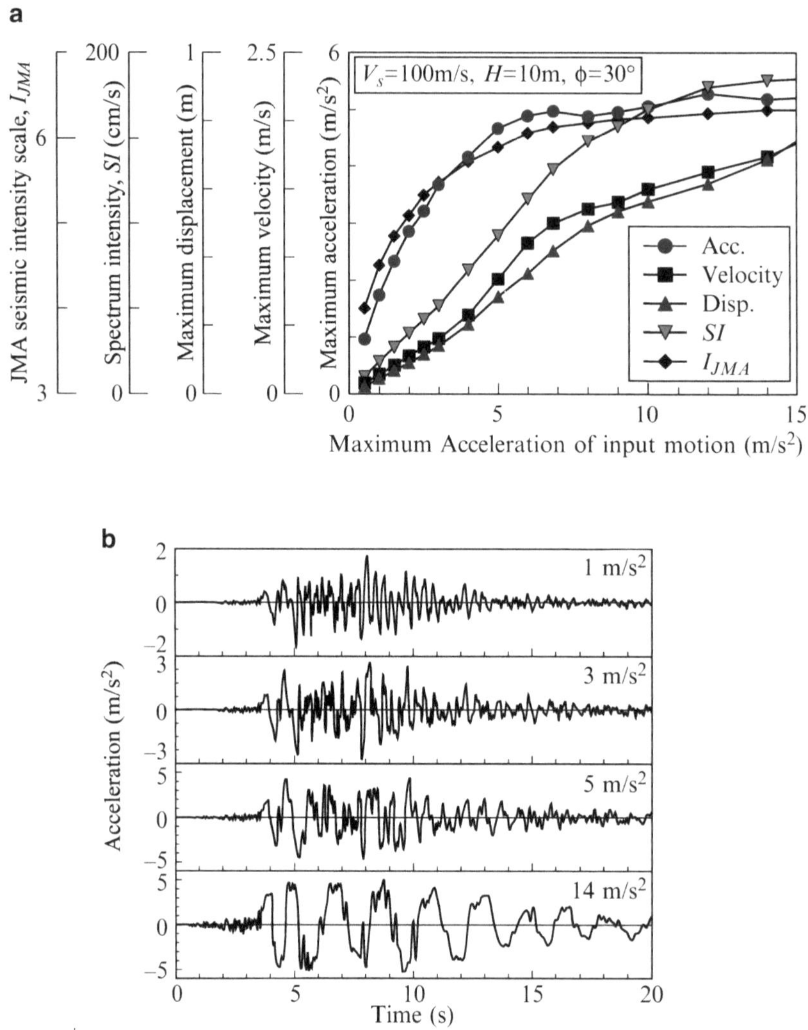

Fig. 1.16 Change of earthquake motion and upper bound. (**a**) Ground motion indices vs. maximum acceleration. (**b**) Acceleration time histories under different input motion

input earthquake, they finally reach upper bound, but this does not have sense in the engineering practice.

It is thought that both *SI* value and maximum velocity have similar characteristics and the JMA instrumental seismic intensity is an average quantity between acceleration and velocity. If it is true, the *SI* value and the JMA seismic intensity should not have upper bound; however the result of the analysis shows different features. Exactly speaking, there are significant differences in the definition of the *SI* value and the JMA seismic intensity from that of the velocity regarding the frequency range. The *SI* value considers period range from 0.1 to 2.5 s or frequency from 0.4 to 10 Hz. JMA seismic intensity is calculated for periods from 0.1 to 2 s or frequency from 0.5 to 10 Hz. Therefore, waves with longer period caused by the nonlinear behavior are not taken into account if their period exceeds the longest period in the definition. The JMA seismic intensity reaches upper bound earlier than the *SI* value since it considers less long period than the *SI* value.

The mechanisms of the upper bound are different for these indices than the maximum acceleration. Longer period component does not have an effect on the damage to the structures; however, they all indicate that the damage also has upper bounds depending on the nonlinear behavior of soil.

A typical example showing the existence of the upper bound was observed during the 1995 Kobe earthquake (Suetomi and Yoshida 1998). A cross section passing the Sannomiya railway station is shown in Fig. 1.17 with soil profiles obtained by the borehole tests. Many seismic response analyses were carried out based on these soil profiles. Additionally, the earthquake motion indices such as the peak acceleration, velocity, and the JMA seismic intensity were evaluated from the acceleration time history recorded on the ground surface. Here, the JMA seismic intensity is calculated by using only one-directional component although it is to be calculated from three components in the original definition. Therefore, the value is smaller than the actual seismic intensity, but it does not affect the discussion here. Earthquake motion indices are shown at the bottom of the profile in Fig. 1.17. Here, damage belt zone is the region where damage to buildings and wooden houses is the

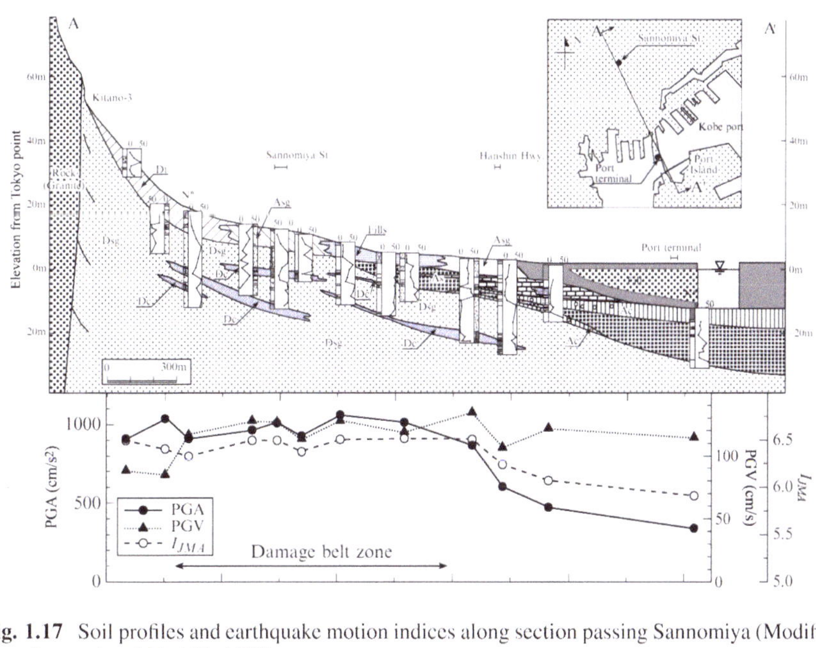

Fig. 1.17 Soil profiles and earthquake motion indices along section passing Sannomiya (Modified from Suetomi and Yoshida 1998)

most significant; collapsed houses ratio reaches several tens percent and sometimes 90 % or more. In the south of the damage belt zone, the seismic intensity as well as the acceleration decreases rapidly, the same as significant decrease of the collapsed houses ratio. However, it is also noted that decrease of the velocity is not observed in this case, too.

From the analysis point of view, decrease of the acceleration is due to the existence of the Holocene clay layer shown as ▩ or Ac in the figure, which is the same layer (Ma13) as in Fig. 1.14. In other words, the south boundary of the damage belt zone is controlled by the Holocene soft clay layer. Actually, it is also reported that this Holocene clay layer is not found under the damage belt zone (Editorial Committee for the Report on the Hanshin-Awaji Earthquake Disaster).

References

AIJ (ed) (1987) Seismic loading – state of the art and future developments. AIJ, Tokyo, 438pp (in Japanese)

Committee of gas facility standard (2000) Design specification of high pressure gas pipeline, Japan Gas Association, Tokyo (in Japanese)

Digital Strong-Motion Seismograph Network KiK-net (2010) National Research Institute for Earth Science and Disaster Prevention, Tsukuba, http://www.kik.bosai.go.jp/kik/index_en.shtml [2010]

Editorial Committee for the Report on the Hanshin-Awaji Earthquake Disaster, Report on the Hanshin-Awaji earthquake disaster, General issue volume 2, Earthquake and strong motions, geological setting and geotechnical condition, Maruzen, Tokyo, 577 pp (in Japanese)

Editorial committee of records of Nagano-ken-seibu earthquake (1986) Happened at Ohtaki, Report of the Nagano-ken-seibu earthquake, Ohtaki village, Nagano (in Japanese)

Housner GW (1965) Intensity of earthquake ground shaking near the causative fault. In: Proceedings of the 3rd WCEE, Auckland and Wellington, vol I, pp III-94–III-115

Idriss IM (1990) Response of soft soil sites during earthquakes. In: Proceedings, H. Bolton Seed memorial symposium, vol 2, Berkeley, California, pp 273–289

Irikura K (1978) Seismic bedrock and earthquake motion. In: Proceedings of the 6th symposium of earthquake ground motion, AIJ, Tokyo, pp 1–8 (in Japanese)

JGS reconnaissance team (2000) Reconnaissance Report on damage during the 2000 Tottoriken-seibu earthquake, JGS, Tokyo (in Japanese)

JSCE Earthquake Committee (2003) Summary of reconnaissance report session on damage during the 2003 Tokachi-oki earthquake. JSCE, Tokyo (in Japanese)

Kameda H (ed) (1990) Loma Prieta earthquake of October 17, 1989 Reconnaissance report, Natural Disaster Research Report, Supported by the Japanese Ministry of Education, Science and Culture (Grant No. 01102044), Japanese Group for the Study of Natural Disaster Science, 347 pp (in Japanese)

Kinoshita S, Ohtake M (eds) (2000) Fundamentals of strong motion, National Research Institute for Earthquake Science and Disaster Prevention. http://www.k-net.bosai.go.jp/k-net/gk/publication/ (in Japanese)

Kobe City Developing Department (1995) Investigation of ground deformation of fill by Hyogoken-nambu earthquake (in Japanese)

Nozu A (2003) What was made clear by strong earthquake motion observation. Found Eng Equip Mon 31(5):42–46 (in Japanese)

Sawada S, Kishimoto T (2001) Approximation of natural period of ground based on reflection-transmission factor method. In: Proceedings of the 36th Japan national conference on geotechnical engineering, pp 2383–2384 (in Japanese)
Si H, Midorikawa S (1999) New attenuation relationships for peak ground acceleration and velocity considering effects of fault type and site condition. J Struct Eng 523:63–70 (in Japanese)
Suetomi I, Yoshida N (1998) Nonlinear behavior of surface deposit during the 1995 Hyogoken-nambu earthquake, soils and foundations, special issue on geotechnical aspects of the January 17 1995 Hyogoken-Nambu earthquake, No. 2, pp 11–22
Toki K (1981) Seismic response analysis of structures, New series of civil engineering, vol 11, Gihodo Shuppan, 250 pp (in Japanese)
Tokyo Prefecture (1990) Reconnaissance report on Loma Prieta earthquake by reconnaissance team of Tokyo Prefecture, 255 pp (in Japanese)
Yoshida N (1995) Earthquake response analysis at Port Island during the 1995 Hyogoken-nanbu earthquake. Tsuchi-to-Kiso 43(10):49–54 (in Japanese)
Yoshida N (1999) Large earthquake motion and ground -nonlinear problem-, Jishin Journal, ADEP (28):66–74 (in Japanese)
Yoshida N, Shinohara H, Sawada S, Nakamura S (2005) Role of engineering seismic base layer on defining design earthquake motion. JSCE J Earthq Eng Symp 28, Paper No. 170 (in Japanese)

Chapter 2
Introduction of Seismic Ground Response Analysis

2.1 Brief History of Seismic Ground Response Analysis

Computer programs for seismic response analysis have been developed mainly in the field of structural analysis. The analysis of ground might be possible by these computer programs. However, they are not used at present. Significant difference exists in the constitutive model or stress–strain relationships. As soil shows nonlinear behavior at very small strains, a simple model such as a bilinear model is not applicable to soil. We need a computer program that is designed for the seismic ground response analysis, especially to consider stress–strain relationships and to take soil particle–water mixture into account.

Method to obtain cyclic shear deformation characteristics was developed in the early 1970s by Seed and Idriss (1970) and Hardin and Drnevich (1972a, b). The hysteretic stress–strain curve is expressed by (secant) shear modulus and damping ratio as a function of shear strain amplitude. This idea is widely used even at present.

At the same time, a computer program "SHAKE" was developed (Schnabel et al. 1972). It was the first computer program that aimed for analyzing the behavior of ground during earthquakes. It solved the equation of motion in a frequency domain by employing a Fourier series expansion and the concept of complex modulus. The latter is necessary to consider nonlinear behavior in a linear system. This method is named an equivalent linear method. SHAKE has got support from the engineers because of its easy handling method, and it has been used long time even today with minor modification. Not a few computer programs have been developed based on the same concept, and now the name "SHAKE" is treated as if it is a common noun of this concept. This book also uses SHAKE as if it was a noun.

Unlike the name "equivalent," SHAKE is an approximate method and has a few limitations, and attempts to overcome them were made. Improvements were mainly made by considering frequency-dependent characteristics in stiffness and/or damping. For example, the computer program FDEL (Sugito et al. 1994) improved

N. Yoshida, *Seismic Ground Response Analysis*, Geotechnical, Geological and Earthquake Engineering 36, DOI 10.1007/978-94-017-9460-2_2

one limitation, and DYNEQ (Yoshida et al. 2002) improved two limitations with theoretical background. Frequency-dependent nature is also considered in Kausel and Assimaki (2002), Assimakia and Kauselb (2002), Satoh et al. (1997) and Nakamura and Yoshida (2002), but some of them do not relate limitations of SHAKE.

Computer programs which can consider nonlinear stress–stress relationships more accurately were developed a little later. These programs evaluate instantaneous or tangential stiffness and solve equation of motion in time domain by a step-by-step time integral scheme. This method is frequently called "truly nonlinear method" in order to distinguish from the equivalent linear method which also deals with nonlinear problem. CHARSOIL (Streeter et al. 1974) and DESRA (Finn et al. 1976) are programs developed in the early time. CHARSOIL uses a Ramberg–Osgood model, and DESRA uses a hyperbolic model for stress–strain relationships. CHARSOIL was hardly used in engineering practice as far as the author knows, but DESRA is one of the frequently used computer programs and improved several times by considering, for example, elastic base (Lee and Finn 1978) and by considering water film (Yoshida and Finn 2000).

Effective stress analysis is also possible in DESRA. Here, effective stress analysis indicates the analysis to consider dilatancy of soil. Stiffness and strength are evaluated as a function of effective stresses following the principle of effective stress. Flow of pore water is considered based on Biot's formulation (Biot 1963). On the contrary, analysis that does not consider dilatancy is called a total stress analysis. Procedures to develop total and effective stress analyses are similar; the only difference is stress–strain model. Therefore, almost all truly nonlinear analysis can make both total and effective stress analyses.

YUSAYUSA (Ishihara and Towhata 1982) is also a computer program based on truly nonlinear method developed in the early time, and it is one of the most frequently used computer programs. YUSAYUSA uses a hyperbolic model for stress–strain relationships and uses a stress-path model for excess pore water pressure generation. It was improved by the author for considering the Ramberg–Osgood model (Yoshida and Towhata 1991).

Many truly nonlinear programs use mathematical equations for the backbone curve and Masing's rule to obtain hysteresis curves. This indicates that perfect agreement is impossible for the backbone curve. In addition, damping characteristics are also not satisfied because actual soil does not follow Masing's rule. This limitation was overcome partly by defining a backbone curve and a hysteresis loop separately (Ishihara et al. 1985) and partly by defining the backbone curve with piecewise interpolation (Yoshida et al. 1990). This model was installed in DYNES 3D (Yoshida 1995) as well as many other stress–strain models. In the old days, one computer program uses only one stress–strain model; DESRA and YUSAYUSA use the hyperbolic model, and CHARSOIL uses the Ramberg–Osgood model; therefore, the name of the computer program and the stress–strain model are one-to-one correspondence. In this meaning, DYNES 3D was a unique program.

Computer programs shown in the preceding were designed for the horizontally layered deposit or one-dimensional problem. Computer programs for two-dimensional analyses were also developed. Examples of these computer programs developed in the early time are FLUSH (Lysmer et al. 1975) and TARA-3 (Finn et al. 1984). FLUSH is based on equivalent linear method the same with SHAKE although finite element method is used in space, whereas TARA-3 is a truly nonlinear analysis program.

At present, many computer programs have been developed, and some of them can deal with not only soil but also structures. Computer programs, in which many stress–strain models are installed, are not uncommon. In addition, multipurpose programs to consider not only dynamic problem but also static and consolidation problems, such as STADAS (Yoshida 1993) developed by the author, have been developed and have been used.

Computer programs that the author developed, DYNEQ, DYNES3D, YUSAYUSA, and STADAS, are open to the public from http://www.civil.tohoku-gakuin.ac.jp/yoshida/computercodes/index.html.

2.2 Procedure of Seismic Ground Response Analysis

Flow of the seismic ground response analysis is schematically shown in Fig. 2.1. Before explaining the detailed procedure in this figure, it may be better to understand the procedure or flow of the analysis roughly. The procedure in the simplest form consists of the following steps: (1) to collect data, (2) to model them for computer programs, (3) to execute computer program, and (4) to interpret the results.

Several input data are required in the seismic ground response analysis. They are classified into four categories:

1. Geological or topological configuration, such as soil profiles and cross-sectional shape
2. Mechanical properties
3. Input earthquake motion
4. Parameters to control the flow of the computer program or the method of the analysis

Among these input data, topological configuration will be explained in Chap. 10, and input earthquake motion will be discussed in Chap. 3. Controlling parameters are deeply related to the method of analysis and will be described in different places in this book, while the most important features are given in Chaps. 11 and 12. Accuracy of the seismic ground response analysis strongly depends on the mechanical properties. Therefore, many pages in this book deal with this issue.

The last step in Fig. 2.1, i.e., engineering judgment, is not well recognized by the practical engineers, but is important, which will be mentioned through case studies in Chap. 15.

In Fig. 2.1, numbers in parenthesis are chapter/section numbers where related descriptions are given. However, one important issue, to choose the computer program for the analysis, is not given in the flow in Fig. 2.1. It is very difficult

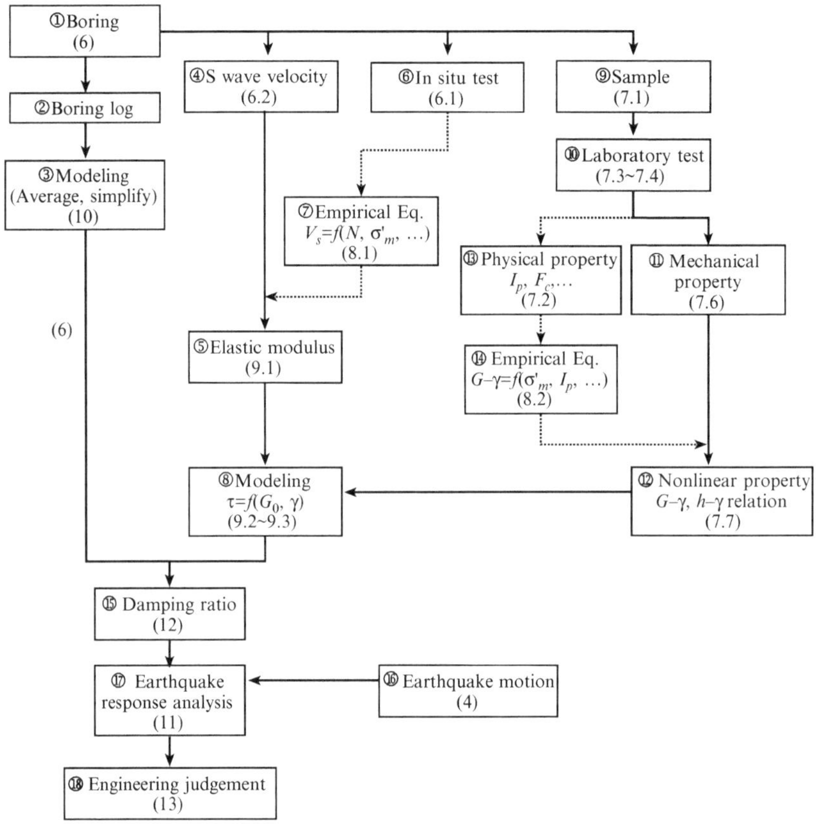

Fig. 2.1 Flow of the seismic ground response analysis

to write it in a figure. If an engineer has only one program, then it comes at the top of the flow. In practice, however, many engineers have more than one program, in which case the engineer may/can change the program when, for example, the result seems far from satisfactory or the engineer found that the program is out of applicable range in the interested problem. This is the reason why it is not given in Fig. 2.1.

In the following, flow is explained by referring circled sequence numbers.

After getting the borehole investigation data (①), modeling of the topographic configuration is required by dividing the profile into layers having the same mechanical properties (③) using the boring logs (②). Lateral boundaries and boundary conditions are to be set in the multidimensional analysis as part of this process. Layering the profile generally indicates to classify the soil into sand, silt, or clay, but sometimes it may need to consider geological age and/or depositional conditions. Subdivision of each layer is also required depending on the method of analysis. Additionally, subdivided layers with the same classification may not

Table 2.1 In situ and laboratory tests for obtaining geotechnical property

Geotechnical parameter \ Test		Existing documents	Field investigation	Borehole investigation	Laboratory test: Physical test	Laboratory test: Unconfined compression test	Laboratory test: Triaxial compression test	Laboratory test: Cyclic shear test	In-situ test: Standard penetration test	In-situ test: Elastic wave exploration	In-situ test: PS logging	In-situ test: Groundwater survey
Soil structure		△	△	○					○	○	△	
Ground water		△	△	○								○
Physical property		△			○							
Mechanical property test	SPT-N	△							○			
	Cohesion	△				○	○		○			
	Internal friction angle	△					○		△			
	Elastic wave velocity	△						○	△	○	○	
	Cyclic characteristics	△						○	△		○	

Modified from Kutsuzawa and Morita 1991)
○: Geotechnical parameter can be directly obtained.
△: Method to deduce geotechnical property or method is under study

have the same mechanical property because the mechanical property depends on the effective confining pressure and/or plasticity index. The shear modulus and the shear strength, for example, change with depth. Properties changing continuously with depth, however, cannot be handled in the computer program; representative values appointed for each layer are used at certain depths. Nonetheless, the SPT N-value varies even in the same layers; means for averaging is thus required. As these operations affect the results of the seismic ground response analysis, the engineer who will conduct analysis should have theoretical background and/or enough experience.

Determining the mechanical properties of soil is the most difficult and important job in the data preparations. Methods to obtain them are schematically shown in Table 2.1. Among these methods, borehole investigation (①) is made at almost all sites where seismic response analyses will be conducted. However, as the results of a borehole investigation are not sufficient to determine the mechanical and the in situ properties, the laboratory tests will be necessary.

There are only two methods to obtain the elastic modulus. One is to measure them on site (④), which is explained in Chap. 6. The other is to use empirical equations based on other field measurements (⑥). There are many empirical equations to obtain the wave velocity from, for example, the SPT N-value (⑦). Some physical quantities such as the void ratio and/or the effective confining stress may be required depending on the empirical equation. Elastic modulus can be evaluated from the wave velocity and the density (⑤).

There are also two methods to obtain the nonlinear soil parameters (⑫). In the first method, they (⑪) are directly obtained through the laboratory tests (⑩) by using undisturbed samples taken from in situ (⑨). Secondly, they can be obtained by using judgment based on past experience or empirical equations (⑭). Soil classification is, of course, important in this case, but other parameters (⑬) such as the plasticity index and the fines contents may be necessary. These are explained in details in Sect.6.2. This process is sometimes a little complicated, and one or two subsequent jobs will be required. Test data is converted into one or a few parameters that represent the material property (⑫) by the empirical equations proposed to evaluate them at first. This procedure may not be necessary depending on the computer program. Next step is to determine the values of parameters that the computer program uses (⑧). Stress–strain relationships are expressed by means of a mathematical formula in many computer programs, and the value of the coefficients has to be determined in this stage. This process may not be necessary, too, depending on the computer programs, which is explained in Chap. 8

Setting on the damping characteristics is also required (⑮), which will be explained in Chap. 12.

The last input data is earthquake motions (⑯) to be used in the analysis. As it is usually given a priori as the analysis condition, it is not explained in this book in detail although some information is given in Chap. 3.

After preparing all these data, the seismic ground response analysis can be executed (⑰). Then the engineer needs to evaluate the results whether they are rational or not considering the result from various points of view (⑱), and the job terminates. The last step, evaluation or engineering judgment, is very important as can be seen in the case studies referred in Chap. 15.

References

Assimakia D, Kauselb E (2002) An equivalent linear algorithm with frequency-and pressure-dependent moduli and damping for the seismic analysis of deep sites. Soil Dyn Earthq Eng 22(9–12):959–965

Biot MA (1963) Theory of stability and consolidation of a porous media under initial stress. J Math Mech 12:521–541

Finn WDL, Byrne PL, Martin GR (1976) Seismic response and liquefaction of sands. J Geotech Eng Div 102(GT8):841–856

Finn WDL, Yogendrakumar M, Yoshida N, Yoshida H (1984) TARA-3, a computer program to compute the response of 2-dimensional embankment and soil-structure interaction systems to seismic loading, soil dynamic group. University of British Columbia, Vancouver

Hardin BO, Drnevich VP (1972a) Shear modulus and damping in soils: measurement and parameter effects. Proc Am Soc Civ Eng 98(SM6):603–624

Hardin BO, Drnevich VP (1972b) Shear modulus and damping in soils: design equations and curves. Proc Am Soc Civ Eng 98(SM7):667–692

Ishihara K, Towhata I (1982) Dynamic response analysis of level ground based on the effective stress method. In: Pande GN, Zienkiewicz OC (eds) Soil mechanics – transient and cyclic loads. Wiley, New York, pp 133–172
Ishihara K, Yoshida N, Tsujino S (1985) Modelling of stress-strain relations of soils in cyclic loading. In: Proc. 5th international conference for numerical method in geomechanics, Nagoya, vol 1, pp 373–380
Kausel E, Assimaki D (2002) Seismic simulation of on elastic soils via frequency-dependent moduli and damping. J Eng Mech 128(1):34–47
Kutsuzawa S, Morita Y (1991) Design specification and soil constant, Tsuchi-to-Kiso. JGS 39(12):63–67
Lee MKW, Finn WDL (1978) DESRA-2C, Dynamic effective stress response analysis of soil deposits with energy transmitting boundary including assessment of liquefaction potential. The University of British Columbia, Faculty of Applied Science, Vancouver
Lysmer J, Udaka T, Tsai CF, Seed HB (1975) FLUSH a computer program for approximate 3-D analysis of soil-structure interaction problems, report no. EERC75-30. University of California, Berkeley
Nakamura S. Yoshida N (2002) Proposal of nonlinear earthquake response analysis in frequency domain considering apparent frequency dependency of soil property. J Geotech Eng (Proc. JSCE, No. 722/III-61): 169–187 (in Japanese)
Satoh T, Horike M, Takeuchi Y, Uetake T, Suzuki H (1997) Nonlinear behavior of scoria soil sediments evaluated from borehole record in eastern Shizuoka prefecture Japan. Earthq Eng Struct Dyn 26:781–795
Schnabel PB, Lysmer J, Seed HB (1972) SHAKE a computer program for earthquake response analysis of horizontally layered sites, report no. EERC72-12. University of California, Berkeley
Seed HB, Idriss IM (1970) Soil moduli and damping factors for dynamic response analyses, report no. EERC70-10, Earthquake Engineering Research Center. University of California, Berkeley, 40p
Streeter VL, Wylie EB, Richart FE (1974) Soil motion computation by characteristic method. J Geotech Eng Div AXE 100:247–263
Sugito M, Goda H, Masuda T (1994) Frequency dependent equi-linearized technique for seismic response analysis of multi-layered ground. J Geotech Eng (Proc. of JSCE, No. 493/III-27) 49–58 (in Japanese)
Yoshida N (1993) STADAS, a computer program for static and dynamic analysis of ground and soil-structure interaction problems, report, soil dynamics group. The University of British Columbia, Vancouver
Yoshida N (1995) DYNES3D, A computer program for dynamic response analysis of level ground by effective stress-nonlinear method, Revised in 2009 (version 2.74), Tohoku Gakuin University; http://www.civil.tohoku-gakuin.ac.jp/yoshida/computercodes/index.html
Yoshida N, Finn WDL (2000) Simulation of liquefaction beneath an impermeable surface layer. Soil Dyn Earthq Eng 19(5):333–338
Yoshida N, Towhata I (1991) YUSAYUSA-2 and SIMMDL-2, theory and practice, revised in 2003 (version 2.1), Tohoku Gakuin University and University of Tokyo; http://www.civil.tohoku-gakuin.ac.jp/yoshida/computercodes/index.html
Yoshida N, Tsujino S, Ishihara K (1990) Stress-strain model for nonlinear analysis of horizontally layered deposit. In: Summaries of the technical papers of annual meeting of AIJ, Chugoku, Vol. B (Structure I), pp. 1639–1640 (in Japanese)
Yoshida N, Kobayashi S, Suetomi I, Miura K (2002) Equivalent linear method considering frequency dependent characteristics of stiffness and damping. Soil Dyn Earthq Eng 22(3): 205–222

Chapter 3
Input Earthquake Motions

As input earthquake motion for the seismic ground response analysis is usually given a priori as an analysis condition, the engineer need not choose them. This book also stands on the same point of view. It may be, however, necessary to know how it is chosen or determined.

Input earthquake motions come from different concepts. The first one is earthquake motions recorded in the past earthquakes and has been used in the past earthquake resistant designs, which is called a historical earthquake motion in this book. The second is artificial earthquake motions developed by assuming the collapse of a particular fault or average feature of the past observed earthquakes. The former earthquake motion is frequently called a site wave in Japan, because it is developed for a particular site. The third one is an earthquake record observed near the site. It is used as input motion of the seismic ground response analysis directly, or deconvoluted to obtain the incident wave at the bedrock where earthquake motions are supposed to be unique in wide area.

3.1 Engineering Seismic Base Layer

In the Japanese practice, almost all earthquake motions to be used in the seismic ground response analysis or earthquake resistant design are defined at the engineering seismic base layer. This definition is not usually used outside Japan, but input motions are defined at the rock or the hard deposit outcrops. As explained later in this section, these definitions are nearly equivalent to each other. Therefore, they are treated as one at this moment. The engineering seismic base layer is sometimes called engineering bedrock which probably comes from its similarity to the seismic bedrock, but this term is not used in this book because the engineering seismic base layer may not be rock.

N. Yoshida, *Seismic Ground Response Analysis*, Geotechnical, Geological and Earthquake Engineering 36, DOI 10.1007/978-94-017-9460-2_3

It is equivalent to define bedrock or a base layer and to set the boundary conditions of the equation of motion. The definition of the boundary condition indicates that behavior outside the boundary does not affect the result of the seismic ground response analysis. However, it is very difficult to define base layer in this very strict definition. In the old days, seismic bedrock explained in Sect. 1.1 had been used as the definition of the engineering seismic base layer (Toki 1981; Japan Society for Natural Disaster Science 2002). After that the depth of the seismic bedrock moved downward according to the improvement of the research. However, it is difficult or impossible in the ordinary earthquake resistant design to make investigations at very deep depths. Then the engineer needs to separate them from the practical requirement. In addition, based on the second definition of the seismic bedrock in Sect. 1.1, the location of the bedrock for the earthquake resistant design need not be very deep depth for almost all structures whose predominant period is not very long.

The background concepts to use the engineering seismic base layer in the design are as follows (Amachi 2009):

1. One of the most important issues on the input earthquake motion is the amplification of the earthquake motion whose period is close to the predominant period of the structure. From this point, the base layer need not be a fixed depth or location, but can vary depending on the predominant period of the structure.
2. In the urban area in Japan, the seismic bedrock is located at very deep depths and observed earthquake motion is very few. On the other hand, there are not a few observed earthquake motions at the engineering seismic base layer. Therefore, it is practical or rational to define the earthquake motion at this depth.
3. In the Japanese practice, information on the basin structure is not well known except the Kanto, Osaka, and Nohbi plains. Therefore, it is very difficult to evaluate amplification characteristics from the seismic bedrock.

It is preferable to define the boundary conditions by which response in the analyzed region is not affected. Seismic bedrock may be a good boundary in this sense. When practical requirement from the design is considered, however, above discussions indicate that very strict definition may not be necessary, which is the reason why the engineering seismic base layer has been used in Japan.

Mechanical meaning of the engineering seismic base layer is sometimes defined to be the layer that has sufficiently large impedance ratio to the surface layer or hard deposit that expands widely in the region. These definitions come from the definition of the seismic bedrock or are introduced to define the base layer as rigid boundary. In the old days, earthquake motion is applied as a base motion, i.e., the sum of the incident and the reflected waves. Since reflected waves are affected by the behavior in the analyzed region, it is impossible to define them a priori. These considerations make the definition of the engineering seismic base layer more difficult, and discussions are continued (The Research Subcommittee on the Earthquake Motion 2002).

On the other hand, the definition of the engineering base layer in the design specifications in Japan is very clear because it is not defined by its mechanical

Table 3.1 Definition of engineering seismic base layer in design specifications

Design specification	V_s (m/s)
Port facility (Ports and Harbours Bureau, Ministry of Transport 1999)	300
Road bridge (Japan Road Association 2002)	300–350
Building and houses (Construction Ministry 2000)	400
Nuclear power plant (Electric Technical Guideline Committee 1987)	700

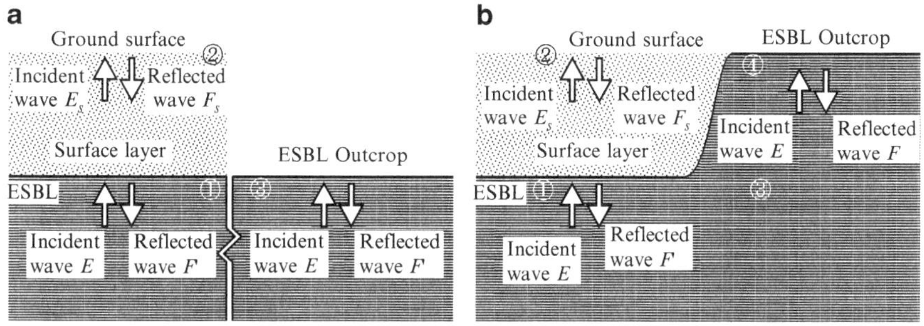

Fig. 3.1 Earthquake motion near the interested site. (**a**) Definition of outcropped base. (**b**) Image frequently used

characteristics, but is defined by the S-wave velocity. Representative definitions in Japan are shown in Table 3.1. The engineering seismic base layer defined for port facilities (Ports and Harbours Bureau, Ministry of Transport 1999) and for road bridge (Japan Road Association 2002) seems to correspond to the layers with SPT N-value of 50. On the other hand, definition in the buildings (Construction Ministry 2000) seems to indicate layers deeper than the $N = 50$ layer because a layer with $N = 50$ appears when V_s is a little larger than 300 m/s in many cases.

Considering the definition of the engineering seismic base layer, a specified earthquake motion at the engineering seismic base layer is preferable not to be affected by the behavior in the subsurface layers. It is possible to define it as an incident wave. This boundary condition is called an elastic base. Reflected wave can go out into the base layer; therefore, this boundary can consider the radiation damping. It is discussed in detail in Sect. 10.4.2.

There are two methods to define the earthquake motion at the elastic base, which is shown in Fig. 3.1a. Left side figure shows an ordinary case where there is subsurface layer (①) and, on the other hand, base outcrops (③) in the right side figure. Among the incident and reflected waves in the left figure, the incident wave is not affected by the behavior in the surface layer. Therefore, definition by the incident wave is one method. On the other hand, as explained in Sect. 1.2.2, incident and reflected waves are the same in the right side figure because of the ground surface outcrops. Therefore, the definition of the design earthquake motion as an outcrop motion is another method. As will be explained in Sect. 11.3, incident wave and reflected wave are usually expressed by variable E and F, respectively. Following this method, the first method defines E and the second method defines $2E$.

Figure 3.1b is frequently used image in North America. Earthquake motion is defined as an outcrop motion at the rock/hard deposit outcrop (④) near the interested site. If the rock or hard deposit is the same between ③ and ④, both incident waves toward the surface layer and those at the outcrop surface have the same waveform although there is some time lag from ③ to ④. Both Japanese and North American definitions are identical under this condition. In addition, if changes of the waveform between ③ and ④ are small, they are again identical in the engineering practice. Therefore, they are treated identical in this book.

The earthquake motion is sometimes defined at the ground surface. As mentioned, the engineering seismic base layer is defined as a layer where incident wave does not change in a wide area. Therefore, the earthquake record observed at the ground surface is sometimes deconvoluted in order to obtain the incident wave at the engineering seismic base layer. Obtained incident wave is used as the input motion at other sites under the assumption that the incident wave is unique regardless of the subsurface layer. Using the hysteretic earthquake motion as design earthquake corresponds to this case. The investigation of the damaged structure by using the earthquake record nearby is also this case.

Here, it is noted that the incident wave at ④ is different from the incident wave at ① if mechanical properties change from ③ to ④, and the situation that wave velocity is constant from ③ to ④ hardly happens. This may cause serious problem when an engineer computes the incident wave at ① from the earthquake record at ④.

Figure 3.2, for example, shows a blind test result (Kobayashi et al. 1992). Response at the ground of soft soil site (KS2) is required to be predicted under the given earthquake record at the rock site (KR1 in Fig. 3.2a). Three-dimensional topology hardly affects and one-dimensional analysis is sufficient in this problem (Sato et al. 1998). There were 44 applicants and 28 applicants used the equivalent

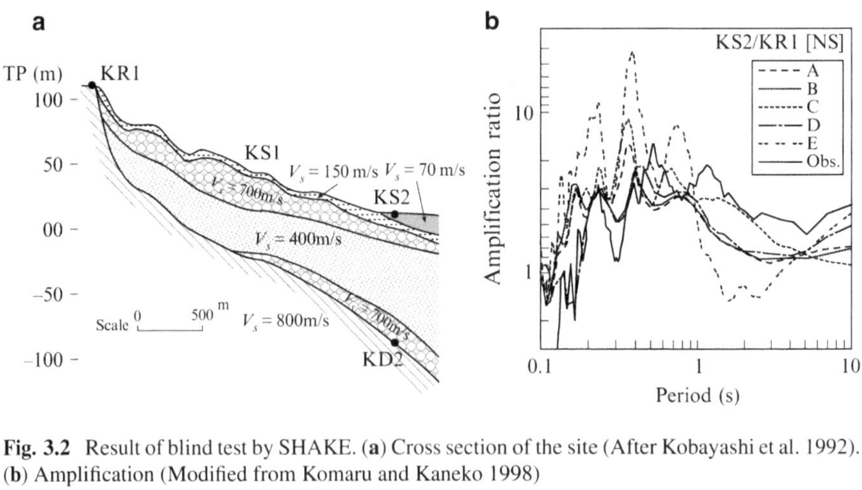

Fig. 3.2 Result of blind test by SHAKE. (**a**) Cross section of the site (After Kobayashi et al. 1992). (**b**) Amplification (Modified from Komaru and Kaneko 1998)

linear 1D analysis. Among them five applicants clearly wrote that they used SHAKE (Schnabel et al. 1972) although the other applicants were also supposed to use SHAKE but not sure because the program name was not required in this test. The amplification ratios by these five applicants are compared in Fig. 3.2b (Komaru and Kaneko 1998), and the result varies very much. Komaru and Kaneko (1998) made many analyses changing the sampling time, number of soil layers, filter frequency, etc., and found that these parameters do not affect the result significantly; error was within 10 %. They concluded that the reason of this scattering is the choice of the input motion at the KS2 site. Among five results, two applicants treated observed record at KR1 as outcrop motion and applied it directly at $V_s = 800$ m/s or $V_s = 1{,}500$ m/s layers, respectively. Other applicants deconvoluted record at KR1 to $V_s = 800$ m/s or $V_s = 1{,}500$ m/s layers to evaluate the incident wave and used it at the corresponding layers.

The similar issue will be discussed later in Sect. 15.8. This example clearly shows that the word "rock outcrop" or "hard deposit outcrop" is not sufficient to define the input earthquake motion. In addition, the definition of these rocks/soils also scatters in researchers or engineers. Figure 3.3 shows variation of the S-wave velocity in each soil classification by Japanese and US researchers (Yoshida and Iai 1998). It is recommended to define the engineering seismic base layer or rock outcrop by a more clear definition such as the S-wave velocity.

The sum of the incident and the reflected waves, which is called a composite wave, is also used as input motion in some case such that observed earthquake record in the underground is used as the input motion. The boundary where composite wave is specified is called a rigid base, where impedance of the base layer is assumed to be infinite. It is also possible in the multiple reflection theory (see Sect. 11.3) although stiffness of the base layer is not infinite and radiation toward the base layer is considered. These two definitions show exactly the same behavior in the subsurface layer.

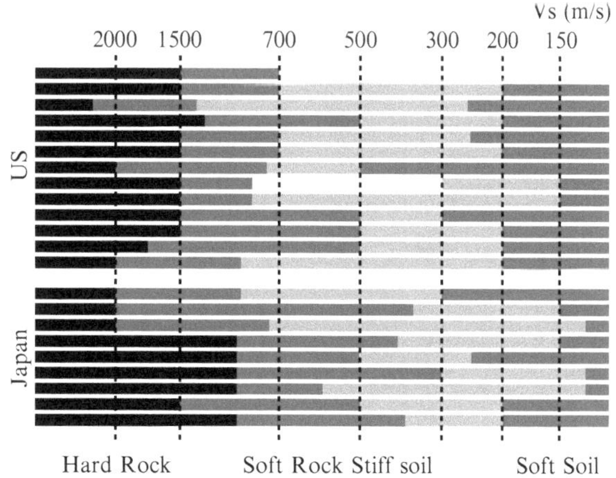

Fig. 3.3 Variation of S-wave velocity in each soil classification by researchers

3.2 Historical Earthquake Motions

Prof. Suehiro, the first president of the Earthquake Research Institute, University of Tokyo, made a presentation titled "Earthquake Engineering" in the 1932 ASCE meeting and emphasized the importance of the earthquake observation. Following his recommendations, several tens of earthquake instruments were installed in California (Ohsaki 1983), and earthquake records were obtained from fairly large earthquakes. These records have been used long time in the earthquake resistant design in the world. They are still used now, partly because it is convenient to compare past experiences or researches.

The acceleration record with maximum acceleration more than 3 m/s^2 was obtained at a town El Centro, located near the boundary of Mexico, during the 1940 Imperial Valley earthquake of $M = 7.1$. This record is called the El Centro wave and has been used as one of the representative earthquake motions to be considered in the earthquake resistant design. Waveforms of the NS component are shown in Fig. 3.4, and soil profiles are shown in Table 3.2. There are many versions in this record depending on the method of digitization and correction because original record is written on the paper. The records by USGS (Seekins et al. 1992) and UC Berkeley (Chopra 2010) are shown in Fig. 3.4; there are differences both in phase and maximum value.

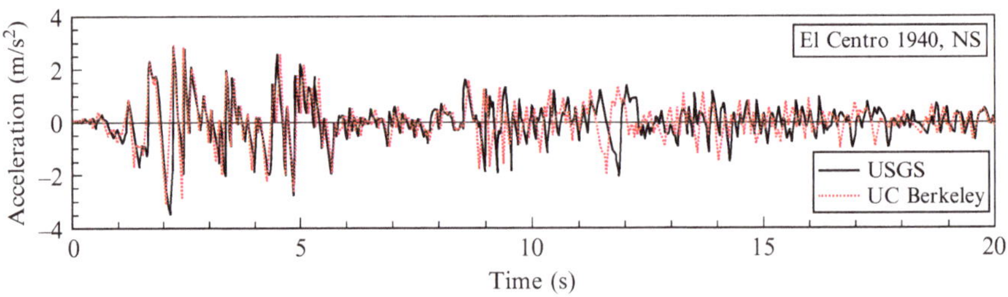

Fig. 3.4 Acceleration waveform of the El Centro 1940 NS wave

Table 3.2 Soil profiles of the sites of strong motion observatory

El Centro			Taft		
Depth (m)	Density (t/m^3)	V_s (m/s)	Depth (m)	Density (t/m^3)	V_s (m/s)
30.5	2.05	159.2	12.2	2.24	163.2
335.5	2.23	857.0	61.0	2.31	731.6
1,250.4	2.23	1,104.0	213.5	2.31	961.2
2,317.8	2.35	1,235.1	1,219.9	2.31	1,170.8
3,385.2	2.35	2,195.8		2.72	2,883.2
	2.76	3,049.7			

After Duke and Leeds (1962)

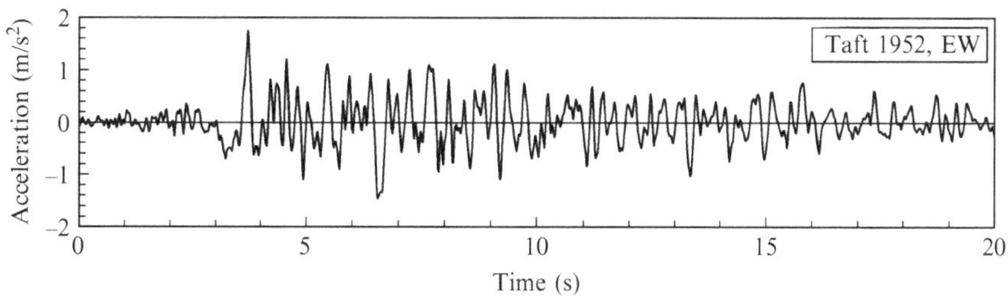

Fig. 3.5 Acceleration waveform of Taft 1952 EW wave

The earthquake motion obtained at Taft during the 1952 Kern County earthquake is called the Taft wave and has been used in the earthquake resistant design same as the El Centro wave. Figure 3.5 shows observed earthquake record, and soil profile is shown in Table 3.2.

These earthquake motions have been used in the earthquake resistant design in Japan, too, probably because large earthquake records were not observed in Japan. A relatively large earthquake motion was obtained at the University of Tokyo during the 1956 north Tokyo Bay earthquake, which is called Tokyo 101. This wave, however, is hardly used at present.

Then many strong motion records began to be observed in Japan, and some of them have been frequently used in the practical design.

During the 1978 Miyagiken-oki earthquake, a record, called the Kaihoku wave, was obtained at the Kaihoku Bridge, Miyagi prefecture, Japan. The acceleration waveform (longitudinal direction) and a soil profile are shown in Fig. 3.6. The sand/silt rock in Mesozoic Triassic period spreads widely in shallow depth near the site, and the seismograph is set on the rock. The record on the rock site is the reason why the Kaihoku wave has been used widely.

Another frequently used earthquake motion was observed at the Hachinohe port during the 1968 Tokachi-oki earthquake and is called the Hachinohe wave. Acceleration waveform is shown in Fig. 3.7, and soil profiles are shown in Fig. 3.8a. This wave contains long period component significantly, which is the reason why this wave has been used in the practical design. One needs some technique to calculate the incident wave at the engineering seismic base layer, which will be shown in Sect. 11.4.

Many earthquake records were obtained during the 1995 Hyogoken-nambu earthquake, which hit the Kobe City and caused serious damage and is frequently called as the Kobe earthquake internationally although the original name Hyogoken-nambu earthquake is usually used in Japan. Because of the significant damage, the disaster caused by this earthquake is called the Hanshin-Awaji earthquake disaster. In the early publication in Japan, the name "the Hanshin-Awaji earthquake" was sometimes used. Among many records, a record observed at the Kobe meteorological observatory has been used as a representative level 2 ground motion, which is shown in Fig. 3.9 with soil profiles in Fig. 3.8b.

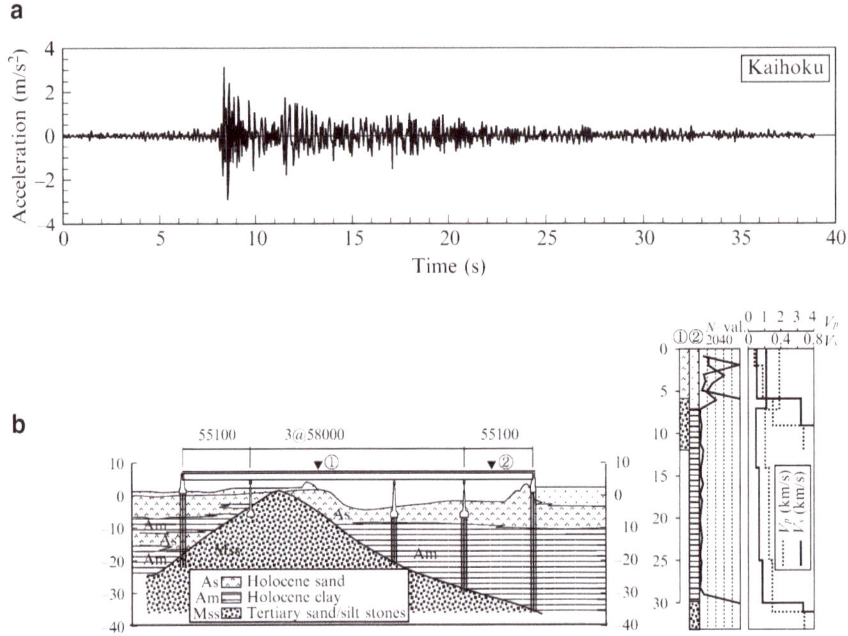

Fig. 3.6 Earthquake record at Kaihoku Bridge. (**a**) Acceleration waveform. (**b**) Cross section in longitudinal direction of bridge axis and soil profiles (Iwasaki et al. 1981)

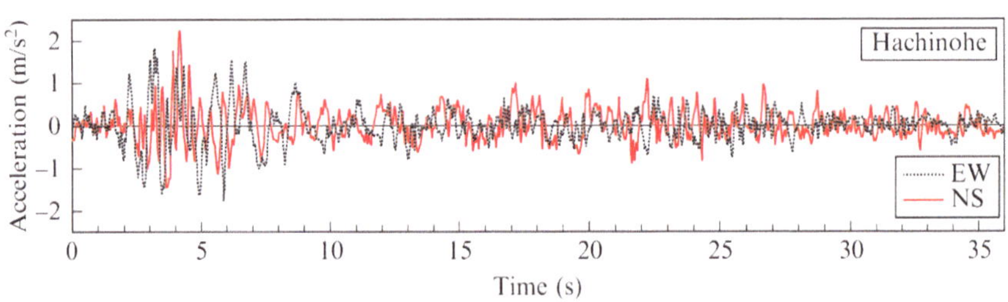

Fig. 3.7 Acceleration waveform

3.3 Intensity of Design Ground Motion

Ground motions during the 1995 Kobe earthquake were much larger than the design earthquake motion on the civil engineering structures, and significant damage occurred in these structures (Editorial Committee for the Repost on the Hanshin-Awaji Earthquake Disaster 2000). After that, design specification began to require two-step design. The first step considers the earthquake motions that will occur with high possibility during the lifetime. The second step considers very large earthquake motions that will occur with very small possibility. The similar concept had been employed for buildings in a Building Standard Law in 1981, in which seismic

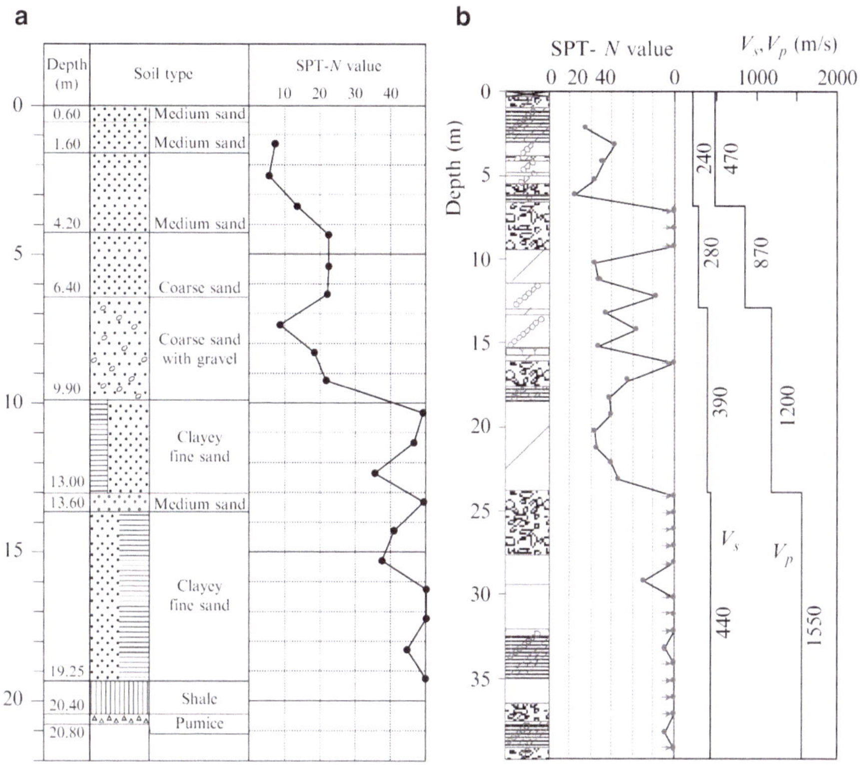

Fig. 3.8 Soil profiles at the strong motion observatory. (**a**) Hachinohe -S (Tsuchida and Uwabe 2008). (**b**) Kobe Marine Observatory (Tamura and Honda 1998)

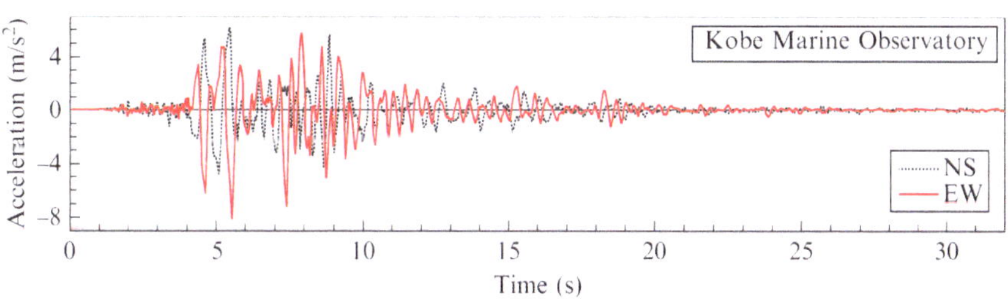

Fig. 3.9 Acceleration waveform at Kobe meteorological observatory

intensity five times larger than conventional one (seismic coefficient =0.2) is used to check the safety under the limit state. Now, the earthquake motion that is considered in the first step design is frequently called as the level 1 earthquake motion and that in the second step design as the level 2 earthquake motion.

The detailed description of these earthquake motions may be somewhat different between design specifications. According to the proposal by Japan Society of Civil Engineers (JSCE) (1996), level 2 ground motion is defined to be the possible

Table 3.3 Vision 2000

	Earthquake performance level			
Earthquake design level	Fully operational	Operational	Life safe	Near collapse
Frequent (43 years)	•	○	○	○
Occasional (72 years)	■	•	○	○
Rare (475 years)	★	■	•	○
Very rare (970 years)		★	■	•

○ Unacceptable performance (for new construction), • Basic objective, ■ Essential/hazardous objective, ★ Safety critical objective

strongest earthquake motion at the site. Then the earthquake motion changes site by site, which is the reason why artificial earthquake motions have been developed. In many design specifications in Japan, the definition is a little looser than that by JSCE such as experienced maximum motions, but essential concept seems to be held.

Intensity of the earthquake motion is to be related to the performance demand or the requirement of structures although actual earthquake is unique. It means that engineering judgment is required to determine the design earthquake motion. As mentioned, two criteria are considered in Japan, whereas more than two criteria are considered in North America such as Vision 2000 (Structural Engineers Association of California Vision 2000 Committee 1995) as shown in Table 3.3.

3.4 Synthesized Earthquake Motions

Historical earthquake records have been used in the earthquake resistant design when the seismic ground response analysis was necessary in the old days. According to the development of the computer and the research, however, synthesized earthquake motions began to be used. There are two types of synthesized earthquake motions. One is a wave that represents certain area, and the other is a wave that represents strong motion for general design. Some examples in Japan are shown here.

Many high-rise buildings were going to be built in the Tokyo Bay waterfront area, in which case it is convenient to make an earthquake motion that represents this area and can be used for all structures in this area. The Rinkai wave was developed for this purpose (Structural Safety Committee in Water Front Area 1992, 1993). In making this wave, past and future earthquakes, such as the 1855 Ansei-Edo earthquake, the 1923 Kanto earthquake, and the future Tokai earthquake, are considered, and a response spectrum that envelops these earthquakes was build.

The Rinkai wave is defined as an outcrop wave at the engineering seismic layer with S-wave velocity $V_s = 300$–500 m/s. The Tokyo gravel layer or the Edogawa formulation corresponds to this layer. This earthquake motion is supposed to be the level 2 ground motion, and that of the level 1 ground motion is defined to be

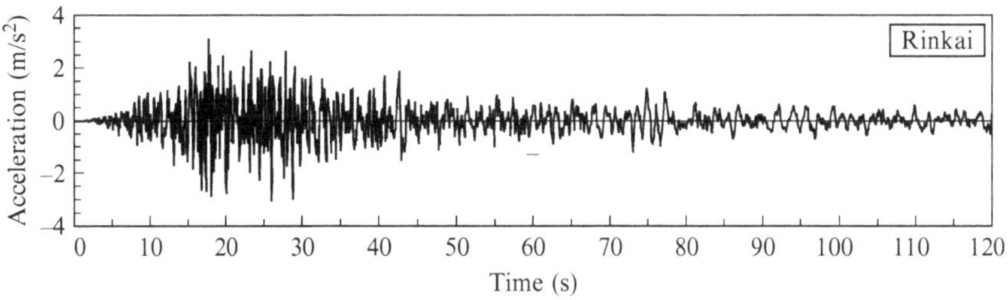

Fig. 3.10 Rinkai, Tokyo Bay waterfront area, wave

a half magnitude. Although only a response spectrum is defined in the report, the acceleration time history is also developed for the discussion of this wave in the committee which is shown in Fig. 3.10, and this wave is usually used in the practical design, too.

Several waves have been built by the similar method. The MM21 wave (Housing and Architecture Bureau of Yokohama City 2010), for example, is developed for the design in Minato Mirai 21 region and Shin-Yokohama area in the Yokohama City, Japan. Four earthquakes, the Minami-Kanto earthquake, the Tokai earthquake, the inland earthquake in Tokyo area, and the inland earthquake in the Yokohama City, are considered, and the wave is made by fitting envelops of the response spectrum developed by using the Kobayashi–Midorikawa method (Midorikawa and Kobayashi 1978) and Irikura method (Irikura 1986).

The Building Center wave is built for the design of high-rise buildings in Japan and is defined at the engineering seismic base layer outcrop wave with $V_s \geq 400$ m/s (The Building Center of Japan 1992, 2010). A velocity response spectrum is first developed at the site with equivalent focal depth of 52.5 km under the $M = 8$ earthquake. Earthquake records in Japan and strong earthquake records outside Japan are used in making the response spectrum. Then the acceleration time history is made. Two earthquake motions, level 1 and level 2, were developed.

3.5 Strong Ground Motion Databases

It was difficult to obtain the earthquake records in the old days, but earthquake waves can be obtained from many sources at present; many organizations release the earthquake records in Japan as well as in the world.

For example, representative observed earthquake records that have been used in the earthquake resistant design can be obtained from the Building Center of Japan (2010). Old earthquake records are published from the Association for Science Documents Information (Association for Science Documents Information 1972). Port and Airport Research Institute distributes earthquake records obtained by Strong-Motion Earthquake Observation Results from 1963 (Port and Airport Research Institute 2010). Vertical array records are distributed from Japan

Association for Earthquake Engineering (JAEE) until 1995 (Association for Earthquake Disaster Prevention 1992; Association for Earthquake Disaster Prevention). National Research Institute for Earthquake Science and Disaster Prevention, Japan, distributes earthquake records by K-NET, F-NET, and KiK-net (http://www.bosai.go.jp/e/index.html 2010).

Many data outside Japan can be obtained from PEER (Pacific Earthquake Engineering Research Center, http://peer.berkeley.edu/), COSMOS (Consortium of Organizations for Strong Motion Observation System, http://www.cosmos-eq.org/), etc. Past earthquake records can also be obtained in these websites. In addition, there are many related links.

References

Amachi F (2009) Concept of bedrock and ground characteristics, Lecture on fundamentals of strong motion geotechnology, Seismological Society of Japan, http://wwwsoc.nii.ac.jp/ssj/publications/KISOKOZA/kisokoza08.html [2009]

Association for Earthquake Disaster Prevention (1992) Report of Strong motion array record database committee in 1991

Association for Earthquake Disaster Prevention. Strong motion array record database, No. 1 (1993), No. 2 (1995) and No. 3 (1998)

Association for Science Documents Information (1972) Digitized strong-motion earthquake acceleration in Japan

Building Center of Japan (1992) Report of committee on design earthquake wave, 349 pp.; Engineering guideline for making design earthquake wave and commentary, 73 pp (in Japanese)

Building Center of Japan (2010) On distribution of representative observed earthquake records (acceleration data). http://www.bcj.or.jp/download/src/point.pdf [2010]

Chopra AK (2010) El Centro, 1940 ground motion data. http://nisee.berkeley.edu/data/strong_motion/a.k.chopra/index.html [2010]

Construction Ministry (2000) On structural calculation standard for structural safely of high rise building, ministerial announcements 1461 (in Japanese)

Duke CM, Leeds DJ (1962) Site characteristics of Southern California strong motion earthquake stations, U.C.L.A. Department of Engineering Report No. 62–55

Editorial Committee for the Repost on the Hanshin-Awaji Earthquake Disaster (2000) Report on the Hanshin-Awaji Earthquake Disaster, General Issue Volume 1, Compendium of earthquake disaster including Stochastic description, Maruzen, 549 pp (in Japanese)

Electric Technical Guideline Committee (1987) Technical guidelines for aseismic design of nuclear power plants, Japan Electric Association

Housing and Architecture Bureau of Yokohama City (2010) Characteristics of synthesized earthquake motion (yoko rock). http://www.city.yokohama.lg.jp/kenchiku/center/kenchiku/kozo/yokonami/yokorock.html [2010]

http://www.bosai.go.jp/e/index.html [2010]

Irikura K (1986) Prediction of strong ground acceleration motions using empirical Green's function. In: Proceedings of the 7th Japan earthquake engineering symposium, pp 151–156

Iwasaki T, Kawashima K, Takagi Y (1981) Analysis of earthquake record at Kaihoku bridge. In: Proceedings of the 18th JSCE earthquake engineering symposium-1989, earthquake Engineering Committee, JSCE, pp 293–296

Japan Road Association (2002) Specifications for highway bridges. Part V, Seismic design, Maruzen, 317 pp

Japan Society for Natural Disaster Science (ed) (2002) Subject-book on disaster prevention, Tsukiji Shobo, 543 pp
JSCE (1996) Proposal on earthquake resistance for civil engineering structures. http://www.jsce.or.jp/committee/earth/propo-e.html [2005]
Kobayashi K, Amaike F, Abe Y (1992) Attenuation characteristics of soil deposits and its formulation. In: Proceedings of the international symposium on the effects of surface geology on seismic motion. Odawara
Komaru Y, Kaneko F (1998) Why the 1D results varied much? -Ashigara Valley test site. In: Proceedings of the 2nd international symposium on the effects of surface geology on seismic motion, vol 1, pp 313–317, Yokosuka, Japan
Midorikawa S, Kobayashi H (1978) Spectral characteristics of incident wave from seismic bedrock due to earthquake, Transactions of AIJ, No. 273, pp 43–54 (in Japanese)
Ohsaki Y (1983) Earthquake and building, Iwanami, 200 pp (in Japanese)
Port and Airport Research Institute (2010) Port area strong motion earthquake records observatory. http://www.mlit.go.jp/kowan/kyosin/eq.htm [2010]
Ports and Harbours Bureau, Ministry of Transport (eds) (1999) Technical standards and commentary of port and harbour facilities in Japan. The Overseas Coastal Area Development Institute of Japan (in Japanese)
Sato K, Higashi S, Yajima H, Sasaki S (1998) Ashigara valley test site, 1D or 2D-3D? In: Proceedings of the 2nd international symposium on the effects of surface geology on seismic motion, vol 1, pp 319–340, Yokosuka, Japan
Schnabel PB, Lysmer J, Seed HB (1972) SHAKE a computer program for earthquake response analysis of horizontally layered sites, Report no. EERC72-12. University of California, Berkeley
Seekins LC, Brady AG, Carpenter C, Brown N (1992) Digitized strong-motion accelerograms of North and Central American earthquakes 1933–1986. U.S. Geological Survey Digital Data Series DDS-7. http://nsmp.wr.usgs.gov/data_sets/ncae.html [2010]
Structural Engineers Association of California Vision 2000 Committee (1995) Vision 2000, performance based seismic engineering of buildings
Structural Safety Committee in Water Front Area (1992) Report on safety of large scaled buildings in water front area, Japan Building Disaster Prevention Association, 160 pp
Structural Safety Committee in Water Front Area (1993) Special issue on safety of buildings in water front area, Building Disaster Prevention, no. 182, pp 2–36
Tamura K, Honda R (1998) Soil survey at Kobe Marine Observatory, PWRI Report No. 3625, Public Work Research Institute, Japan (in Japanese)
The Building Center of Japan (2010) Synthesized base layer wave of the Japan Building Center of Japan, BCJ L1 and BCJ-L2. http://www.bcj.or.jp/download/wave.html [2010] (in Japanese)
The Research Subcommittee on the Earthquake Motion (2002) Proceedings of the 30th symposium of earthquake ground motion -can design earthquake motion be defined at engineering seismic base layer, AIJ
Toki K (1981) Seismic response analysis of structures. New series of civil engineering, vol 11, Gihodo Shuppan, 250 pp (in Japanese)
Tsuchida H, Uwabe T (2008) Characteristics of earthquake motion at bees layer estimated from earthquake records at ground surface, Technical note of the Port and Airport Research Institute, No. 11, 270 pp (in Japanese)
Yoshida N, Iai S (1998) Nonlinear site response and its evaluation and prediction. In: Proceedings of the 2nd international symposium on the effect of surface geology on seismic motion, vol 1, pp 71–90, Yokosuka, Japan

Chapter 4
Fundamentals of Soil Mechanics

Knowledge on fundamental mechanics that are required in the seismic ground response analysis is introduced in this chapter. Practical models will be explained in Chaps. 7 and 8.

4.1 Stress and Strain

4.1.1 Positive Directions of Stress and Strain

Stress and strain are expressed as

$$\text{stress} : \left\{ \sigma_x \ \sigma_y \ \sigma_z \ \tau_{xy} \ \tau_{yz} \ \tau_{zx} \right\}$$

$$\text{strain} : \left\{ \varepsilon_x \ \varepsilon_y \ \varepsilon_z \ \gamma_{xy} \ \gamma_{yz} \ \gamma_{zx} \right\}$$

where σ and τ denote normal and shear stresses, respectively; ε and γ denote axial and shear strains, respectively; and subscripts indicate corresponding coordinate axis.

There are two definitions on the shear strain; the one is a tensor strain, and the other is an engineering strain. The tensor strain is used for the tensor expression and is usually used in the textbook on mechanics. In order to distinguish the tensor strain from the engineering strain, the notation ε is used; hence

$$\text{Tensor strain} : \left\{ \varepsilon_x \ \varepsilon_y \ \varepsilon_z \ \varepsilon_{xy} \ \varepsilon_{yz} \ \varepsilon_{zx} \right\}.$$

In the geotechnical engineering practice, however, engineering strain is frequently used, and this book also uses engineering strain. The engineering strain is two

N. Yoshida, *Seismic Ground Response Analysis*, Geotechnical, Geological and Earthquake Engineering 36, DOI 10.1007/978-94-017-9460-2_4

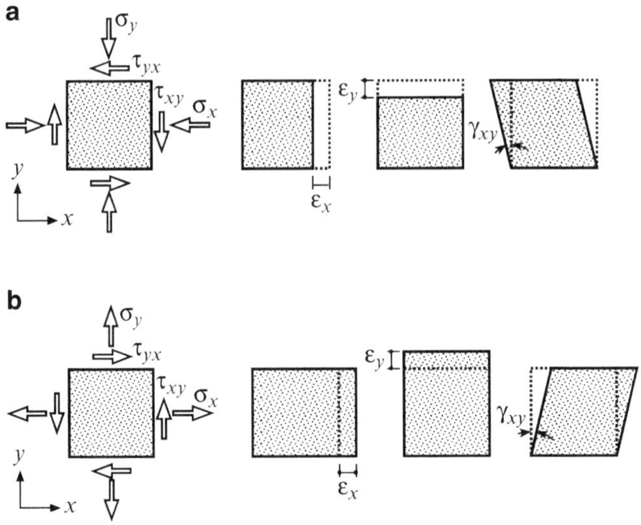

Fig. 4.1 Definition of stress and strain. (**a**) Positive in compression. (**b**) Positive in tension

times larger than the tensor strain such as $\gamma_{xy} = 2\varepsilon_{xy}$. When engineering strain is employed, the concept of Mohr's circle cannot be used. Even so, the engineering strain is more convenient than the tensor strain as the name "engineering" indicates. For example, stress–strain relationships on the shear deformation can be written by introducing the elastic shear modulus G as

$$\tau = G\gamma \tag{4.1}$$

The coefficient becomes $2G$ if tensor strain is used, which is obviously inconvenient in the daily use.

The definition of a stress and a strain is shown in Fig. 4.1, in which two different definitions are shown. Figure 4.1b is the definition used in general mechanics; normal stresses such as σ_x and σ_y are positive for tension. This definition is called definition of positive in the following. On the other hand, soil hardly resists under tensile force, and strain of soil is usually in compression. Therefore, almost all numbers are to be written with minus symbol (negative value) when definition in Fig. 4.1b is used, which is obviously very inconvenient. Therefore, in the soil mechanics, normal stress is defined to be positive under the compression, which is called definition of negative in the following. Correspondingly, positive direction changes for shear stress as shown in Fig. 4.1a.

In some places, in addition, positive direction for normal stress follows Fig. 4.1a and that for shear stress follows Fig. 4.1b. This definition brings inconveniency such that Mohr's circle cannot be used, but may be practical. For example, let us consider the case that a small element in Fig. 4.1a is subjected to force toward x-direction at the top with the bottom fixed. This force is positive because the direction coincides with the direction of the coordinate axis. Then the deformation becomes such as in the right side of Fig. 4.1b. Therefore, resulting stress and strain are negative value. Response by negative value under the positive direction of loading may not be accepted easily although it is theoretically correct. This inconveniency can be avoided by considering the third definition.

All these three definitions are used in the field of the seismic response of ground. For example, multipurpose software that deals with both structure and ground uses the definition of positive. On the other hand, the definition of negative is used in the computer program that mainly deals with ground. Even in these programs, the definition of positive may be employed for structure. In the one-dimensional analysis, the definition of positive is usually used. In many cases, positive direction may not be a big issue, but it may become important in particular cases.

4.1.2 Effective Stress Principle

As soil is composed of soil particles, the deformation of soil does not come from the deformation of the soil particles but comes from the change of the arrangement of the soil particles that is called a soil skeleton. The deformation of the soil particles is much less than that of the soil skeleton and is usually neglected. Relative movement of the soil particle is mainly controlled by a friction force between the neighboring soil particles. In other words, friction force between soil particles is to be considered in the behavior of the soil. It can be done by considering an effective stress and is known as the effective stress principle. According to Lambe and Whitman (1969), for example, the effective stress principle is explained as:

1. The effective stress is equal to the total stress minus the pore pressure.
2. The effective stress controls certain aspects of soil behavior, notably compression and strength.

The first statement is schematically shown in Fig. 4.2. Here the term "total stress" is the same as "stress" in the ordinary mechanics; it acts the entire area under consideration. In the soil mechanics, however, "total stress" is used in order

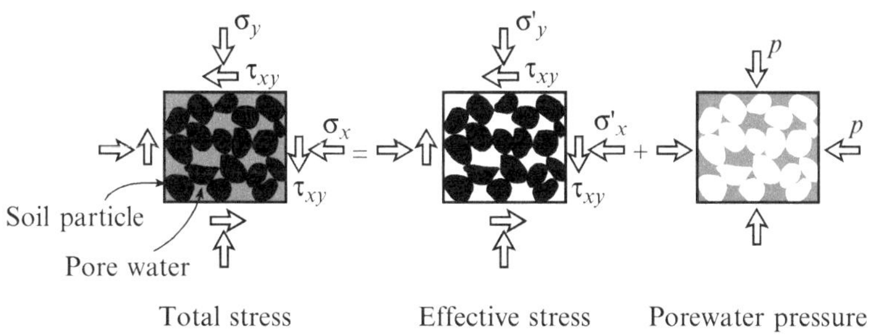

Fig. 4.2 Total stress and effective stress

to distinguish "effective stress." The effective stress is frequently indicated by prime such as σ'_x as also shown in the figure. Since water does not resist under the shear stress, total and effective are distinguished only for the normal stresses.

In practice, the elastic modulus and the shear strength are affected by the effective stress, which will be explained in Chap. 7.

The concept on the effective stress is very important especially in the analysis of liquefaction. However, it is not strongly interested in the total stress analysis because the change of the mechanical property caused by the change of the effective stress during an earthquake is frequently neglected in the total stress analysis although the initial elastic modulus and the shear strength are calculated based on the effective stress.

4.2 Characteristics of Soil Behavior

Young's modulus and the shear modulus are usually used to express the elastic moduli of the structural material such as steel and concrete. This expression is strongly related to the method they are used. Young's modulus is, for example, a coefficient between the axial stress and the axial strain when they are modeled into a lineal member. Soil is not, however, modeled like this, which means that Young's modulus is not a convenient modulus for the soil mechanics, because the elastic moduli that correspond to the frequently appeared deformations are convenient to be used in the practical engineering.

Young's modulus and the yield stress are sufficient to express the behavior of the steel, for example. On the other hand, behavior of the soil is very complicated, and it is almost impossible to express the behavior by small numbers of indices. Therefore, the representative value is measured and used only in the field of interest. For example, volume change is a major interest in the consolidation analysis, whereas shear strength is interested in the bearing capacity problem. In the field of the earthquake geotechnical engineering, the most important behavior is a shear deformation. Volumetric change characteristics are also important especially in the effective stress analysis.

Considering these situations, the behavior of soil is classified into shear deformation, volume change, and dilatancy in this book.

4.2.1 Volume Change

Volume change behavior is characterized by using a bulk modulus K as

$$\sigma_m = K\varepsilon_v \tag{4.2}$$

where σ_m and ε_v are a confining stress (average stress) and a volumetric strain, respectively, and are defined to be

$$
\begin{aligned}
\sigma_m &= \left(\sigma_x + \sigma_y + \sigma_z/3\right) \\
\varepsilon_v &= \varepsilon_x + \varepsilon_y + \varepsilon_z
\end{aligned}
\tag{4.3}
$$

If prime is put on stresses in Eqs. (4.2) and (4.3), they denote the effective stress. However, since this book deals with the total stress analysis in which case the soil particle and the porewater are considered to behave as one body, the prime may be left out when it is clear to express the effective stress.

4.2.2 Shear Deformation

Constitutive relations against the shear deformation are already explained in Eq. (4.1), which was

$$\tau = G\gamma \tag{4.1}$$

In addition, the shear stress appears under a deviator stress or difference between normal stresses. Therefore, the deviatoric stress s_{ij} and the equivalent stress σ_e are used in expressing the shear behavior under multidimension. Corresponding to these stresses, the deviatoric strain e_{ij} and the equivalent strain ε_e are used for strain. They are defined as

$$s_{ij} = \sigma_{ij} - \delta_{ij}\sigma_m, \quad e_{ij} = \varepsilon_{ij} - \frac{1}{3}\delta_{ij}\varepsilon_v, \quad \sigma_e = \sqrt{\frac{3}{2}s_{ij}s_{ij}}, \quad \text{and} \quad \varepsilon_e = \sqrt{\frac{2}{3}e_{ij}e_{ij}} \tag{4.4}$$

where σ_{ij} and ε_{ij} are stress and strain tensors, respectively. The δ_{ij} is a Kronecker delta which takes value 1 when $i=j$ and 0 for other combinations of i and j. Both the equivalent stress and the equivalent strain are representative values to express the shear deformation by one number. In this definition coefficients 3/2 and 2/3 are used, but different values may be used in other book. They must be an inverse number as the product of the deviatoric stress, and the deviatoric strain must become energy. The constitutive relationships yield

$$s_{ij} = 2Ge_{ij} \quad (i \neq j) \tag{4.5}$$

4.2.3 Other Parameters

Young's modulus stands for a lineal structure when there is no stress in the lateral surface, although this fact is sometimes overlooked. This stress state hardly appears in the soil, which is the reason why Young's modulus is hardly used in the soil mechanics. Conversely, one-dimensional behavior with no lateral strain change frequently appears such as one-dimensional consolidation. The coefficient corresponding to this behavior is called a one-dimensional rebound modulus and is denoted by B. Then

$$\sigma_x = B\varepsilon_x \left(\varepsilon_y = \varepsilon_z = 0\right), \quad B = K + \frac{4}{3}G \tag{4.6}$$

This relation is used when dealing with P-wave propagation in the earthquake geotechnical engineering.

In addition to these elastic moduli, Poisson's ratio ν is frequently used in the engineering practice. It is not an elastic modulus, but treated similar to the elastic modulus. In the homogeneous media, only two elastic moduli are independent and other moduli can be calculated from them. Various conversion equations are summarized in Table 4.1a. It may be noted that the soil is usually an anisotropic material, but it is usually dealt with an isotropic media in the engineering practice.

4.2.4 Dilatancy

Another important behavior of soil is dilatancy, which is shown in Fig. 4.3 schematically. Soil is composed of soil particles, and due to the presence of void, soil

Table 4.1 Conversion between elastic moduli

(a) Three-dimensional state (ordinary case)

	E, G	E, ν	E, K	G, ν	G, K	ν, K
E	–	–	–	$2G(1+\nu)$	$\dfrac{9KG}{3K+G}$	$3K(1-2\nu)$
G	–	$\dfrac{E}{2(1+\nu)}$	$\dfrac{3KE}{9K-E}$	–	–	$\dfrac{3K(1-2\nu)}{2(1+\nu)}$
K	$\dfrac{GE}{3(3G-E)}$	$\dfrac{E}{3(1-2\nu)}$	–	$\dfrac{2G(1+\nu)}{3(1-2\nu)}$	–	–
ν	$\dfrac{E-2G}{2G}$	–	$\dfrac{3K-E}{6K}$	–	$\dfrac{3K-2G}{2(3K+G)}$	–
B					$K+4G/3$	

(b) Simple two-dimensional state

	E, G	$E, \tilde{\nu}$	$E, \tilde{K}$	$G, \tilde{\nu}$	$G, \tilde{K}$	$\tilde{\nu}, \tilde{K}$
E	–	–	–	$2G(1+\tilde{\nu})$	$\dfrac{4\tilde{K}G}{\tilde{K}+G}$	$2\tilde{K}(1-\tilde{\nu})$
G	–	$\dfrac{E}{2(1+\tilde{\nu})}$	$\dfrac{\tilde{K}E}{4\tilde{K}-E}$	–	–	$\dfrac{\tilde{K}(1-\tilde{\nu})}{1+\tilde{\nu}}$
$\tilde{K}$	$\dfrac{GE}{4G-E}$	$\dfrac{E}{2(1-\tilde{\nu})}$	–	$\dfrac{G(1+\tilde{\nu})}{1-\tilde{\nu}}$	–	–
$\tilde{\nu}Z$	$\dfrac{E-2G}{2G}$	–	$\dfrac{2\tilde{K}-E}{2\tilde{K}}$	–	$\dfrac{\tilde{K}-G}{\tilde{K}+G}$	–
B					$\tilde{K}+G$	

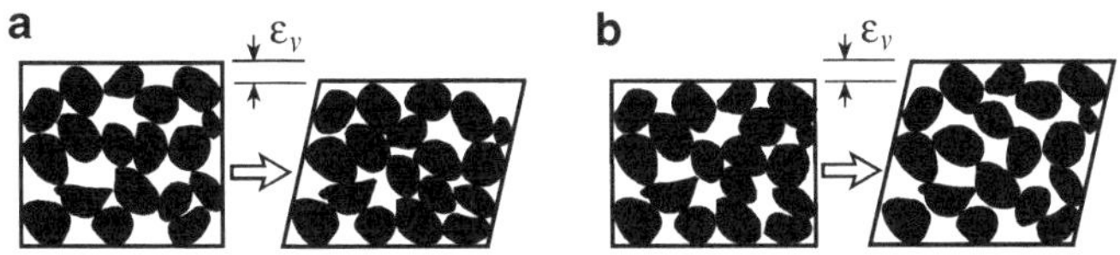

Fig. 4.3 Schematic figure showing mechanism of dilatancy. (**a**) Negative dilatancy. (**b**) Positive dilatancy

particle may drop into the pore when subjected to shear force as shown in Fig. 4.3a. While this is true for loose sand, dense soil behaves differently as the soil particles must ride up to the neighboring soil particle to move horizontally under shear and as a result volume increases. Then volume change is also associated with shear deformation. This behavior is called dilatancy and is one of the characteristic behaviors in granular material; dilatancy is not observed in the ordinary structural material such as steel and concrete. This behavior makes analysis of soil very complicated.

There are several methods to consider dilatancy. The stress–dilatancy model

$$\frac{d\varepsilon_v^p}{d\gamma^p} = \mu_m - \frac{\sigma_e}{\sigma_m'} \tag{4.7}$$

is frequently used in the plasticity theory (see Sect. 8.3.2 for further details). However, a more simple empirical formula such as the Martin–Finn–Seed model (Martin et al. 1975) or the bowl model (Fukutake 1997) may be used. In addition, stress path in the shear stress-effective mean stress path is used to express the dilatancy behavior under the undrained condition such as in Ishihara and Towhata (1982) and Towhata and Ishihara (1985).

Consideration of the dilatancy is essential in the effective stress analysis, but it is hardly considered in the total stress analysis. Therefore, this book does not deal with dilatancy, hereafter.

4.2.5 Constitutive Relations for Elastic Behavior

The bulk modulus K and the shear modulus G are usually used as elastic moduli in the soil mechanics. Stress–strain relationships are written as follows by using these moduli:

$$\begin{Bmatrix} \sigma_x \\ \sigma_y \\ \sigma_z \\ \tau_{xy} \\ \tau_{yz} \\ \tau_{zx} \end{Bmatrix} = \begin{bmatrix} K+4G/3 & K-2G/3 & K-2G/3 & 0 & 0 & 0 \\ K-2G/3 & K+4G/3 & K-2G/3 & 0 & 0 & 0 \\ K-2G/3 & K-2G/3 & K+4G/3 & 0 & 0 & 0 \\ 0 & 0 & 0 & G & 0 & 0 \\ 0 & 0 & 0 & 0 & G & 0 \\ 0 & 0 & 0 & 0 & 0 & G \end{bmatrix} \begin{Bmatrix} \varepsilon_x \\ \varepsilon_y \\ \varepsilon_z \\ \gamma_{xy} \\ \gamma_{yz} \\ \gamma_{zx} \end{Bmatrix} \tag{4.8}$$

However, three-dimensional analysis is hardly made in the earthquake geotechnical engineering partly because of the lack of the computer power and partly because of the difficulty to obtain geotechnical data for three-dimensional analysis. One-dimensional analysis is the most popular, and two-dimensional analysis is sometimes used. Formulation for the three-dimensional analysis can easily be used in the one-dimensional analysis because only shear deformation is considered. On the other hand, formulation in two dimensional may require special formulation.

The plane stress or the plane strain condition is assumed in the ordinary two-dimensional analysis, among which plane strain condition is relevant in the two-dimensional analysis because displacement perpendicular to the analyzed plane is assumed not to occur.

Strains ε_z, ε_{yz}, and ε_{zx} are zero in the plane strain formulation. Stresses τ_{yz} and τ_{zx} become zero under this assumption. Therefore stress and strain used in the plane strain analysis are $\{\sigma_x \sigma_y \sigma_z \tau_{xy}\}$ and $\{\varepsilon_x \varepsilon_y \gamma_{xy}\}$. Then constitutive relation yields

$$\begin{Bmatrix} \sigma_x \\ \sigma_y \\ \sigma_z \\ \tau_{xy} \end{Bmatrix} = \begin{bmatrix} K + 4G/3 & K - 2G/3 & 0 \\ K - 2G/3 & K + 4G/3 & 0 \\ K - 2G/3 & K - 2G/3 & 0 \\ 0 & 0 & G \end{bmatrix} \begin{Bmatrix} \varepsilon_x \\ \varepsilon_y \\ \gamma_{xy} \end{Bmatrix} \tag{4.9}$$

Among the four stresses, σ_z may not be output in some computer programs as it is not so important in the engineering practice.

This formulation just reduces the number of nonzero stresses and strains. Therefore, the plane strain formulation does not imply that three-dimensional constitutive model is not necessary. Stress σ_z is necessary to calculate the confining pressure σ_m, which is necessary to evaluate the stiffness and the strength. Three-dimensional constitutive model is, however, sometimes too complicated to use in practice. Therefore, constitutive models that do not consider σ_z are also used. In this model, volume change behavior is expressed as

$$\sigma_m = \frac{\sigma_x + \sigma_y}{2} = \tilde{K}\varepsilon_v = \tilde{K}\left(\varepsilon_x + \varepsilon_y\right) \tag{4.10}$$

where $\tilde{K}$ is a bulk modulus in this formulation. Here, the volumetric strain is the same with three-dimensional case because $\varepsilon_z = 0$ under the plane strain condition. However, K evaluated in three-dimensional field does not stand in this formulation since σ_m is different from the definition in the three-dimensional field. It can be easily understood from Eq. 4.9 as

$$\frac{\sigma_x + \sigma_y}{2} = \left(K + \frac{1}{3}G\right)\left(\varepsilon_x + \varepsilon_y\right) \quad \text{or} \quad \tilde{K} = K + \frac{1}{3}G \tag{4.11}$$

By introducing a new bulk modulus $\tilde{K}$, the constitutive relation is written as

$$\begin{Bmatrix} \sigma_x \\ \sigma_y \\ \tau_{xy} \end{Bmatrix} = \begin{bmatrix} \tilde{K}+G & \tilde{K}-G & 0 \\ \tilde{K}-G & \tilde{K}+G & 0 \\ 0 & 0 & G \end{bmatrix} \begin{Bmatrix} \varepsilon_x \\ \varepsilon_y \\ \gamma_{xy} \end{Bmatrix} \tag{4.12}$$

This condition is called a simple two-dimensional formulation (Yoshida and Ohya 2012) in this book. Conversion of the elastic moduli under this formulation is shown in Table 4.1b, where quantities "∼" are not the same values with the corresponding quantities in the three-dimensional plane. This conversion is made by assuming that G and E are the same in both dimensional planes, which is necessary to keep V_s and V_p the same. Since the bulk moduli in three-dimensional and the simple two-dimensional formulation are different to each other, several differences appear between other moduli. For example, Poisson's ratio is 0.5 under no volume change condition in the three-dimensional formulation, whereas it is 1.0 in the simple two-dimensional formulation. In addition, the one-dimensional rebound modulus B, which relates P-wave velocity, is treated with care. B is calculated as

Three dimensional: $B = K + \frac{4}{3}G$. Two dimensional: $\tilde{B} = \tilde{K} + G = B$

Both B's are the same value from Eq. 4.11, which means that the P-wave velocity is kept even under the simple two-dimensional analysis. The bulk modulus K is, however, frequently used instead of $\tilde{K}$ in Eq. 4.11, in which case the P-waves become different. Finally, relationship between Poisson's ratio and the coefficient of earth pressure at rest K_0 is as follows:

Three dimensional: $K_0 = \nu/(1-\nu)$. Simple two dimensional: $K_0 = \tilde{\nu}$

Poisson's ratio is frequently set 1/3 in the switch-on-gravity analysis to obtain the initial stress because coefficient of earth pressure at rest K_0 becomes 0.5. However, in the simple two-dimensional analyses, it must be 0.5 to make $K_0 = 0.5$.

In the engineering practice of the seismic ground response analysis, this difference does not affect the result significantly, partly because the behavior is predominantly controlled by the shear deformation and partly because the bulk modulus is much larger than the shear modulus in the saturated sand. However, in the case that volume change behavior becomes important or analysis of vertical wave propagation, this difference may become effective.

4.2.6 Confining Stress Dependency

Effective confining stress dependency frequently appears in the geomaterial because of the effective stress principle. Since this book deals with the total stress analysis, the effective stress and the total stress are not clearly distinguished, although they must be distinguished in theory. There are, of course, places where they must be distinguished. The effective stress must be evaluated, for example, by considering the water table when the initial stress is calculated.

The definition of the effective confining stress is shown in Eq. (4.3). Here, let us replace subscript y into v by setting the y-axis as vertical direction and subscripts x and z into h as they are horizontal directions. Then Eq. (4.3) yields

$$\sigma_m = \frac{\sigma_v + 2\sigma_h}{3} \tag{4.13}$$

In addition, let us introduce a coefficient of earth pressure at rest K_0; then

$$\sigma_m = \frac{1 + 2K_0}{3}\sigma_v \tag{4.14}$$

where σ_v is an overburden stress. As the value of K_0 is very difficult to measure in the actual situation, it is frequently assumed that $K_0 = 0.5$, which result in

$$\sigma_m = \frac{2}{3}\sigma_v \tag{4.15}$$

Therefore, the confining stress dependency is frequently replaced by the overburden stress dependency, especially in the one-dimensional analysis. It is noted that this replacement is justified only when K_0 is constant in the whole analysis; it is supposed to change during an analysis. If excess porewater generates, for example, both σ_v and σ_h becomes small with the same amount (generated porewater pressure), which will cause decrease of K_0.

If effective confining stress dependency is replaced into the effective overburden stress dependency, the bulk modulus $\overline{K}$ for one-dimensional analysis becomes

$$\overline{K} = K + \frac{4}{3}G \tag{4.16}$$

from Eq. (4.9). This bulk modulus is exactly the same as one-dimensional rebound modulus defined in Eq. (4.6). Therefore, B is usually used instead of $\overline{K}$.

4.3 Nonlinear Characteristics

Soil shows nonlinear behavior from very small strain, which is one of the characteristics different from the other structural material such as steel and concrete. Figure 4.4 schematically shows various features of the soil regarding the strain dependency. Abscissa is a log scale because it becomes impossible to see small strain behavior in a linear axis. Ordinate shows various informations.

Mechanical properties of the rock and the soil related to strain are written at the top row. Hard material such as hard rock and soft rock is written in the upper part, and soft material such as clay and sand is written in the lower part. All material shows elastic behavior at very small strains the same as other materials. When strain

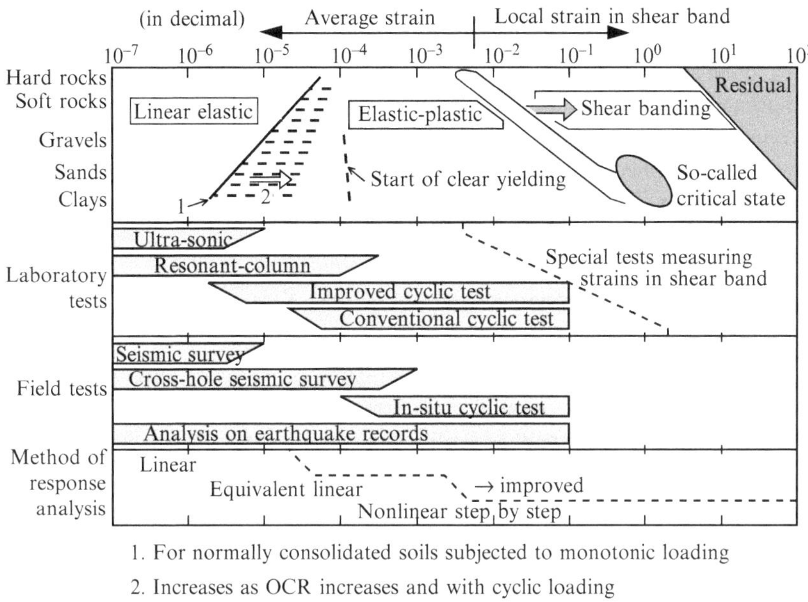

Fig. 4.4 Nonlinear characteristics of soil and its measurement, seismic ground response analysis (Modified from Tatsuoka and Shibuya (1992) and Ishihara (1982))

becomes larger than 10^{-6}, soft material such as clay begins to show nonlinear behavior. Ordinary geomaterial interested in the seismic ground response analysis shows clear nonlinear behavior at strain 10^{-4}, which is less than 1/10 of the yield strain of the steel.

When the strain becomes very large, deformation is not uniform but localizes. Deformation concentrates along a particular plane called a shear band in the hard material. On the other hand, soft material reaches a failure state called a critical state. Fortunately, as maximum strains in the seismic ground response analysis are less than several percent in many cases, the localization of deformation is needless to consider and we can treat geomaterial as continuum.

Laboratory tests are written in the next row. It is generally a difficult task to measure mechanical properties from very small strain to very large strain by one test apparatus, and different test methods have been used depending on the interested strain range. Recent improvement of the test apparatus enables to measure a wide range of strain behaviors and it is already in practice.

Field tests are written in the next row. As will be explained in Sect. 6.5, the elastic modulus is to be measured in situ, not in laboratory. Seismic survey such as the PS logging, described in Sect. 5.2, is the most frequently used method. The cross-hole method is hardly used in Japan partly because of the complicated ground configuration in Japan. In situ cyclic test is also hardly used in the practice of the seismic ground response analysis because the evaluation of the strain is difficult.

Back analysis from the earthquake records can be applied in wide range of strain theoretically. However, there are difficult problems in each strain range. This also relates to the accuracy of the seismic ground response analysis.

Methods of the seismic ground response analysis are written at the bottom row. There are three types of analysis, linear, equivalent linear, and (truly) nonlinear analyses. The applicable strain ranges are also shown in the figure, which will be discussed later in Sect. 11.4 and Chap. 14.

4.3.1 Nonlinear Characteristics Against Shear

Nonlinear behavior against the shear deformation is the most important behavior in the total stress seismic ground response analysis. Therefore, when the term "nonlinear" appears without other explanations, it indicates nonlinear against shear deformation in this book.

Because of the dilatancy, the behavior of soil against shear deformation cannot be discussed by only shear behavior, but behavior of void or porewater is also to be considered. Two extreme conditions are usually used. The one is a drained condition in which water can move freely, and this behavior is similar to dry soil. The other is an undrained test in which porewater cannot come out of the soil element. Behaviors are quite different between two conditions. A typical example is shown in Fig. 4.5. Strain hardening is seen in the drained condition test, which is caused as the soil becomes dense under the cyclic loading. On the other hand, stress–strain curve deteriorates as loading because of the decrease of the effective mean stress.

Detailed behavior is described in Chap. 6.

4.3.2 Nonlinear Characteristics Under Volumetric Change

Figure 4.6 shows an example of the relationships between the void ratio e and the effective stress σ'_m for both sand and clay. These behaviors are usually modeled as in

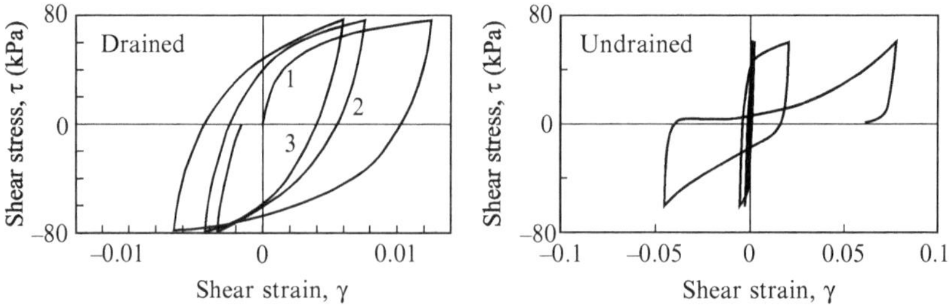

Fig. 4.5 Stress–strain relationships under drained and undrained conditions (After Towhata (1989))

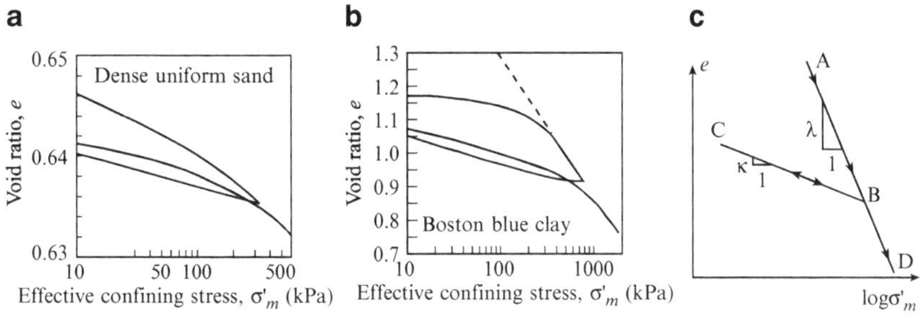

Fig. 4.6 Volume change characteristics and its modeling. (**a**) Sand (After Yoshikuni 1982). (**b**) Clay (After Yoshikuni 1982). (**c**) Modeling

Fig. 4.6c. Here, path A → B → D is an irreversible path called a normal consolidation, where path B → C → B is reversible path called a rebound behavior. Both the normal consolidation and the rebound behaviors are considered in the consolidation analysis as the effective confining stress usually increases. On the other hand, σ_m' does not increase in the seismic ground response analysis, and it decreases in the liquefaction problem. Therefore, only rebound path is usually considered.

Relationship between the volumetric strain ε_v and the void ratio e is expressed as

$$d\varepsilon_v = -\frac{de}{1+e} \tag{4.17}$$

Negative sign is put in the right-hand side as the volumetric strain is positive for compression. Linear relation in $e - \log\sigma_m'$ plane in Fig. 4.6c is written as

$$de = -\mu d\left(\log\sigma_m'\right) = \frac{-\mu d}{\sigma_m'} \cdot \frac{d\sigma_m'}{\ln(10)} \approx \frac{-\mu d}{\sigma_m'} \cdot \frac{d\sigma_m'}{2.3} \tag{4.18}$$

Here, μ is a swelling index (the slope of the line in Fig. 4.6). Substitution of this equation into Eq. (4.17) yields

$$d\varepsilon_v \approx \frac{\mu d\sigma_m'}{2.3\,(1+e)\,\sigma_m'} \quad \text{or} \quad d\sigma_m' \approx \frac{2.3\,(1+e)\,\sigma_m'}{\mu} d\varepsilon_v \tag{4.19}$$

Therefore, effective stress increment is proportional to the volumetric strain increment. This equation has three limitations.

The first limitation is that the void ratio increases to infinity as effective stress tends to zero. The second limitation is that plastic volumetric strain caused by, for example, dilatancy is not included.

These limitations may become significant problem in the effective stress analysis. The effective stress cannot become zero, and the volume change after the excess porewater pressure dissipation is significantly underestimated as the volume change caused by the dilatancy is not accounted for the dissipation analysis although it is considered in the process of the excess porewater pressure generation

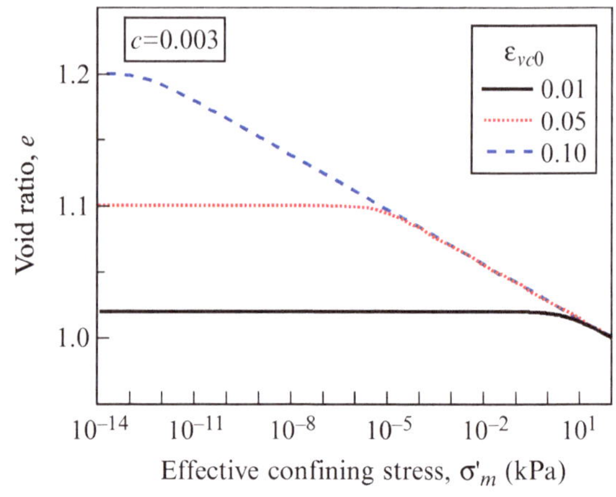

Fig. 4.7 Volumetric change characteristics model that shows upper limit of void ratio (Ohya et al. 2009)

(Yoshida 2003). According to the test result (Unno et al. 2006), the void ratio comes close to the minimum void ratio after the excess porewater pressure dissipation, but it is not considered in Eq. (4.19).

These two limitations are improved by setting the minimum possible void ratio and considering the plastic volumetric change. Ohya et al. (2009) showed, for example, that

$$d\sigma'_m = \left\{ \frac{\sigma'_m}{c} + \frac{\sigma'_{m0}/c}{\exp\left[\left(\varepsilon_{v0c} + \varepsilon_{vd}\right)/c - 1\right]} \right\} d\varepsilon_{vc} \tag{4.20}$$

where ε_{vc0} denotes a volumetric strain due to compression from $\sigma'_m = \sigma'_{m0}$ to $\sigma'_m = 0$, ε_{vd} denotes a volumetric strain due to dilatancy, and c is almost equal to the swelling index μ. A numerical example is shown Fig. 4.7; the relationship is almost linear at the ordinary confining stress same as in Eq. (4.19), but the void ratio reaches upper limit as σ_m tends to zero.

This book deals with the total stress analysis. Therefore, the case that the effective confining stress reaches zero hardly occurs. In this sense, Eq. (4.19) can be used with sufficient accuracy.

The third limitation is that Poisson's ratio depends on the effective stress because powers of the effective stress dependency are different between the shear modulus and the bulk modulus. Therefore, in some cases, power of Eq. (4.19) is set equal to that of the shear modulus, in which case Poisson's ratio becomes constant. This will be discussed in Sect. 8.1.2 again.

References

Fukutake K (1997) Study of three dimensional liquefaction analysis of soil-structure system considering multi-directional nature of shear deformation, Ohsaki Research Report, No. 97–03, Shimizu Corp. (in Japanese)

Ishihara K (1982) Evaluation of soil properties for use in earthquake response analysis. In: Proceedings of the international symposium on numerical models in geomechanics, Zurich, pp 237–259

Ishihara K, Towhata I (1982) Dynamic response analysis of level ground based on the effective stress method. In: Pande GN, Zienkiewicz OC (eds) Soil mechanics – transient and cyclic loads. Wiley, New York, pp 133–172

Lambe TW, Whitman RV (1969) Soil mechanics, SI version. Wiley, New York

Martin GR, Finn WDL, Seed HB (1975) Fundamentals of liquefaction under cyclic loading. FED ASCE 101(GT5):423–438

Ohya Y, Yoshida N, Sugano T (2009) Volume change characteristics of sand considering loading history. J Struct Eng AIJ 55A:405–414 (in Japanese)

Tatsuoka F, Shibuya S (1992) Deformation characteristics of soils and rocks from field and laboratory tests. Key note lectures, 9th Asian regional conference on soil mechanics and foundation engineering, Bangkok, Report of the Institute of Industrial Science the University of Tokyo, vol 37, No. 1, pp 1–136

Towhata I (1989) Models for cyclic loading, mechanics of granular materials, Report of ISSMFE technical committee on mechanics of granular materials, ISSMFE, pp 80–90

Towhata I, Ishihara K (1985) Shear work and pore water pressure in undrained shear. Soils Found 25(3):73–84

Unno T, Kazama M, Uzuoka R, Sento N (2006) Relation of volumetric compression of sand between under drained cyclic shear and reconsolidation after undrained cyclic shear. Doboku Gakkai Ronbunshu C 62:757–766

Yoshida N (2003) Dynamic property of soil and analysis of ground for performance based design. Tsuchi-to-Kiso 51(1):2–7 (in Japanese)

Yoshida N, Ohya Y (2012) Simple two-dimensional formulation for liquefaction analysis. In: Proceedings of the 9th international conference on urban earthquake engineering (9CUEE) and 4th Asia conference on earthquake engineering (4ACEE), pp 457–482

Yoshikuni H (1982) Compression and consolidation of soil, Handbook of soil engineering. JGS, Tokyo, pp 147–221 (in Japanese)

Chapter 5
In Situ Soil Testing

Mechanical properties of the geomaterial frequently change during the sampling and the traveling as will be described in Sect. 6.5. Therefore, it is recommended to obtain the mechanical properties in situ as possible. As shown in Fig. 4.4, however, an in situ test is difficult to obtain the behavior at large strain. The elastic modulus is to be measured in situ at minimum.

In situ tests frequently used to obtain the mechanical properties are as follows:

1. Standard penetration test (SPT)
2. Cone penetration test (CPT)
3. PS logging
4. Surface wave exploration
5. MWD (measurement while drilling) method

Among them, the SPT is a standard method in almost all sites in Japan and a many other countries. The CPT is also frequently used outside Japan. The PS logging becomes to be used in addition to the SPT/CPT to measure wave velocities. The surface wave exploration and the MWD method are hardly used alone to obtain the in situ data required for the seismic ground response analysis. They are usually used to grasp change of the material in space because they are supposed to be less reliable to grasp the mechanical properties compared with the first three methods mainly because of the lack of experiment, but they are cheaper.

5.1 Standard Penetration Test

A standard penetration test is the most common method in Japan. The number of blow count, SPT -value , is measured to drive the sampler by 30 cm by repeated falling of the weight with 63.5 kg at 75 cm high. One of the advantages of this

N. Yoshida, *Seismic Ground Response Analysis*, Geotechnical, Geological and Earthquake Engineering 36, DOI 10.1007/978-94-017-9460-2_5

method is that the soil sample remains in the sampler. Since soil conditions are generally complicated in Japan, this is a big advantage and this becomes one of the reasons why SPT is a standard method in Japan.

At the same time, there are limitations of this method. One of the big issues is the meaning of the SPT N-value. It is easily guessed that it relates to the failure state, and the failure strength has some correlation with the elastic modulus (Sato et al. 1990). Therefore, SPT N-value relates to the strength and the elastic modulus. From the practical point of view, lots of empirical equations have been proposed based on the huge experiences. It is, however, noted that error included in the empirical equations is sometimes large. Therefore, the SPT N-value is to be used with care although it is convenient to be able to obtain mechanical properties through empirical equation.

The result of the SPT is called an SPT N-value in this book, but the name "blow count" is also used in practice.

5.1.1 Energy Correction

The practical method of the standard penetration test is not standardized although its principle is clear. Various methods are used in the world; several methods are sometimes used even in the same country. Differences of the methods appear as the difference of the energy of free fall transferred into the rod. In addition to the effectiveness of the methods, the sharpness of the sampler head also affects blow count significantly.

Table 5.1 summarizes rod energy ratios in several countries (Seed et al. 1985). The prevalent method in the USA uses safety, and hammer release is made by rope and pulley. The energy ratio is 60 %. A method that uses donut and hammer release by the rope is prevalent in Japan. In addition, an automatic mechanical release method began to be used. These differences indicate that the engineer must be careful when using the empirical equations developed in other countries or different energy ratios. If one applies an empirical equation developed in USA for

Table 5.1 Rod energy ratio for SPT procedures

Country	Hammer type	Hammer release	Estimated rod energy (%)
Japan	Donut	Free fall	78
	Donut[a]	Rope and pulley with special throw release	67
United States	Safety[a]	Rope and pulley	60
	Donut	Rope and pulley	45
Argentina	Donut[a]	Rope and pulley	45
China	Donut[a]	Free fall	60
	Donut[a]	Rope and pulley	50

After Seed et al. (1985)
[a]Prevalent method in this country today

site characterization in Japan, the SPT N-value must be increased by 67/60 $\approx$ 1.12 times in Japan. However, this correction is hardly made in the engineering practice partly because using the small value is conservative side and partly because the method or energy ratio is not so relevant. In addition, error of 10 % seems within the allowable range considering the method of SPT.

Energy ratio is smaller than that shown in Table 5.1 in some countries. The SPT N-value in the Philippines and Turkey, for example, is about twice as large as that in Japan (Kiso-jiban Consultants 1990; Kiku et al. 2002). This does not indicate that energy ratio is always small in these countries because they are just one case, but it is also true that small energy ratio was found. This may not be a problem of the energy absorption but relates to the sharpness of the installer. In addition, it is also frequently said that blow count depends on the existence of the supervisor at the test. If the SPT N-value is two times larger, for example, shear strength and elastic modulus will be evaluated significantly large, which may result in a critical side design.

5.1.2 Effective Confining Stress Dependency

The SPT N-value increases with the increase of the effective confining stress even in the same material. Therefore, observed SPT N-value should be corrected when necessary. Three methods have been used in Japan. Correction is usually made to evaluate the liquefaction potential, for example, but correction is hardly made in the field of total stress, partly because many empirical equations are developed without considering the confining stress dependency:

1. Meyerhof's method
   Meyerhof showed a relationship between the SPT N-value and the relative density D_r (Meyerhof 1957)

$$N = 1.7D_r^2\left(\sigma'_m + 10\right)\left[\mathrm{lb/in^2}\right] \tag{5.1}$$

where σ'_m denotes an effective confining stress. This equation stands in lb/in.^{2} unit. It is converted into the SI unit system as

$$N = \frac{D_r^2\left(\sigma'_m/98 + 0.7\right)}{441}, \text{ or } D_r = 21\sqrt{\frac{N}{\sigma'_m/98 + 0.7}} \tag{5.2}$$

On the other hand, the liquefaction strength R is related to D_r. Tatsuoka et al. (1978), for example, proposed

$$R = 0.0042D_r \tag{5.3}$$

Equations 5.2 and 5.3 indicate that liquefaction strength depends on the effective confining stress. Then the following conversion equation is derived under the reference confining stress of 98 kPa:

$$N_c = \frac{1.7N}{\sigma'_m/98 + 0.7} \tag{5.4}$$

where N_c denotes a corrected SPT N-value under the effective overburden stress of 98 kPa. Here, as it is very difficult to obtain the in situ confining stress, the confining stress dependency is frequently modified into the overburden stress σ'_v dependency because overburden stress can be obtained with ease and certain accuracy. Then,

$$N_c = \frac{1.7N}{\sigma'_v/98 + 0.7} \tag{5.5}$$

It is noted that the relationship between σ'_m and σ'_v in Eq. 4.14 is not considered in this expression.
The value 98, which corresponds to ten times of acceleration of gravity in meter, is frequently replaced into 100 in the engineering practice. This equation is used in the design specification of road bridges (Specifications for Highway Bridges 2002).

2. Equation by Liao et al.
Liao et al. gathered several empirical equations to consider the effective confining stress dependency and proposed the following equation that is simple and has nearly the same accuracy with other empirical equations (Liao and Whitman 1986):

$$N_c = \sqrt{\frac{98}{\sigma'_v}} \cdot N \tag{5.6}$$

This equation evaluates SPT N-value at $\sigma'_v = 98\,\text{kPa}$. It is employed in the design specification of foundation of building (AIJ 2001).

3. Equation by Iai et al.
Iai et al. gathered data on the confining pressure dependency in the port areas in Japan. They selected the data of liquefiable soil (Technical standards and commentary of port and Harbour facilities in Japan 1989) for port facilities and proposed (Iai et al. 1986) that

$$N|_{65} = \frac{N - 0.019\left(\sigma'_v - 65\right)}{0.0041\left(\sigma'_v - 65\right) + 1.0} \tag{5.7}$$

Here SPT N-value is corrected at the effective overburden pressure of 65 kPa.

5.2 PS Logging

The PS logging is one of the best methods to measure the wave velocities as wave velocities are directly measured. Elastic moduli are calculated from wave velocities as

$$\begin{aligned} G &= \rho V_s^2 \\ B &= K + \tfrac{4}{3}G + \tfrac{K_w}{n} = \rho V_p^2 \end{aligned} \tag{5.8}$$

where ρ denotes a density (sum of the soil particle and the porewater), G denotes the elastic shear modulus, K denotes the bulk modulus of soil skeleton, K_w denotes the bulk modulus of water, n denotes porosity, and B denotes the one-dimensional rebound modulus corresponding to the total stress. If there is no water, the term K_w/n is omitted, and B becomes the same as shown in Table 4.1. The existence of the water is considered in V_p, but is not considered in V_s because the porewater resists against the volume change but not against the shear deformation.

Poisson's ratio ν is calculated to be

$$\nu = \left\{1 - 2\left(\frac{V_s}{V_p}\right)^2\right\} \Big/ \left\{2 - 2\left(\frac{V_s}{V_p}\right)^2\right\} \tag{5.9}$$

Poisson's ratio again corresponds to the mixture of the soil particle and the porewater for the soil under the water table, and generally it is greater than 0.45 (Report of Task force on countermeasures against Tokai earthquake). If it becomes less than 0.45, soil is a significant unsaturated state.

Three methods shown in Fig. 5.1 are frequently used in the PS logging depending on the location of the vibration source and receiver. Among them the downhole method and the suspension method are usually used, but the cross-hole method is hardly used in Japan partly because Japanese ground is complicated in general and partly because it is costly.

The result of the downhole method is schematically shown in Fig. 5.2. The vibration is caused at the ground surface, and it is received at the underground receiver. Measurements are carried out by changing the depths of the receiver. Then, waveforms are obtained as shown in Fig. 5.2a. In the actual situation, several

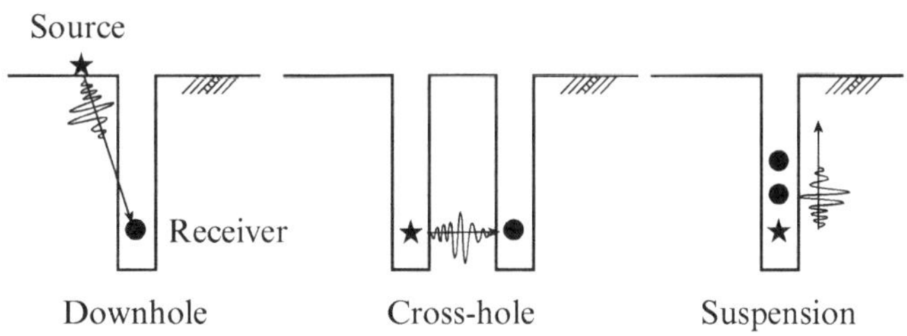

Fig. 5.1 Schematic figure showing methods of PS logging

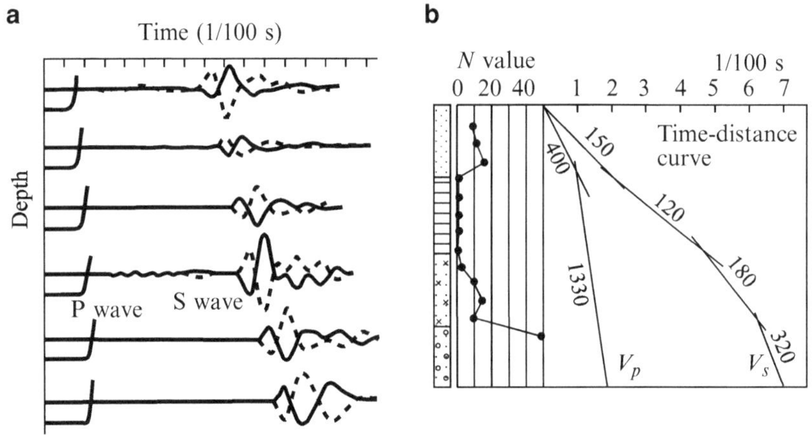

Fig. 5.2 Schematic figure showing downhole method in PS logging (Modified from Sakai 1985). (**a**) Waveforms. (**b**) Time–distance curve

measurements are made in one depth in order to increase the signal-to-noise ratio, which is called the stacking. Then the arrival time is measured and relationships between the arrival time and depths are drawn as time–distance relationships as shown in Fig. 5.2b. The wave velocity is obtained from the slope of the line.

In the practice, small disturbance may be neglected as a linear line is drawn by hand. In other words, wave velocities are set constant in certain thickness, and the depths that the velocity changes may not fit the layer boundary based on the soil classification.

A suspension sonde in which vibration source and two receivers are installed is used in the suspension method. High-frequency pulse waves (several hundred Hz) are generated from the source. The wave velocity is calculated from the delay of the arrival time by two receivers. This method is available only under the water, and error becomes large when side wall of the borehole is disturbed.

Comparison between the two methods is shown in Fig. 5.3. Wave velocities scatter by the suspension method, whereas it is constant in large thickness by the downhole method. The wave velocities by the downhole method seem to be an average of those by the suspension method. Another example is shown in Fig. 5.5 later, which also has the same feature.

The wave velocity distribution by the suspension method in Fig. 5.3 is similar to a linear line drawn by hand, SPT *N*-value distribution, which indicated that the suspension method catches local change of the wave velocity well. Recently, however, average wave velocity in the layer becomes to be output as a result of the suspension method, in which case detailed information is lost.

Since there are differences between two methods, a question "which method is better" may arise, but it is difficult to answer it. The suspension method may be accurate at the borehole point, but obtained structure may not spread in the horizontal direction. The soil profile used in the seismic ground response analysis is to be a representative one at the site. In this sense, downhole method has advantages,

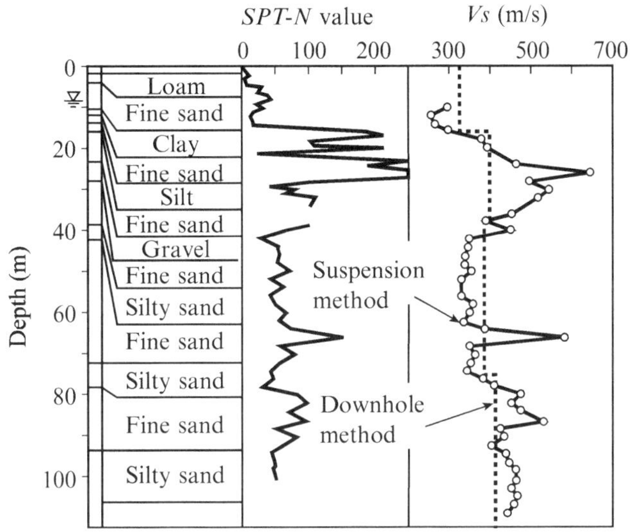

Fig. 5.3 Comparison between downhole method and suspension method (After Kokusho 1992)

but it may miss the existence of weak layer which affects the response very much as the upper bound acceleration. A good example of how these limitations appear in the practical calculation will be shown in Sect. 14.4. Through this case study, it is important to modify the soil profiles and the wave velocities so that it becomes a representative one can be recognized.

It is important to obtain the elastic moduli from the wave velocities by the PS logging. It can be obtained from the SPT *N*-value through empirical equations shown in Sect. 7.1 and from the laboratory test, too, but it is not suggested because there may be significant error. They will be discussed in Sect. 6.5.

5.3 Other Methods

The standard penetration test is a popular method in Japan, and the cone penetration test is hardly made in the engineering practice.

In addition to the standard penetration test, PS logging, and laboratory test by means of an undisturbed sample, a new method, called MWD (measurement while drilling), began to be used. Resistance of the soil during the boring is measured, and mechanical properties are estimated from the resistance. Material properties can be measured continuously along the depth in MWD. Since these methods are cheaper than the standard penetration test, it is used to obtain the ground configuration or cross section in a wide area. Detailed or accurate data is obtained by, for example, SPT, and data by MWD are used to interpolate them in space. It is hardly used alone to obtain the mechanical property for the seismic ground response analysis. The surface wave exploration is also used in the same purpose.

Empirical equations will also be developed against the MWD. Here, there are cases that the S-wave velocity is directly evaluated from resistance (Robertson et al. 1986) and that the SPT N-value is first evaluated (Yamada et al. 1992; Inada 1960) and the empirical equations from SPT N-value to V_s are used. In the latter case, error from the MWD to the SPT and that from the SPT to the wave velocity may be cumulated. In addition, an empirical equation may be derived so that SPT N-value becomes in the conservative side, but not an average value, in which case error may become more. Therefore, the engineer should be careful in using these methods because a smaller evaluation does not result in a safe side and vice versa in the seismic ground response analysis.

5.4 Geological Age of Soil

The subsurface layers considered in the seismic response analysis are frequently classified into the alluvial and the diluvial layers in Japan. These terms came from Germany and are used very frequently such as As (alluvial sand), Dc (diluvium clay), and Ag (alluvial gravel). This classification originally came from the disposal condition: alluvial for marine deposit and diluvial for fluvial deposit. These classifications may fit the geological age in Europe but does not fit in Japan and probably in many countries outside Europe. Therefore, for the international common acceptance, it is better to use a geologic age for the classification of soils.

Comparison between geologic ages is shown in Fig. 5.4. Generally speaking, the alluvial deposit corresponds to the Holocene period, and the diluvial deposit corresponds to the Pleistocene period. However, there is some difference between them because the alluvial deposit starts at the end of recent ice age (18,000 years ago), whereas the Holocene period starts 10,000 years ago. It is also noted that some

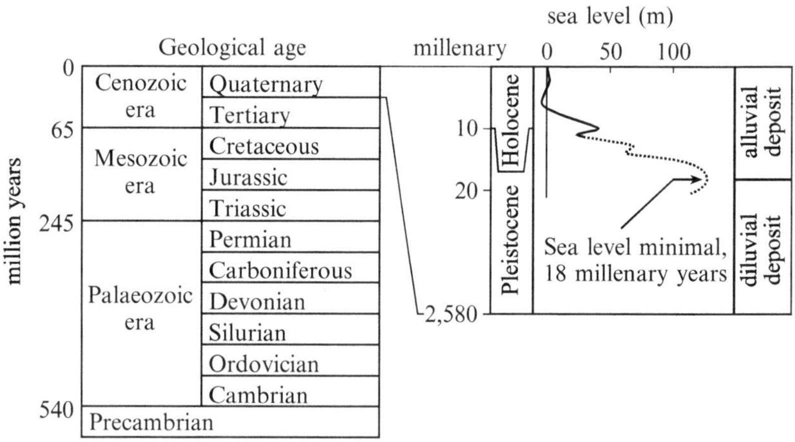

Fig. 5.4 Geologic age and classification based on depositional condition (After JGS 1998; National Astronomical Observatory of Japan 2002)

researchers in Japan separate alluvial and diluvial the same as the Holocene and the Pleistocene periods, i.e., at 10,000 years ago (JGS 1998).

5.5 Continuity of Soil Layers

It is a fairly difficult task to obtain the accurate soil profile of the ground. Layers may not be horizontally deposited. Lens shape layer may exist. Boundary between the layers may not be clear. The engineer cannot know detailed soil profile because, in many cases, soil profile is evaluated from the limited number of borehole data. A result of the detailed field and laboratory investigations (Masuda et al. 2001) is introduced in this section to show how material changes in the ground.

Result of the soil survey is summarized in Fig. 5.5. The site is filled in the Tokyo Bay area, Japan. Several borehole investigations are carried out in short distance.

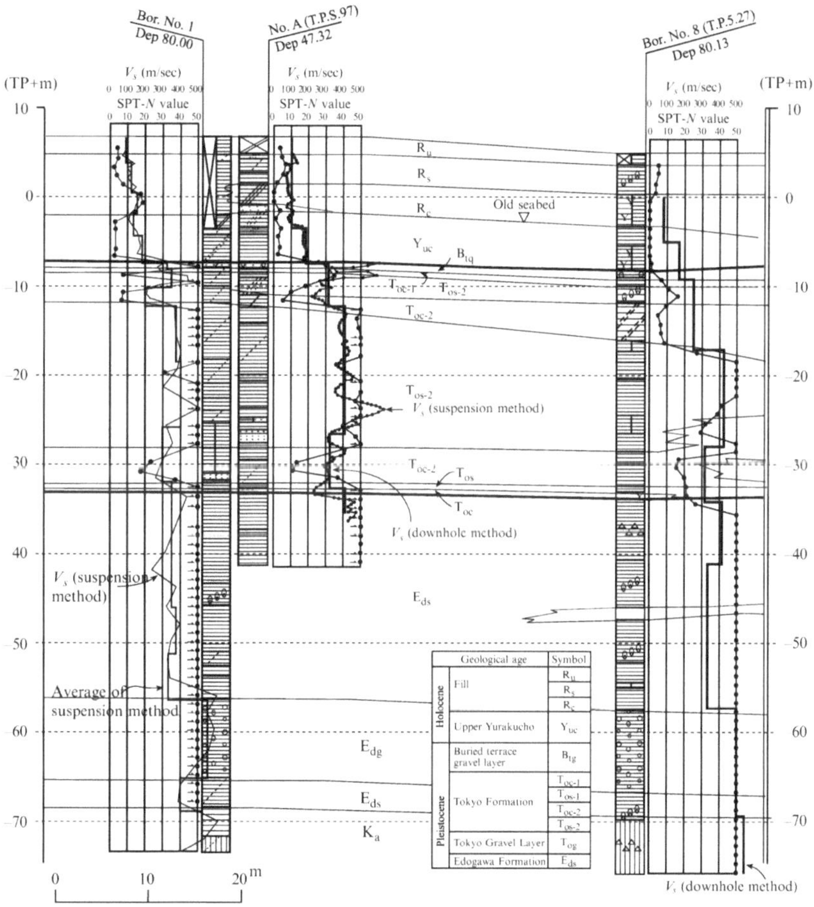

Fig. 5.5 Soil profiles of the site

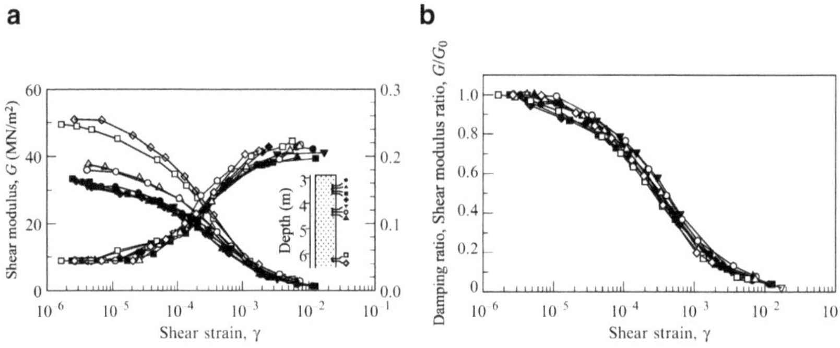

Fig. 5.6 Change of fill material as depths. (**a**) Cyclic shear deformation characteristics. (**b**) Normalized shear modulus ratio

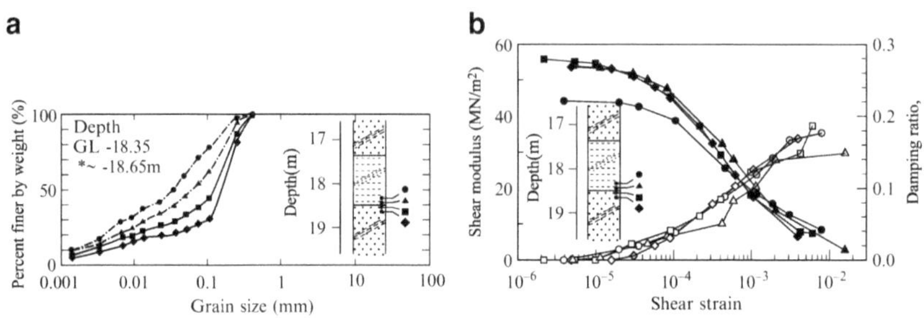

Fig. 5.7 Boundary between Pleistocene sand and silt. (**a**) Distribution of fines. (**b**) Cyclic shear deformation characteristics

There are layers with the same material at deep depths (deeper than old seabed), but these layers are not horizontally layered deposit. Soils are fairly different above the old seabed as this is fill ground.

Soils were sampled continuously up to GL-20 m, and the cyclic shear deformation characteristics tests were carried out. Some result of the tests is shown in Fig. 5.6 (see Chap. 6.3 as well). The shear modulus increases with depth, whereas the damping characteristics are almost the same. The shear modulus is normalized by the maximum shear moduli and is shown in Fig. 5.6b. All the data lie in one line, which suggests that the difference of the shear modulus in Fig. 5.6a comes from the confining stress dependency of the elastic modulus and the G/G_0-γ curve does not depend on the confining stress.

The percent finer by weight and cyclic shear deformation characteristics of the soil layers between the boundary of the Pleistocene sand and the Pleistocene silt layers are compared in Fig. 5.7. The percent finer changes gradually and no clear boundary is observed. The same feature can also be seen in the cyclic shear deformation characteristics. This indicates that sedimentation occurred continuously by changing little by little. However, a boundary is required by considering the difference between the upper and the lower material in the practical situation. Therefore the depth of boundary is somewhat accidental; the boundary can be a little upper or lower than that in Fig. 5.7. However, the engineer usually treats this boundary a truly boundary by which the material clearly changes.

There are, of course, cases that a clear boundary exists. For example, mechanical properties are shown to change significantly between the old seabed and the fill in the same report.

References

AIJ (2001) Recommendations for design of building foundations, 2001 Revision, AIJ, Tokyo, 486 pp (in Japanese)

Iai S, Koizumi K, Tsuchida H (1986) A new criterion for assessing liquefaction potential using grain size accumulation curve and N-value. Report of the Port and Harbour Research Institute, Yokosuka, No. 11, pp 125–234 (in Japanese)

Inada M (1960) On test result of Swedish sounding test. Tsuchi-to-Kiso 8(1):13–18 (in Japanese)

JGS (1998) Diluvial layer, Geotech note 8, 98pp (in Japanese)

Kiku H, Sawada S, Yasuda S, Yoshida N (2002) In-situ penetration tests in Adapazari damaged by 1999 Kocaeli earthquake in Turkey. In: Proc. 11th Japan symposium of earthquake engineering, Tokyo, pp 43–46 (in Japanese)

Kiso-jiban Consultants: investigation report of Luzon earthquake of 16, July 1990, 87 pp (in Japanese)

Kokusho T (1992) Dynamic characteristics of ground, lecture: analytical method of the interaction between ground and structure. Tsuchi-to-Kiso 40(No. 4):76–84 (in Japanese)

Liao SSC, Whitman RV (1986) Overburden correlation factors for SPT in sand. J Geotech Eng ASCE 112(3):373–377

Masuda T, Yasuda S, Yoshida N, Sato M (2001) Field investigations and laboratory soil tests on heterogeneous nature of alluvial deposits. Soils Found 41(No. 4):1–16

Meyerhof GG (1957) Discussion. In: Proceedings of the 4th international conference on soil mechanics and foundation engineering, Zurich, vol 3, p 110

National Astronomical Observatory of Japan (2002): Chronological scientific tables, Maruzen, 2009 (in Japanese)

Report of Task force on countermeasures against Tokai earthquake, Central Disaster Prevention Council

Robertson PK, Campanella RG, Gillespie D, Rice A (1986) Seismic CPT to measure in situ shear wave velocity. J Geotech Eng 112(8):791–803

Sakai K (1985) Plan and management of ground survey for earthquake resistant design, Kajima Institute Publishing, Tokyo, 109pp (in Japanese)

Sato T, Kim Y-S, Shibuya, S, Tatsuoka F (1990) Correlation of deformation modulus to the shear strength for various geotechnical engineering materials. In: Proceedings of the 25th Japan National Conference on SMFE, JSSMFE, Okayama, vol I, pp 718–718 (in Japanese)

Seed HB, Tokimatsu K, Harder LF, Chung RM (1985) Influence of SPT procedure in soil liquefaction resistance evaluations. J Geotech Eng III(12):1425–1435, ASCE

Specifications for highway bridges, Part V, seismic design, Japan Road Association, 2002 (in Japanese)

Tatsuoka F, Iwasaki T, Tokida K, Yasuda S, Hirose M, Imai T, Kon-no M (1978) A method for estimating undrained cyclic strength of sandy soils using standard penetration resistances. Soils Found 18(No. 3):43–58

Technical standards and commentary of port and Harbour facilities in Japan, The Overseas Coastal Area Development Institute of Japan, 1989

Yamada K, Kamao S, Yoshino H, Masuda Y (1992) Correlations between N-value and cone index. Tsuchi-to-Kiso 40(8):5–10 (in Japanese)

Chapter 6
Laboratory Test and Assemble of Test Result

In situ test as a tool to obtain the elastic modulus was discussed in the previous chapter. In addition to the elastic soil properties, it is necessary to determine the nonlinear characteristics of soil for the seismic response analysis. These characteristics can only be evaluated by the laboratory tests.

Laboratory tests consist of physical tests to obtain void ratio, plasticity index, liquid limit, grain-size distribution, etc., and mechanical tests to obtain stress–strain relationships and strength characteristics such as internal friction angle and cohesion. The procedure to obtain the stress–strain relationships for the seismic response analysis is described throughout this chapter. It is quite different from the one used in the ordinary structural material such as steel and concrete. As previously discussed, the soil behavior is very complicated; it is impossible to depict the whole behavior by a single test. The behavior cannot be represented by a few parameters, which is also a different feature from the ordinary structural elements. For example, Young's modulus and yield stress may be sufficient for steel, but many parameters related to consolidation, load carrying capacity, liquefaction, etc., are necessary to understand the whole behavior or soil. In the field of geotechnical engineering, therefore, tests are carried in order to grasp the important characteristics that are of interest from a geotechnical standpoint. Volume change characteristics, for example, are the most important in the consolidation analysis, whereas the shear strength is of great interest in the stability analysis. In the same manner, in the seismic ground response analysis, cyclic shear deformation characteristics are the most important behavior.

6.1 Soil Sampling

Various methods have been used for sampling of soils. They are explained in detail by JGS (1995), for example, among which methods those appear in the practice of the earthquake geotechnical engineering are focused on in this section.

N. Yoshida, *Seismic Ground Response Analysis*, Geotechnical, Geological and Earthquake Engineering 36, DOI 10.1007/978-94-017-9460-2_6

Proper soil sampling is very important to obtain correct mechanical property of soil as will be explained in Sect. 6.5. Several methods have been used in practice depending on the type and conditions of soils.

The simplest sampling procedure is to use a thin-walled sampler. The thin-walled sampler is a circular hollow cylinder with a thin-walled thickness. It is penetrated into the soil quasi-statically without any rotation. This method is relevant for soft clay or sand with large amount of fines. If grain size of soil becomes large or ground becomes hard, thin-walled sampler becomes difficult to penetrate. A double-tube or triple-tube sampler is preferred for soils of this type. The double-tube sampler is used in the stiff clay, in which an external tube is rotated in order to cut away the soil and an inner tube is penetrated in the double-tube sampler. On the other hand, a liner is prepared in an inner tube in addition to the apparatus of the double-tube sampler, and the soil sample is stored in this liner in the triple-tube sampler, which is used for hard clay and sand.

These methods are frequently specified as tube sampling methods, and the samples obtained from these procedures are described as tube sample since soil sample is stored in the tube. Tube samples are treated as undisturbed in many engineering practice, but, as explained in Sect. 6.5, these samples are usually disturbed to an extent.

A frozen sampling method is one of the most highly undisturbed soil sampling methods (Tokimatsu and Oh-hara 1990). The ground is frozen by pumping liquid nitrogen down the vertical pipes placed in the sampling area by extruding the existing water on the location to exteriors. Unfortunately, it hardly is employed in the engineering practice, because it is limited for use with less fine sands and it is very costly.

The rather inexpensive methods are also proposed in the stage of development to get samples less disturbed than the tube samples, such as GP sampler (Sakai et al. 2006) and GS sampler (Research committee on early effective use of wasted material fill 1991). These methods have the advantage to be applied on a variety of soils such as soils with large fines contents and unsaturated soil.

A block sampling method is used to take samples near the ground surface. Soil is excavated up to the upper depths of sampling, and a sampler is driven statically. A large size sample can be taken by this method.

6.2 Physical Tests

General properties of soil are evaluated by the physical test. Results of a grain-size analysis, a liquid limit test, and a plastic limit test are frequently used in practice in the seismic ground response analysis.

The particle-size distribution of soil is obtained by the grain-size analysis, and the result is used to separate soil particles into size range such as clay, silt, sand, and gravel. Percentage of clay and silt in soil sample is called the fines fraction content or simply fines content and is usually denoted as F_c. Fines content affects liquefaction strength significantly. It is also used to classify soil. When the fines content is less than 50 % and sand is prone, the sample is termed sandy soil. On the other hand, when $F_c \geq 50$ % and clay is prone, the sample is termed clayey soil. Sandy and clayey soils are frequently simply called as sand and clay, respectively.

Another commonly used parameter while describing the cyclic shear deformation characteristics of soil is the plasticity index. The liquid limit w_L is the water content at the boundary between the liquid and the plastic state, and the plastic limit w_P is the water content at the boundary between the plastic and the semisolid state. The plasticity index I_p is then defined to be the difference between those two limits as

$$I_p = w_L - w_P \tag{6.1}$$

Characteristics as sand become predominant when I_p is small, whereas those as clay become predominant when I_p is large. The boundary between two cases is approximately at $I_p = 30$–40.

6.3 Cyclic Shear Deformation Characteristics Test

Behavior of soils under cyclic loading such as seismic loads is expressed by the cyclic shear deformation characteristics. Test to obtain cyclic shear deformation characteristics use a very complicated loading procedure compared to a test under a static loading because it should express behaviors including repetitive unloading and reloading states.

In Japan, the cyclic deformation characteristics of soil under the shear deformation are obtained through the laboratory test explained in JGS (2009). Each test is named based on the test apparatus adopted for that test, e.g., cyclic triaxial test and cyclic torsional shear test. In engineering practice, the general name "cyclic shear deformation characteristics test" is preferred for describing each of these tests. The classical name "dynamic deformation characteristics tests" is also frequently used in practice. Here, it is noted that the term "dynamic" does not indicate dynamic loading but indicates cyclic loading. Conversely, the term "static" is commonly used for monotonic loading.

The most frequently used loading method in Japan is schematically shown in Fig. 6.1. Constant amplitude cyclic shear stress is applied for 11 cycles as shown in Fig. 6.1a under the undrained loading condition with a 0.1 Hz loading speed, and the hysteresis loop at the tenth cycle of loading is used to evaluate the nonlinear characteristics or the cyclic shear deformation characteristics, because the hysteresis loop usually becomes stable under tenth cycle of loading. Then shear modulus G is calculated as the secant modulus, i.e., the slope of the line connecting the two unloading points of the hysteresis loop in Fig. 6.1b. The area of the hysteresis

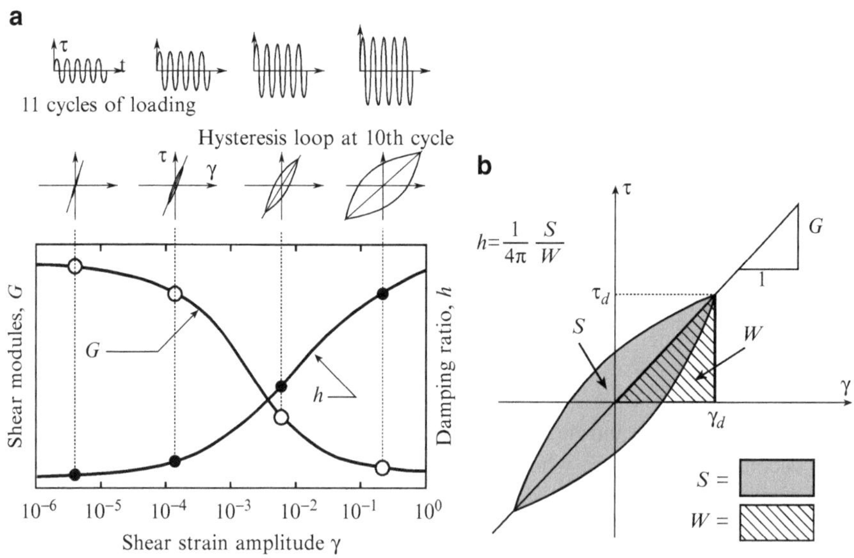

Fig. 6.1 Schematic figure showing cyclic shear deformation characteristics test (**a**) Test method and expression. (**b**) Shear modulus and damping ratio

loop S, which is equal to the energy dissipated by the nonlinear or the plastic behavior, is divided by the area of triangle (shaded area) indicating maximum strain energy $W = \tau_d \gamma_d/2$ in order to obtain damping characteristics, where τ_d and γ_d are amplitude of the shear stress and the shear strain, respectively. Then damping ratio h becomes

$$h = \frac{1}{4\pi}\frac{S}{W} \tag{6.2}$$

The damping ratio is often termed as the damping constant. The factor $1/(4\pi)$ is necessary so that the damping ratio has the same meaning with the damping constant as velocity proportional damping. Single-degree-of-freedom systems with the velocity proportional damping and with the hysteretic damping show the same energy absorption under the cyclic loading if their damping ratios are same. Note that damping coefficient is a coefficient to be multiplied with the velocity in order to evaluate the velocity proportional force, and it is a different quantity from the damping ratio or the damping constant

The cyclic shear deformation characteristics tests initiate with small amplitude shear stress levels. Then the amplitude of shear stress is increased in little increments, and this procedure is repeated until certain shear strain is usually in the order of 1 %. This multistage testing is named a stage test. It is convenient for practice because it uses only one test specimen.

Obtained shear modulus G and damping ratio h are plotted as a function of shear strain amplitude γ as illustrated in Fig. 6.1a. It is also possible to view G and h in separate figures, but conventionally these two characteristics are frequently drawn in one figure as in Fig. 6.1a, in which case the ordinate on the left-hand side denotes shear modulus and ordinate of the right-hand side denotes damping ratio. The abscissa is the shear strain amplitude, which is calculated through dividing the difference between the maximum and the minimum strains in the hysteresis loop by 2. The abscissa is plotted in logarithmic scale because the behavior at small strains cannot be well expressed in a linear scale. Relationships in Fig. 6.1a are occasionally named the G–γ and the h–γ relationships (or curves). The G–γ curve is also termed as the shear modulus reduction curve, but this term is hardly used in Japan as it does not include any information of the damping characteristics.

Generally speaking, the shear modulus decreases and the damping ratio increases with increasing shear strain amplitudes, because, as schematically shown in Fig. 6.1a, soils deteriorate and the area of the hysteresis loop increases by the nonlinear behavior. The maximum possible damping ratio is $2/\pi(\approx 63.7\,\%)$ which occurs in the rigid–perfectly plastic model.

Precisely speaking, terms "nonlinear" and "plastic" refer to different notions. The G–γ curve shows nonlinear characteristics, while the h–γ curve results from the plastic, not nonlinear, behavior. For simplicity, however, the term "nonlinear" is used to express both behaviors in this book because nonlinear behavior is in fact associated with plastic soil behavior.

Excess porewater pressure may develop under the undrained loading conditions. This generated excess porewater pressure is dissipated between the consecutive stages in the stage test. This process indicates that obtained mechanical property is affected by the past loading histories. The stage test has an advantage that it uses only one test specimen, which is important in engineering practice, but it has limitation that loading histories accumulate. A test that uses a new test specimen at all stages is called a fresh test or a single-stage test, but this is not performed in the engineering practice because it requires many test specimens. Difference between the stage and the fresh tests is further discussed in Sect. 6.7.2.

Excess porewater pressure is generated not only in the sandy soils but also in clayey soils. It takes a long time for clayey soil to entirely drain the excess porewater pressure. Therefore, in practice, dissipation is terminated in a certain state before moving the next stage.

6.4 Test Apparatus

Test apparatuses for the cyclic shear deformation characteristics tests are explained in this section. Among them, the cyclic triaxial test apparatus is the most widely adopted apparatus in practice, and the cyclic torsional shear test apparatus is frequently used in the geotechnical research field.

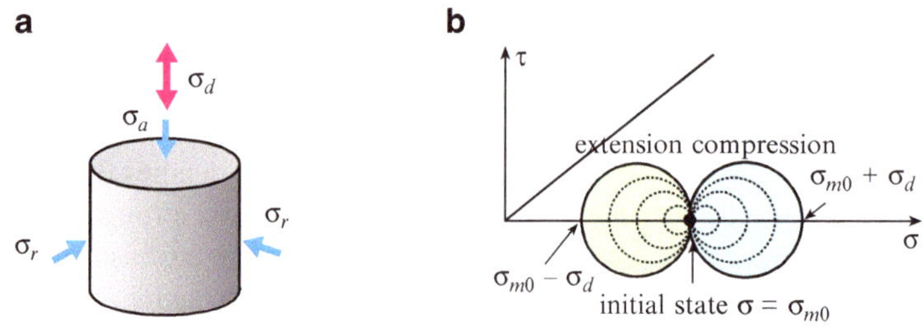

Fig. 6.2 Loading method of triaxial test. (**a**) Method of loading. (**b**) Change of stress state (*dotted circles* denote transient stress states)

6.4.1 Cyclic Triaxial Test

In practice, cyclic shear deformation characteristics test is usually carried out using the triaxial test apparatus. The cylindrical test specimen shown in Fig. 6.2a is used in this test. A soil sample is saturated with the water, and it is covered with a rubber membrane sleeve so as to prevent uncontrolled water flow from the specimen. Both the axial stress σ_a and the lateral stress σ_r are set equal (isotropic stress state) as initial test condition. They are set same as in situ overburden stress in many cases, but test with $\sigma_m^{'} = 100\,\text{kPa}$ (or 98 kPa = 1 kgf/cm^2 in the older test) as a generic value is also seldom used. Then cyclic axial stress σ_d is applied keeping σ_r constant. Figure 6.2b shows the change of stress state during the test by means of Mohr's circle. At the beginning of the test, stress state is expressed as a point since $\sigma_r{=}\sigma_a$. When axial stress σ_d is applied, the size of Mohr's circle increases; the diameter of the circle equals to the deviator stress (difference between the axial and lateral stresses σ_d).

The location of Mohr's circle moves to right (compression side) or left (elongation side) depending on the positive or negative σ_d value. Therefore, Mohr's circle moves point symmetrically with respect to the initial stress point. Notations $\sigma_m^{'}$ and σ_d are used to ensure consistency throughout this book, but, in many textbooks, they are denoted by p and q.

Only the lateral and the axial stresses are applied during testing shear stress work in the test specimen as radius of Mohr's circle is not zero. Maximum shear stress occurs 45° from the principal axis or axial direction of loading σ_d. Therefore, applied shear stress τ_d is calculated as

$$\tau_d = \sigma_d/2 \tag{6.3}$$

Determination of shear strain involves a little more complicated calculation. Let us consider that the axial strain ε_a and the lateral strain ε_r result from loading. Then volumetric strain becomes $\varepsilon_v = \varepsilon_a + 2\varepsilon_r$. Volume change hardly occurs because the bulk modulus of water is very large under the undrained condition test. Therefore, the lateral strain is evaluated under the assumption of $\varepsilon_v = 0$ as

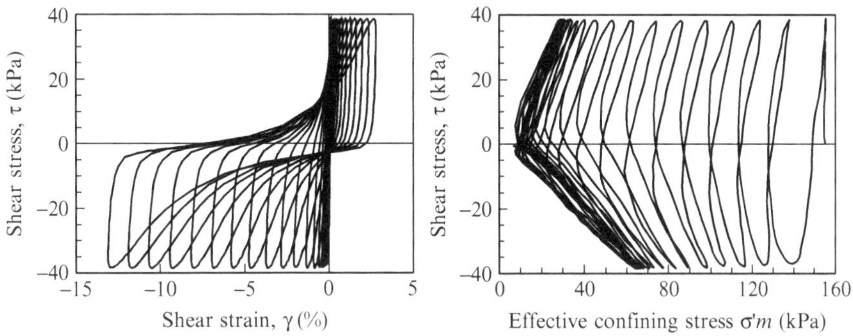

Fig. 6.3 Stress–strain curve and stress path in triaxial test

$$\varepsilon_r = \frac{(\varepsilon_v - \varepsilon_a)}{2} \approx \frac{-\varepsilon_a}{2} \tag{6.4}$$

Shear strain is then obtained as the difference of axial strain and lateral strain, yielding

$$\gamma = \varepsilon_a - \varepsilon_r \approx 1.5\varepsilon_a \tag{6.5}$$

It should be note that tensor shear strain is half of the radius of Mohr's circle by strain and that engineering strain is double of the tensor strain in the calculation of the engineering shear strain.

In cyclic triaxial test, confining stress changes as the shear stress changes with changing shear stresses. This results in changing of mechanical properties such as the stiffness and the strength as a matter of the effective stress principle. In this sense, triaxial test does not reflect the cyclic behavior of soils in a realistic manner. This stress state may occur in certain real-life cases, for example, under the heavy foundation subjected to rocking behavior, but this is an exceptional situation.

When considering a loading cycle, effects of the compression and the elongation sides are presumed to cancel each other, even if nonsymmetrical behaviors are observed in both stress–strain curve and stress trajectory as seen in Fig. 6.3. In those instances, expressions for G–γ and h–γ relationships are justified. This is further discussed in Sect. 6.7.

Instead of these test apparatuses, torsional shear test explained in the next section began to be used especially in the research field.

6.4.2 *Cyclic Direct Simple Shear Test*

When the horizontally layered soil layers are excited under S-waves, the resulting shear deformation is the same as in Fig. 6.4a. The soil layers deform under constant overburden stress σ_v and the normal strain in horizontal direction $\varepsilon_h = 0$. This state is called the simple shear state.

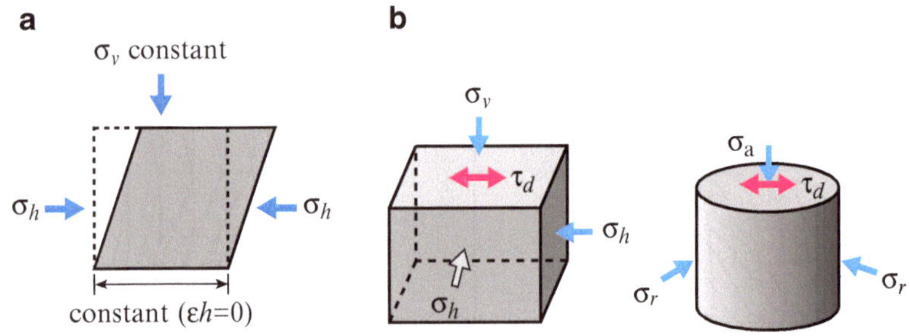

Fig. 6.4 Direct simple shear test. (**a**) Behavior during quake. (**b**) Method of simple shear test

This type of condition can be reproduced using a cyclic direct simple shear test. Typical loading types are shown in Fig. 6.4b. In the rectangular-type solid test specimen, four corners of the test apparatus are connected by hinges so that parallelogram deformation is possible. In the cylindrical shaped specimen, test apparatus is composed of thin donut-shaped frictionless rings. Since it is difficult to obtain uniform deformation in whole specimen and since measurement of the lateral stress is difficult, the direct simple shear test is hardly carried out in the practice.

6.4.3 Cyclic Torsional Shear Test

A hollow cylindrical test specimen as shown in Fig. 6.5a is used in the cyclic torsional shear test. Shear stresses are applied by twisting the specimen while keeping both axial and lateral stresses constant. Since the axial stress is kept constant, the cyclic torsional shear test can reproduce the in situ stress state of soil much better than the triaxial test. Because of this reason, the cyclic torsional shear test is frequently preferred in the research field. On the other hand, it is seldom used in practice partly because of difficulties in preparation of hollow cylinder specimens and the higher costs involved. Thickness of the hollow cylinder is recommended to be made as thin as possible in order to attain the shear strain (and shear stress) uniform in the specimen. Sample preparation becomes more difficult for smaller thickness specimen. Therefore, thickness of a test specimen is usually determined considering both effects. The external and internal diameters are suggested to be greater than or equal to 7 cm and 3 cm, respectively, for sand and 5 and 2 cm for clay (Japanese Geotechnical 2000). In practice, however, former dimensions are frequently used for undisturbed samples of both sand and clay, and a pair of 10 and 6 cm for exterior and interior diameter is used for reconstituted sands. The height of the specimen is typically taken to be 7, 10, or 14 cm for undisturbed soil samples and 10 or 20 cm for reconstituted sand samples. Thus the height is usually double of the diameter in the triaxial test.

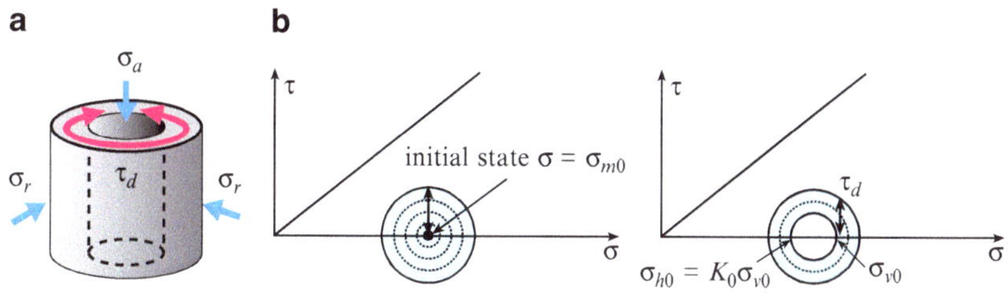

Fig. 6.5 Torsional shear test. (**a**) Loading. (**b**) Change of stress state from isotropic and anisotropic stress states

In cyclic triaxial test, the volume change is assumed not to occur as the bulk modulus of the water is very large. In the case of torsional shear test, however, no lateral strain appears under the assumption of no volume change and zero axial strain. However, it is impossible to set two boundary conditions (constant stress and constant strain) on the lateral surface at the same time. Therefore, lateral strain cannot be not zero and cannot be controlled, which is different from the actual ground behavior under shear waves. Even so, generated stress state is much closer to the in situ condition than that induced by cyclic triaxial test.

Other test methods such as a cyclic direct shear test or a shear box test are hardly used to obtain the cyclic shear deformation characteristics of soil because the stress state developed in these procedures hardly resembles the actual loading conditions shown in Fig. 6.4.

6.5 Effect of Sample Disturbance During Sampling and Traveling

Elastic shear modulus obtained in situ (in situ modulus G_{0F}) and that obtained in laboratory conditions (laboratory modulus G_{0L}) are compared in Fig. 6.6. Here, the ordinate denotes the ratio of moduli $G_{0L}/\ G_{0F}$, and the abscissa denotes in situ modulus G_{0F}. If elastic moduli obtained through both methods are identical, data appear along the line $G_{0L}/G_{0F} = 1$, but data scatters significantly. Generally, the laboratory shear modulus is smaller as the in situ shear modulus becomes larger and vice versa. This mainly is a consequence of soil sample disturbance during sampling and transportation. In general, the shear modulus is small for loose soils as will be shown later in Eq. 7.12 and Table7.4. Samples of loose soils are densified during sampling and transportation process, which increases the laboratory shear modulus. On the other hand, sample of dense soils has large shear modulus and is hardly densified during transportation. Therefore, the effect of the disturbance caused during sampling process governs the test result, resulting in smaller laboratory elastic modulus. In addition, the stress release formed during soil sampling is a contributing effect to sample disturbance. Tendency similar to Fig. 6.6 is also reported by Kokusho (1987).

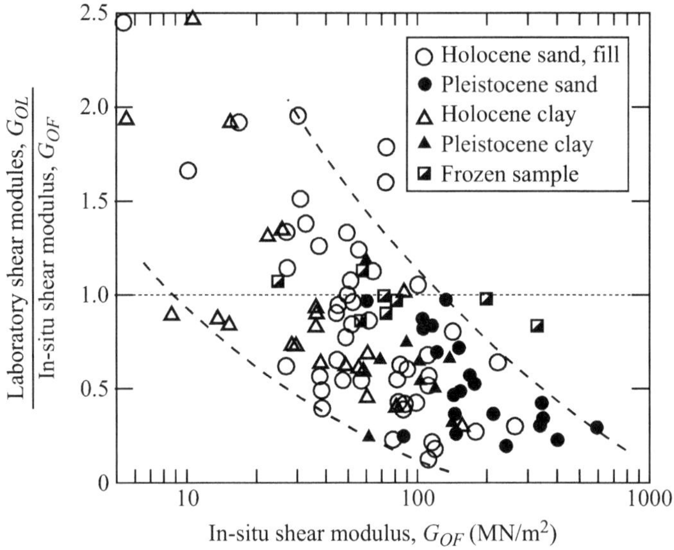

Fig. 6.6 Comparison of in situ and laboratory shear modulus (Modified from Yasuda and Yamaguchi 1984; Tokimatsu 1989)

Data on frozen samples, which are known to be highly undisturbed but costly sample, are indicated with the symbol ◪ in Fig. 6.6. These data show that G_{0F}/G_{0L} is nearly equal to 1.0 in the wide range, which ensures that scatter occurs because of the disturbance and densification during sampling and transportation.

In many engineering practices, empirical equations are developed, which shows average characteristics. However, in this case, empirical equation may be useless because the scattering in Fig. 6.6 is very large.

A similar tendency is also shown in Fig. 6.7. Not a few clay data reside around the line that shows the in situ shear moduli more than twice of the laboratory shear modulus. On the other hand, results that are less than half of the laboratory test are few. This indicates that densification hardly occurs in clay as a consequence of sampling.

If elastic moduli of soils are affected by sampling and transportation process, then it is natural to consider that the nonlinear characteristics of soils are also affected by these processes. Figure 6.8a shows a comparison of the cyclic shear modulus characteristics obtained from a frozen sample, a sample prepared by air pluviation (Tokimatsu and Hosaka 1986), and the in situ measurement. The in situ shear modulus agrees well with maximum laboratory shear modulus of the frozen sample (laboratory shear modulus). On the other hand, the shear moduli acquired for the tube sample and the reconstituted sample is much smaller than that attained by freeze sampling. If shear moduli are normalized by the maximum moduli in each test, however, all test data align on a single curve. Then in situ shear modulus G_F and laboratory shear modulus G_L are related such that

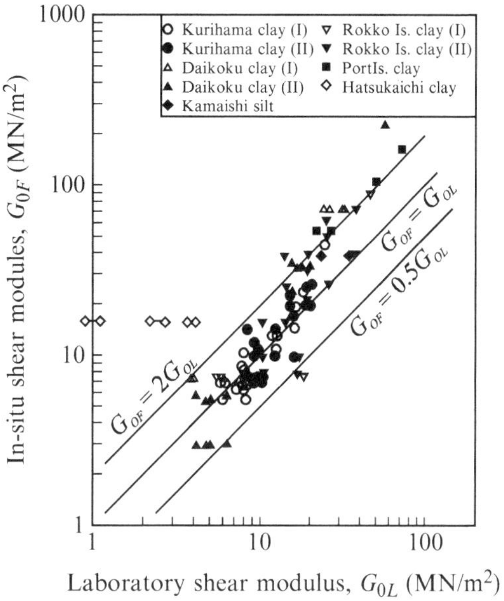

Fig. 6.7 Comparison of in situ and laboratory shear modulus of clay (After Zen 1987)

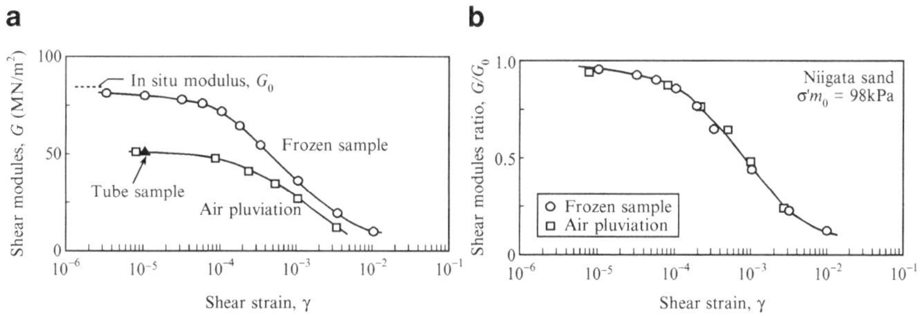

Fig. 6.8 Correction of laboratory test result (After Tokimatsu and Hosaka 1986). (**a**) G–γ relationships. (**b**) Normalized by G_0

$$G_F = \frac{G_{0F}}{G_{0L}} G_L, \quad \text{or} \frac{G_F}{G_{0F}} = \frac{G_L}{G_{0L}} \tag{6.6}$$

Hence, the ordinate G_L is usually normalized by the maximum shear modulus G_{0L}, by which the highest value of the ordinate becomes 1.0. Then in situ shear moduli are simply calculated from G/G_{0L} by multiplying with the in situ elastic modulus. This method is also demonstrated in Iwasaki et al. (1980a). In practice, subscript L is not necessary because nonlinear characteristics are always obtained by the laboratory test. Therefore, this relationship is simply denoted as G/G_0–γ relationships, but, for simplicity, it is also sometimes indicated as G–γ relationships, too.

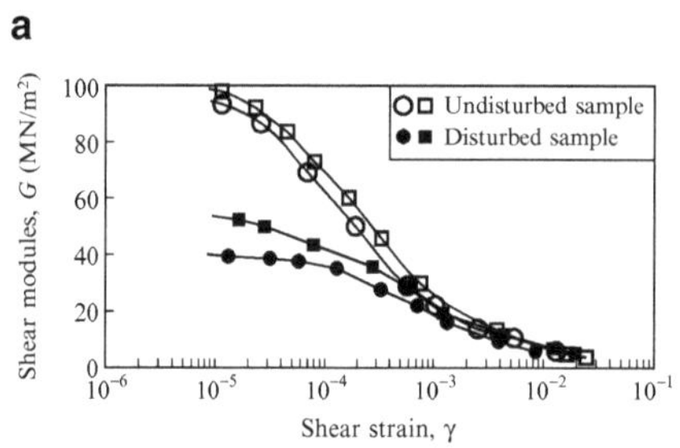

Fig. 6.9 Correction of laboratory test result (After Katayama et al. 1986). (**a**) G–γ relationships. (**b**) Normalized by G_0

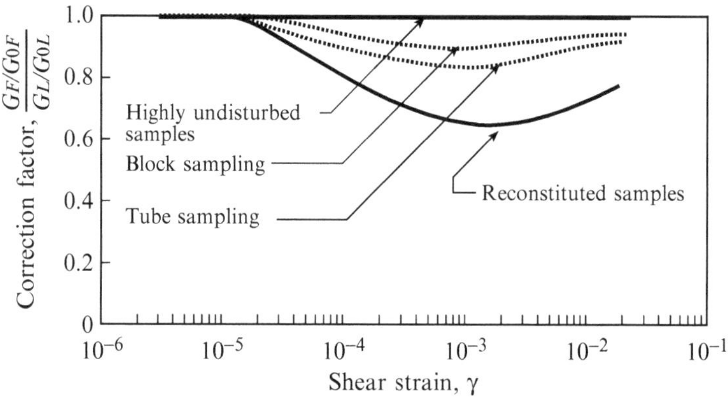

Fig. 6.10 Differences of correction factor depending on sampling method (After Ishihara 1996)

It should be noted that the relationship in Eq. 6.6 may not always hold true as shown by Katayama et al. (1986) and Hatanaka et al. (1988). An example is shown in Fig. 6.9, which shows the effect of sampling method on the cyclic shear deformation characteristics of soils. Kokusho (1987) and Ishihara (1996) demonstrated correction factors, for example, disturbance as summarized in Fig. 6.10. However, correction curves in Fig. 6.10 are not complete. Correction factors are different even by the same tube sampled soils as can be seen by the comparison of Figs. 6.8 and 6.9 for the tube sampler; these correction factors do not lie on one line but lie an interval. Thus, from a practicing engineer point of view, it is almost impossible to estimate a correct correction factor to use in an analysis.

It is also clear from Fig. 6.6 that the effect of the sample disturbance also exits for clays, but research on it is not many. A report (Committee on dynamic property of clay 1995) showed that the laboratory tests result as good agreement with the in situ tests. In the same report, however, the laboratory modulus for the Pleistocene clay is shown to become smaller than that of in situ with increasing shear modulus which is clearly seen in Fig. 6.6. The existence of the disturbance is also reported by JGS (1995).

The difference between in situ and laboratory moduli might also appear from the bedding error, an error caused by the loosened area near the edge of a sample in laboratory testing. Including this effect, many test data on stiffness is compiled by Tatsuoka and Shibuya (1992).

There is a good amount of report on the effect of the disturbance on the shear modulus as shown in this section. On the other hand, research on the damping ratio is few. This may come from the consideration that correction on damping ratio is not necessary if the shear modulus is corrected relevantly or from the difficulty to measure them at the field because cyclic loading test is required.

In summary, sample disturbance affects the mechanical properties of soils, such as the elastic shear modulus and the nonlinear characteristics. Effects to the nonlinear characteristics depend on the sampling method and other factors such as the bedding error. The effect of sample disturbance on elastic moduli can be avoided by using the in situ data. However, the correction of nonlinear characteristics is very difficult because measuring the sample disturbance is nearly impossible. In engineering practice, therefore, the correction of nonlinear characteristics is merely considered by Eq. 6.6.

In summary, sample disturbance affects the mechanical properties of soils such as the elastic shear modulus and the nonlinear characteristics. Effect to the nonlinear characteristics depends on the sampling method and other factors such as the bedding error. In the engineering practice, however, effect to the elastic modulus is avoided by using the in situ data, and effect to nonlinear characteristics is considered only by Eq. 6.6.

6.6 Compilation of Test Results

The G–γ and h–γ relationships are obtained from cyclic shear deformation characteristics test. These relationships are a sequence of points, and it is denoted as a table-type expression in this book. Although the table-type expression is an exact expression of the test result for a particular soil sample, they are not convenient to grasp the general feature of soil behavior. It is more practical if these relationships can be expressed by limited numbers of parameters. This is achieved by mathematical models developed for this purpose. In addition, the use of mathematical models has another advantage that the engineer can imagine the mechanical property from the values of parameters.

6.6.1 Hardin–Drnevich Model

Hardin and Drnevich (1972) showed that the hyperbolic model (Kondner 1963)

$$\tau = \frac{\gamma}{\frac{1}{G_0} + \frac{\gamma}{\tau_f}} = \frac{G_0 \gamma}{1 + \frac{\gamma}{\gamma_r}} \tag{6.7}$$

is applied to a variety of soils from clay to sand, where G_0 denotes the elastic shear modulus and τ_f denotes the shear strength. The shear stress approaches τ_f in an asymptotical fashion. The strain value at the intersection between the initial modulus and the shear strength

$$\gamma_r = \frac{\tau_f}{G_0} \tag{6.8}$$

is called the reference strain, which is one of the good indices to grasp the nonlinear characteristics.

Division of Eq. 6.7 by $G_0\gamma$ yields

$$\frac{G}{G_0} = \frac{1}{1 + \gamma/\gamma_r} \tag{6.9}$$

Equations 6.7 and 6.9 result in as follows by substitution of $\gamma = \gamma_r$:

$$\tau = \frac{\tau_f}{2}, \quad \frac{G}{G_0} = 0.5 \tag{6.10}$$

This indicates that the reference strain can be read from the G/G_0–γ curve as the strain at $G/G_0 = 0.5$. Therefore reference strain is sometimes denoted as $\gamma_{0.5}$.

They also proposed the equation

$$h = h_{\max}\left(1 - \frac{G}{G_0}\right) = h_{\max}\left(\frac{\gamma/\gamma_r}{1 + \gamma/\gamma_r}\right) \tag{6.11}$$

for the damping characteristics, where $h_{\max}$ denotes the maximum damping ratio. This equation is again a hyperbolic equation that approaches to $h_{\max}$ asymptotically as γ increases.

This set of equations adopts only two parameters, γ_r and $h_{\max}$. Because of this simplicity, this model is frequently used to compile the test data and is called a Hardin–Drnevich model (H-D model).

Damping ratio at small strains is zero in Eq. 6.11, but test result usually indicates small but nonzero damping ratio as discussed later in Sect. 6.7.4. Then Eq. 6.11 is modified so that the damping ratio tends to $h_{\min}$ at small strains. Two modifications are possible:

$$h = \max\left(h_{\max}\left(1 - \frac{G}{G_0}\right), \ h_{\min}\right) \tag{6.12}$$

$$h = (h_{\max} - h_{\min})\left(1 - \frac{G}{G_0}\right) + h_{\min} \tag{6.13}$$

In Eq. 6.12, h takes the value $h_{\min}$ when it is less than $h_{\min}$ while keeping the original equation unchanged. On the other hand, the linear relationships between h and G/G_0 hold in a wide range of strains in Eq. 6.13.

Parameters are frequently determined by the least square method. A linear relationship is more convenient to apply the least square method. Equation 6.9 is then modified to

$$\frac{G}{G_0} = 1 - \frac{G}{G_0}\frac{\gamma}{\gamma_r} = 1 - \frac{1}{\gamma_r} \cdot \frac{G\gamma}{G_0} \tag{6.14}$$

Then relationships between G/G_0 and $G\gamma/G_0$ are linear. Subsequently, reference strain is calculated by applying the least square method as

$$\gamma_r = \frac{\sum (G\gamma/G_0)^2}{\sum (G\gamma/G_0)\,(1 - G/G_0)} \tag{6.15}$$

Here, Σ is the summation symbol and denotes summation to all data. In practice, however, not all data points need to be used, but the data that lie on the linear relation range in G/G_0–$G\gamma/G_0$ are adequate for calculation.

In the same manner, maximum damping ratio $h_{\max}$ is obtained from Eq. 6.11 as

$$h_{\max} = \frac{\sum h\,(1 - G/G_0)}{\sum (1 - G/G_0)^2} \tag{6.16}$$

In the practice, either Eq. 6.12 or Eq. 6.13 can be used by looking at h–G/G_0 relationships, but Eq. 6.12 seems to be widely preferred.

Alternatively, substitution of $\tau = G\gamma$ and $\tau_f = G_0\gamma_r$ into Eq. 6.14 yields

$$\frac{G}{G_0} = 1 - \frac{\tau}{\tau_f} \tag{6.17}$$

This is also a linear expression of the hyperbolic model.

6.6.2 GHE Model

Ueno et al. (1991) gathered monotonically loaded test data on cemented hard and soft rocks and test data on gravel, sand, and Kaolin clay reconstituted in laboratory conditions. They examined the applicability of the hyperbolic model (Kondner 1963)

$$\frac{E_{\sec}}{E_{\max}} = \frac{1}{1/c_1 + (1/c_2)\,\varepsilon/\varepsilon_r} \tag{6.18}$$

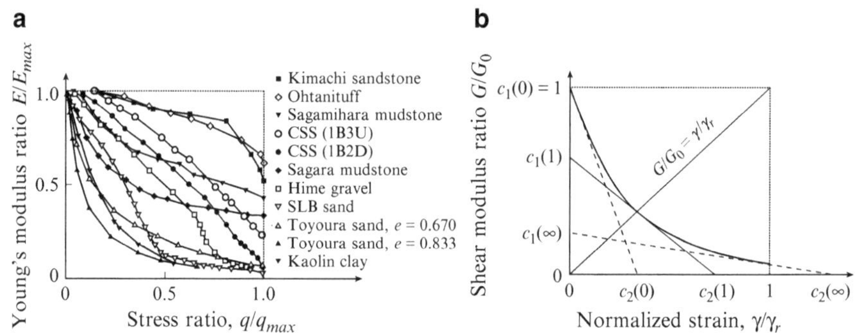

Fig. 6.11 GHE model. (**a**) Monotonically loaded test result. (**b**) Parameters in GHE model

by drawing the $E_{sec}/E_{max} - \varepsilon/\varepsilon_r$ relationships, where E_{sec} denotes the secant Young's modulus, E_{max} denotes the elastic Young's modulus, ε denotes the deviatoric strain, and ε_r is the reference strain (strain at $E_{sec}/E_{max} = 0.5$). This equation is a little modified from the original hyperbolic model shown in Eq. 6.7 by introducing two new parameters c_1 and c_2; $c_1 = c_2 = 1$ results in the original equation (Eq. 6.7). This equation can be rewritten by introducing deviatoric stress q and its maximum q_{max}, following the same procedure from Eqs. 6.7 to 6.17, as

$$\frac{E_{sec}}{E_{max}} = c_1 - \frac{c_1}{c_2}\frac{q}{q_{max}} \tag{6.19}$$

Linear relation is then obtained between E_{sec}/E_{max} and q/q_{max}. The test results they gathered are summarized in Fig. 6.11a. They pointed out that the original hyperbolic model ($c_1 = c_2 = 1$) cannot fit almost all data. Especially, agreement at small strains (less than about 0.1 %) is not good.

The stress–strain relationships is defined from Eq. 6.19 as

$$\frac{q}{q_{max}} = \frac{\varepsilon/\varepsilon_r}{1/c_1 + (1/c_2)\,\varepsilon/\varepsilon_r} \tag{6.20}$$

Tatsuoka and Shibuya (1992) pointed out that this equation cannot be applied for a wide range of strains under constants c_1 and c_2 and proposed an improvement to express c_1 and c_2 as a function of strain. This improved equation is named the GHE model (generalized hyperbolic equation model) although the equation is now substantially different from the hyperbolic equation.

When above equations are rearranged for shear deformation, one obtains

$$\frac{G}{G_0} = c_1 - \frac{c_1}{c_2}\frac{\tau}{\tau_f}, \quad \frac{\tau}{\tau_f} = \frac{\gamma/\gamma_r}{1/c_1 + (1/c_2)\,\gamma/\gamma_r} \tag{6.21}$$

Following the proposal by Tatsuoka and Shibuya (1992), c_1 and c_2 are expressed as a function of normalized strain $\xi = \gamma/\gamma_r$ as

$$c_1(x) = \frac{c_1(0) + c_1(\infty)}{2} + \frac{c_1(0) - c_1(\infty)}{2} \cos\left\{\frac{\pi}{\alpha/\xi + 1}\right\}$$
$$c_2(x) = \frac{c_2(0) + c_2(\infty)}{2} + \frac{c_2(0) - c_2(\infty)}{2} \cos\left\{\frac{\pi}{\beta/\xi + 1}\right\} \tag{6.22}$$

Six additional parameters $c_1(0), c_1(\infty), c_2(0), c_2(\infty), \alpha,$ and β are required in this model. Since the original hyperbolic model used two parameters, eight parameters are used in total to express the shear modulus reduction curve or the G–γ curve. These extra six parameters can be determined as follows:

1. Values of $c_1(0), c_1(\infty), c_2(0),$ and $c_2(\infty)$ are obtained as intersection of the tangential lines at the end as shown in Fig. 6.11b.
2. Values $c_1(1)$ and $c_2(1)$ are read as intersection of the tangential line that paths $G/G_0 = \tau/\tau$. Values α and β are calculated from Eq. 6.22 by substituting $\xi = 1$ and by solving for α and β, as

$$\alpha = \frac{\pi}{\cos^{-1}\frac{c_1(1) - \frac{c_1(0)+c_1(\infty)}{2}}{\frac{c_1(0)-c_1(\infty)}{2}}} - 1, \quad \beta = \frac{\pi}{\cos^{-1}\frac{c_2(1) - \frac{c_2(0)+c_2(\infty)}{2}}{\frac{c_2(0)-c_2(\infty)}{2}}} - 1 \tag{6.23}$$

Since the GHE model is developed using the monotonic stress–strain curve, damping characteristics are not defined. The H-D model can be used, for example, on the damping characteristics.

Physical meaning of these parameters is not well investigated as the purpose of this model is to obtain a mathematical equation that properly fits with the test results. Therefore, it is difficult to grasp the nature of the mechanical properties from the determined values of these parameters. At the moment, the amount of available data is inadequate for elaborating on this aspect of the GHE model. In addition, as discussed later in Sect. 7.3, it is difficult to determine the shear strength for dynamic loading problems. Because of these reasons, the applicability of this equation is very limited, but it has the potential to set a good agreement with the test results, which is the reason why it is included in this book.

Shear strength τ_f is to be determined in order to use this model. Using Mohr–Coulomb failure criteria is suggested to be used for this purpose (Murono 1990), but it is not so simple. Under the drained conditions where change of the confining stress is small, stress–stress curve is a monotonically increasing function. Therefore, τ_f can be obtained from the Mohr–Coulomb criteria because shear strength has a maximum allowable stress. On the other hand, under the undrained conditions where the confining stress may change significantly because of the excess porewater pressure generation, determination of shear strength becomes difficult. As shown later in Sect. 7.33, the stress state eventually satisfies Mohr–Coulomb yield criteria, but maximum stress can occur beforehand.

6.6.3 Comparison of H-D Model and GHE Model

When the H-D model was introduced, it was suggested to agree with the test result well, but, as shown in the previous section, this is hardly the case. This confliction results can be recognized by considering the strain range they considered.

As explained in Sect. 6.7, applicable strain range in the conventional cyclic shear deformation characteristics test is up to 10^{-3} or a little larger. On the other hand, Ueno et al. (1991) discussed that agreement at this strain range is not well defined. This indicates that strain ranges considered by two models are quite different.

Although the relationship between initial modulus and the shear strength (i.e., reference strain) is shown in Eq. 6.8, this shear strength is not the actual shear strength of the material. In other words, the H-D model is applicable to a wide range of strains such as the GHE model. This is a very important issue, especially when dealing with very strong ground motions because shear strength is firmly related to upper bound acceleration.

Another difference arises from the type of test methods. The GHE model deals with monotonically loaded test results, whereas G–γ curve is obtained by cyclic loading test. Stress–strain curve derived from the G–γ curve is not identical with the stress–strain curve under monotonically loading, which is discussed in more detail in Sect. 6.7.

6.6.4 Double Hyperbolic Model

As discussed in the previous sections, the applicability of the hyperbolic model is not well at large strains. This limitation may be solved by the GHE model, but this model uses many parameters and physical properties of many parameters are not clear and not well investigated. The double hyperbolic model is a model that uses only well-known parameters.

The double hyperbolic model uses two hyperbolic models. Ordinary hyperbolic model is used at small strains because it is well applicable at small strains as discussed in Sect. 8.2.3 later and as discussed above. Another hyperbolic model is used at large strains, whose parameters are determined focusing on shear strength.

The ordinary hyperbolic model is already shown in Eq. 6.7 as

$$\tau = \frac{G_0 \gamma}{1 + \gamma/\gamma_r} \tag{6.7}$$

The second hyperbolic model is defined as

$$\tau = \frac{G_0 (\gamma - \gamma_0)}{A + G_0 (\gamma - \gamma_0)/\tau_f} \tag{6.24}$$

where A and γ_0 are unknown parameters and shear strength τ_f is a specified value. These two parameters are calculated by setting two conditions as

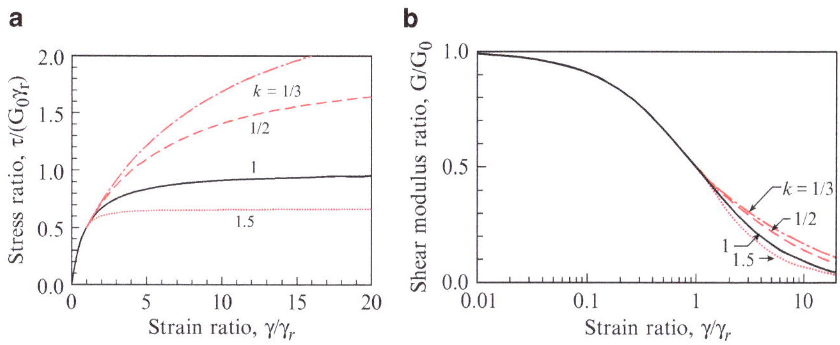

Fig. 6.12 Double hyperbolic model. (**a**) Stress–strain curve. (**b**) Modulus reduction curve

(a) Two models are continuous at $\gamma = \gamma_r$.
(b) Tangent slope is continuous at $\gamma = \gamma_r$.

Then, two unknowns are evaluated, and Eq. 6.24 yields

$$\tau = \frac{G_0\gamma_r\left(\frac{\gamma}{\gamma_r} + 1 - k\right)}{(2-k)^2 + k\left(\frac{\gamma}{\gamma_r} + 1 - k\right)} \tag{6.25}$$

where $k = G_0\gamma_r/\tau_f$ are shear strength ratio of two hyperbolic models.

This model uses two parameters, reference strain and shear strength, whose mechanical properties are well known. Results of parametric study varying k are shown in Fig. 6.12 in two different expressions.

6.6.5 Confining Stress Dependency

The elastic shear modulus is proportional to the square root of the effective confining stress as described later in Sect. 7.1.6. The shear strength of sand is proportional to the effective confining stress, whereas it is constant for clay. These observations indicate that cyclic shear deformation characteristics depend on effective confining stress.

As shown in Sect. 6.6.1, the H-D model is frequently used to represent G–γ and h–γ relationships. Then the change of the cyclic shear deformation characteristics associated with confining stress can be determined as (Yoshida et al. 1990)

$$\frac{G}{G_{\max}} = \left(\frac{G}{G_{\max}}\right)_0 \frac{B + \gamma A\sigma_{vo}'^{m-n}}{B + \gamma A\sigma_v'^{m-n}},\ h = h_0\frac{1 - \frac{G}{G_{\max}}}{1 - \left(\frac{G}{G_{\max}}\right)_0} \tag{6.26}$$

where subscript 0 denotes data before correction. Four parameters, coefficients A and B and powers m and n, are used in this equation. They are coefficients used to express confining stress dependency of the shear modulus and the shear strength, i.e.,

$$\begin{aligned} G_{\max} &= A\sigma_v'^m \\ \tau_f &= B\sigma_v'^n \end{aligned} \tag{6.27}$$

As shown in 7.1.6, m is approximately equal to 0.5. On the other hand, since shear strength is defined as per the Mohr–Coulomb failure criteria, $n = 1$ for sand and $n = 0$ for clay.

6.7 Applicability and Limitations of Cyclic Shear Test

The cyclic shear deformation characteristics test has been the main tool for obtaining the deformation characteristics of soil used in the seismic response analysis. Limitation of this test method has been hardly discussed when input earthquake motion was not so significant. After the 1995 Kobe earthquake in Japan, however, the intensity of design earthquake motion becomes large. In that situation, discussion on the limitation of the test method appeared to be necessary because the maximum strain frequently becomes several percent or more. A new method may be necessary and there are researches considering it. However, in practice, changing a widely adopted standard test is problematic, and it takes time even if a new, better suited test is developed. Therefore, for an engineer, it is necessary to recognize applicability and limitations of the conventional test methods well.

6.7.1 Strain Range and Accuracy of Test

It is important to understand the attainable maximum strain and test accuracies for a testing method and to model it relevantly. Among them, strain ranges and test accuracies are discussed in this subsection, and behavior at large strains is described in Sects. 6.7.5 and 6.7.6.

Figure 6.13 summarizes the result of the questionnaire on the topic of accuracy or the reliable minimum strain (Proc et al. 1994) of the conventional cyclic shear deformation test procedures. Since this survey was carried out in 1994, the data is rather old from the present point of view, but it still gives some ideas. The survey was sent to nearly all organizations that possess test equipment, and return rate was more than half.

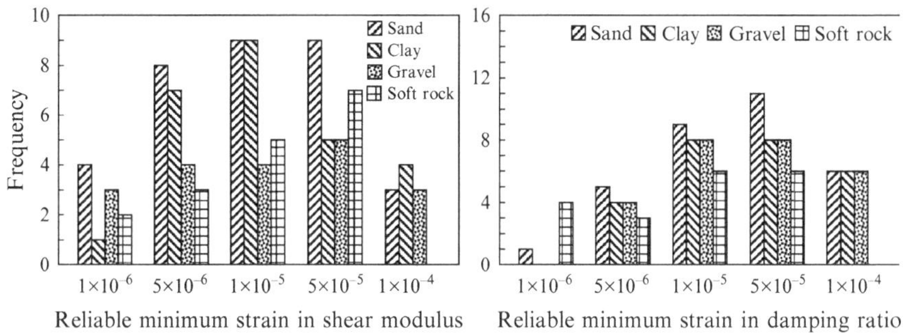

Fig. 6.13 Reliable minimum strain

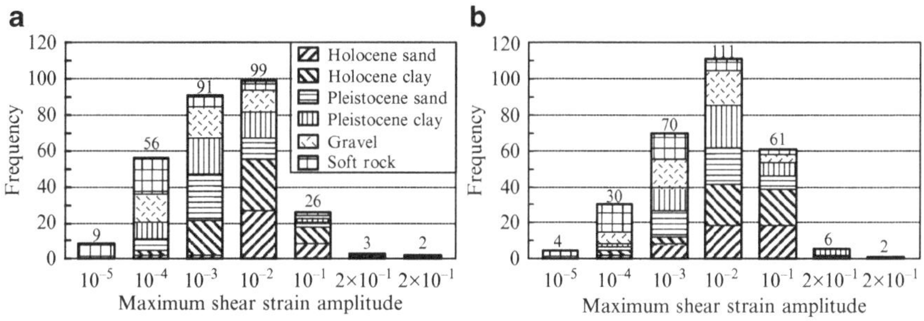

Fig. 6.14 Maximum strain experienced and expected in analyses. Same number 2×10^{-1} appears in the abscissa, which is exactly what was written in the original paper. The right number is supposed to be 5×10^{-1}. (**a**) Past experience. (**b**) Future requirement

It may be surprising that reliable accurately measureable strain expands double figure between 10^{-6} and 10^{-4} both for shear modulus and for damping ratio. However, the majority of values lie in the range between 5×10^{-6} and 5×10^{-5}. Roughly speaking, therefore, reliable minimum measureable strain is approximately 1×10^{-5}. There are methods to measure small strain more accurately such as in Chiba et al. (1994), but these methods are not employed in the engineering practice.

Figure 6.14 shows the questionnaire result on the maximum strain levels. The maximum experienced reliable strains are shown in Fig. 6.14a, whereas the maximum strain levels expected in future testing procedure are shown in Fig. 6.14b. The maximum strain in both questionnaires is less than 1 %. However, it is also noted that there is a data with maximum strain up to 20 %. As mentioned, this survey was made before the Kobe earthquake, and design earthquake motion levels increased compared with that time. Therefore, current required maximum strain levels are much larger than expressed in Fig. 6.14. Actually, maximum strain of several percent frequently appears according to the author's experiences.

6.7.2 Effect of Excess Porewater Pressure Generation

The cyclic shear deformation characteristics tests were carried out under drained conditions in the past. At present, however, they are carried out under undrained conditions partly because undrained test conditions are supposed to be similar condition during earthquakes as the duration of the earthquake is too short for porewater pressure to dissipate and partly because drained conditions require longer test duration and hence are not practical. As shown in Fig. 4.5, behaviors under drained and undrained conditions are quite different at large strains. Research on liquefaction of soils may be of some help for determining the strain at which these differences appear. According to JGS (Japanese Geotechnical 2004), strain levels at which generation of excess porewater initiates are approximately 3×10^{-4}. This strain value is almost on the same order as the reference strain of sand.

Difference associated with the drained condition can be also seen in Fig. 6.15 (Kokusho 1980). The term "reverse" in the figure denotes that the test started in the elongation side; in the ordinary triaxial test, loading starts in the contraction side. Difference between the opposite loading directions is small and may be negligible.

There is not a significant difference of the G/G_0–γ curves between the drained and the undrained conditions. However, a contract appears in the damping characteristics at large strain; a damping ratio increases as the strain increases under the drained condition, whereas it seems to have an upper limit under the undrained condition. As shown later in this section, this upper limit is caused by excess porewater pressure generation in the soil specimen.

As described in Sect. 6.3, the excess porewater is dissipated between the stages even under the undrained test conditions in the stage test. This drainage causes densification of the soil sample. Therefore, differences in the G–γ curves appear between tests under the monotonic and cyclic loading conditions as shown later in Fig. 6.19. However, in engineering practice, the G–γ curves obtained by both loading cases are treated to be the same.

Another problem with stage test is that the past loading history affects the test result as the shear strain increases or the excess porewater pressure begins

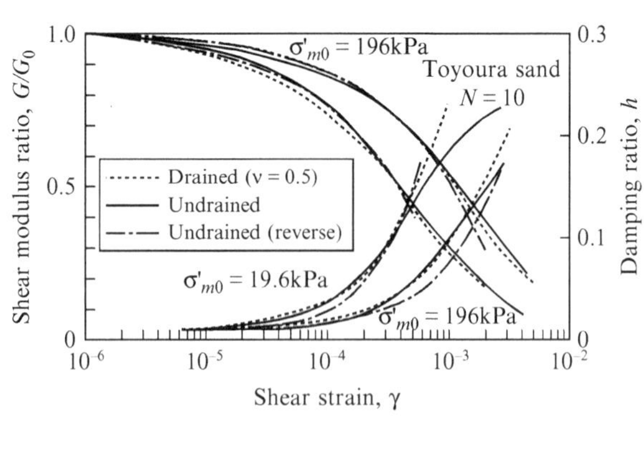

Fig. 6.15 Comparison of cyclic shear deformation characteristics between drained and undrained condition (After Kokusho 1980)

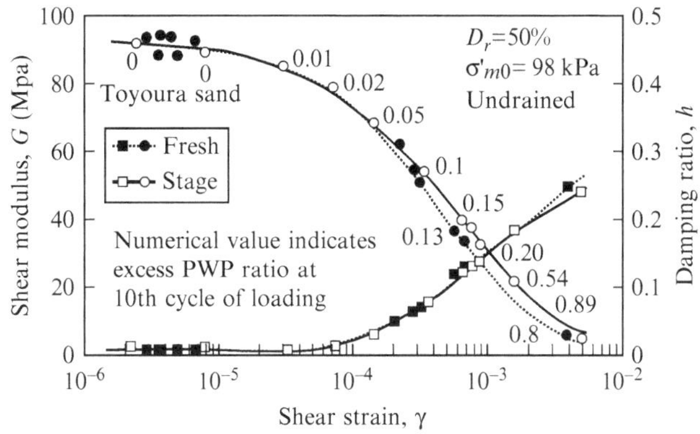

Fig. 6.16 Comparison between fresh test and stage test (After Yasuda et al. 1994)

to generate. A comparison between the stage test and the fresh test is shown in Fig. 6.16 (Yasuda et al. 1994). Numbers indicated on the figure for each strain are excess porewater pressure ratio generated in that specific cycle. No significant differences are observed in the damping characteristics. Difference of the shear modulus is also small until excess porewater pressure generates. Difference appears at the strain where excess porewater pressure begins to generate, and this strain value agrees with the aforementioned strain. Shear moduli by the stage test become larger after this strain than that by the fresh test because the sample becomes denser as excess porewater pressure is dissipated.

The effect of the densification on cyclic shear deformation characteristics is clearly observed by comparing the cyclic deformation characteristics test results and the liquefaction strength test result. In liquefaction strength test, the number of cycles causing liquefaction is plotted against shear stress amplitude ratio. The onset of liquefaction is usually identified when a double amplitude axial strain of 5 % (i.e., 3.75 % single amplitude shear strain) or a residual effective mean stress of 5 % is recorded. Accordingly, the strain in each cycle of loading is reported up to this strain. Shear strain in each loading cycle is also recorded in the cyclic shear deformation test. Therefore, both results can be written in one graph such as that in Fig. 6.17. Equi-strain contours of the cyclic shear deformation characteristics test are located considerably higher than those obtained from the liquefaction strength test. This means that samples by the cyclic shear deformation characteristics test show much larger liquefaction strength than that by the liquefaction strength test although the identical samples are adopted. Clearly, this difference is a consequence of the densification of sand due to the excess porewater pressure dissipation between the stages.

A similar behavior is shown in Fig. 6.19 where stress–strain curves obtained by monotonic and cyclic loading tests are compared (Yamashita 1992). The shear stress in cyclic loading test is larger than that by monotonic loading test. Sand becomes dense at each stage because of the excess porewater pressure dissipation, which results in increase of shear strength.

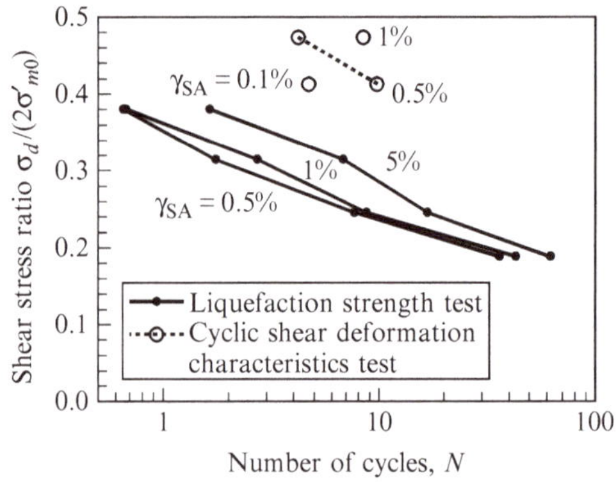

Fig. 6.17 Comparison between cyclic shear deformation characteristics test and liquefaction strength test (After Yoshida and Mikami 2010)

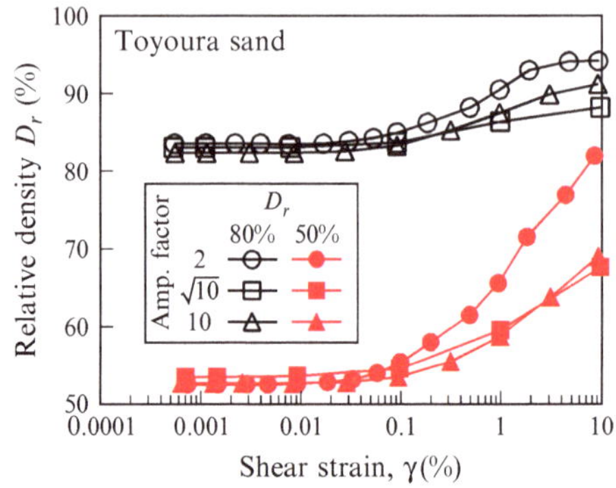

Fig. 6.18 Soil densification in stage test

This densification behavior can also be clearly observed in Fig. 6.18, which is retrieved from the test results by Yoshida el al. (2005). In this test scheme, the shear strain amplitude at each stage is set through multiplying the shear strain amplitude of the previous loading stage by each "amp factor" denoted in the figure; the factor of 2 is the value suggested by JGS (2009). The relative density begins to increase around 5×10^{-4}, and it reaches more than 80 % in the sand with $D_r = 50$ %. Increase of the relative density is not so significant in the sand with $D_r = 80$ % because the specimen is initially dense, but it reaches more than 90 %.

This phenomenon does not occur if excess porewater pressure does not generate. Therefore, it does not become problem when design earthquake motion is small. However, the intensity of design earthquake motion is already large enough to cause significant nonlinear behavior at present, especially in Japan. Thus this issue cannot be neglected.

Next, let us consider the observed damping characteristics at large strain. As typically seen in Figs. 6.15 and 6.16, the maximum damping ratio is about 25 % at maximum in undrained tests, whereas it can become more than 30 % in drained tests. This can be well understood by looking at Fig. 6.20 where the cyclic shear deformation characteristics, and the hysteresis loops normalized by the maximum stress–strain, observed in four stages are drawn.

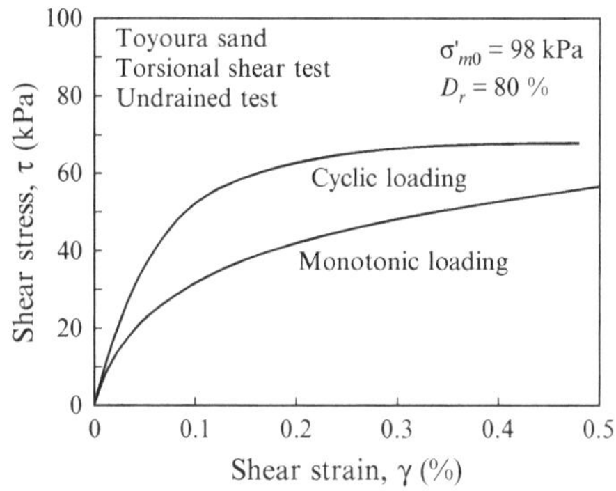

Fig. 6.19 A comparison between monotonic and cyclic loading tests (After Yamashita 1992)

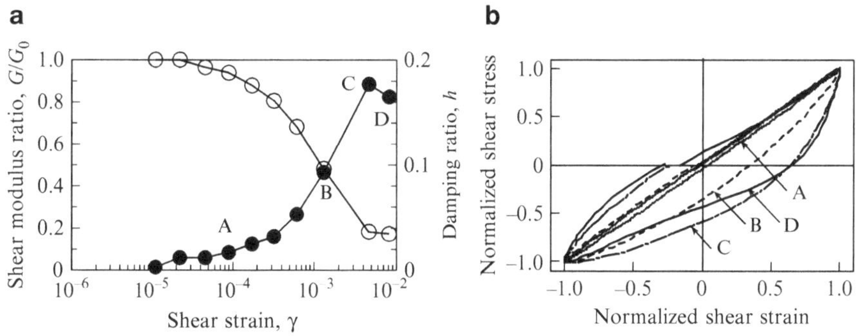

Fig. 6.20 Decrease of damping ratio at large strains under undrained conditions. (**a**) Cyclic shear deformation characteristics. (**b**) Stress–strain curve

The shape of the hysteresis loop in drained test resembles a spindle shape as can be seen in Fig. 4.5, for example. The area of the hysteresis loop (damping ratio) increases as strains increase. Behavior under undrained conditions is similar to that under drained conditions when shear strain is small, as the effect of the dilatancy is limited; however, the hysteresis loop transforms the spindle shape into a flipped S-shape as the strain increases as seen in Fig. 6.19b. Then the damping ratio becomes smaller as shown from point C to point D in Fig. 6.20a.

Another problem occurs if excess porewater pressure develops at large strains. As excess porewater generates in each loading cyclic, the shear strain gradually increases, and the hysteresis loop becomes unstable. For instance, Fig. 6.21a shows stress–strain curves in loading stage D in Fig. 6.20, where the hysteresis loop moves to elongation side gradually with loading. Evidently, this behavior is caused by the degradation of mechanical properties of soils, which can be recognized by comparing the hysteresis loops with that in Fig. 4.5.

Precisely speaking, degradation begins a little prior to stage D. The stress–strain curves at stages C and B are shown in Fig. 6.21b, c, respectively. Behavior similar to stage D is clearly observed in stage C. On the other hand, an increase in the shear strain amplitude is hardly observed in stage B, even though the shift of the hysteresis loop to the elongation side is observed.

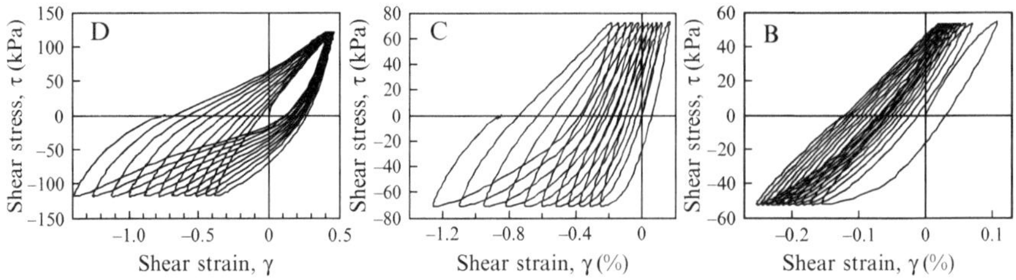

Fig. 6.21 Stress–strain curves at three loading stages in Fig. 6.20

Back to the original concept of the test methods, the hysteresis loops are supposed to become stable in 10 cycles of loading even if transient behavior is observed at the beginning of loading. If strain amplitude increases with loading cycles, then obtained behavior depends on the number of cycles of loading from which G and h are calculated. Thus, this test cannot be objective or a test to output indices for the nonlinear properties. In this sense, the applicability of the conventional tests is limited to the point at which behavior similar to the behavior at stage C is observed. This strain amplitude is a little larger than 10^{-3}; it may extend between stages B and C, but not more than stage C. If the drift of the hysteresis loop becomes is not good (such that as seen in Fig. 6.21c), the applicability of the test is limited to strains smaller than 10^{-3}. Based on the preceding discussion, clearly, accumulation in shear strain occurs with loading cycles as a result of excess porewater pressure generation. Therefore, the applicability of conventional cyclic shear deformation characteristics tests is limited to the stage when the effect of the excess porewater pressure appears in the stress–strain curve, and it is about 10^{-3}(0.1 %) or a little more in the best case, which is a little larger than the strain where excess porewater pressure begins to generate.

It is possible to obtain cyclic shear deformation characteristics even in the strain range exceeding the applicability limit discussed above or even when hysteresis loop does not become stable. One can change the number of cycles in one stage from 10 to 3, for example, or one can use every hysteresis loop in the stage to calculate the cyclic shear deformation characteristics because these methods are also permitted in JGS (2009) as a part of the cyclic shear deformation characteristics test. It is difficult to distinguish these modifications in the G–γ curve, but the effect of the latter method can be clearly recognized in the τ–γ curve in Fig. 6.35 shown later.

Discussion here is mainly focused on the sandy soil. Test is applicable to larger strain for clay because the excess porewater generates much smaller than that of sand.

6.7.3 *Effect of Loading Speed*

Another problem may also arise on the loading speed. The cyclic shear deformation characteristics test is usually performed at a loading speed of 0.1 Hz in stress-controlled methods, but it may be insufficient to get proper properties.

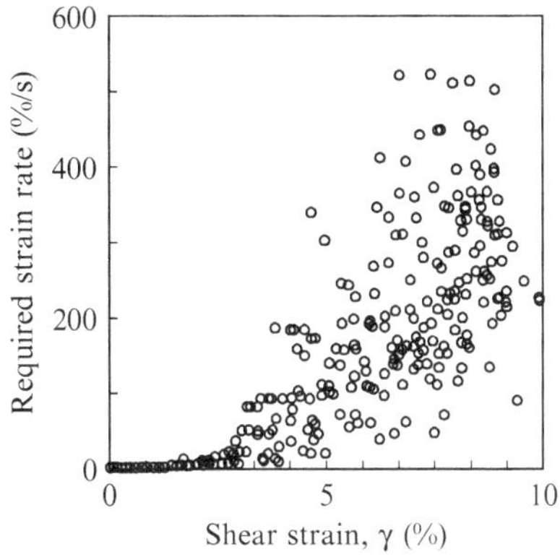

Fig. 6.22 Relationship between double amplitude shear strain and required strain rate (After Momose et al. 1996)

The cyclic shear deformation characteristics test is usually carried out under the stress control condition in which target stress increment produced by a signal processor is transmitted to the test apparatus at a constant time interval. When the stiffness becomes small at large strains, the shear strain increment for reaching the desired shear stress increment becomes large, and the required strain rate may exceed the limits of the test apparatus (Momose et al. 1996). Figure 6.22 shows the required strain rate under different shear strain amplitudes from many test results. Scattering of the data implies that stiffness is not a unique function of the shear strain amplitude.

The signal processor usually transmits sinusoidal wave signal without a feedback from the related physical mechanism in the apparatus. Then if power of the pomp to transmit oil or air to the test device to cause strain is not sufficient, a strain reversal occurs before reaching the target stress level. As a result, both the applied shear stress and the induced shear strain amplitudes become smaller than desired.

These effects can be seen in the liquefaction strength test more easily because applied strain amplitude is larger and resulting stiffness becomes smaller in the liquefaction strength test than the cyclic shear deformation characteristics test. An example of the first effect can be seen in Fig. 6.23 where stress–strain curves as well as the shear stress waveforms at the first and the last cycles of loading are shown. Waveform appears in a sinusoidal shape same as controlling signal at the first cycle, but it is distorted at the last cycle where strain changes considerably.

Another example is shown in Fig. 6.24 where results of liquefaction strength tests for identical samples obtained from different test organizations are shown. Although the same soil sample is used, differences between liquefaction strengths are significant. In this case, the test result indicating the lowest liquefaction strength is the most accurate.

It has been traditionally considered that the stress–strain curves are not affected by the strain rate. As explained in Sect. 8.5, complex modulus is derived based on this assumption. A proof is seen in Fig. 6.25, in which shear modulus is measured under the loading frequencies ranging from 0.01 to 10 Hz. The shear modulus does not appear to be affected by the loading frequency. It should be noted, however,

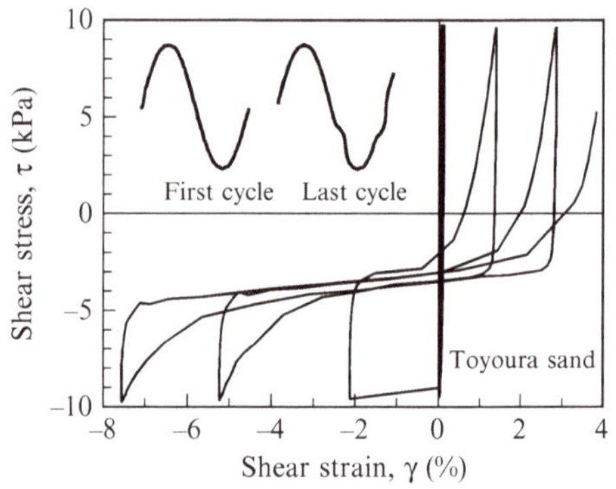

Fig. 6.23 Example of disturbed waveform

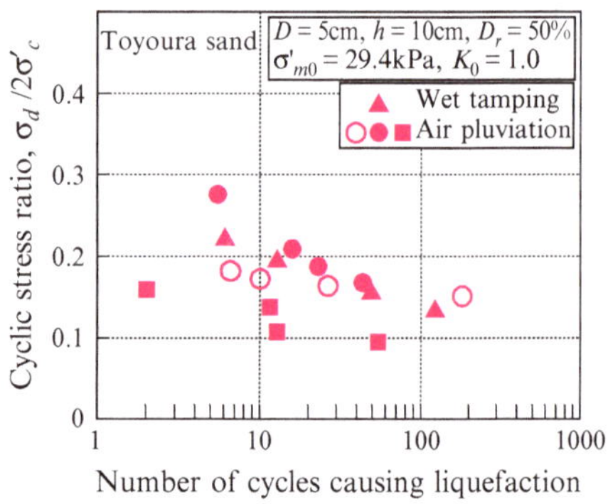

Fig. 6.24 Liquefaction strength test result by different test organizations (After Suzuki et al. 2008)

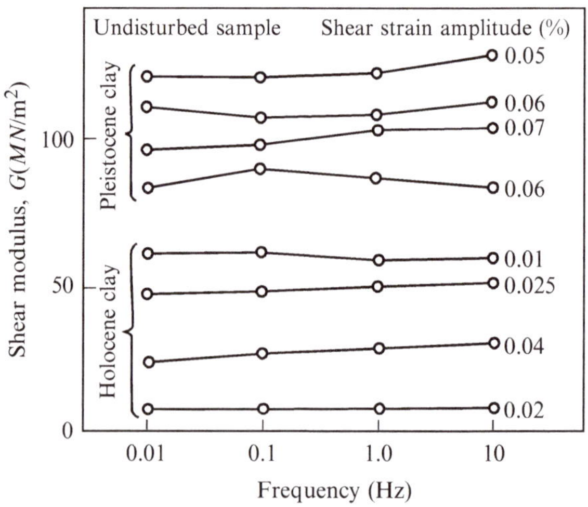

Fig. 6.25 Loading frequency dependency of shear modulus (After Hara 1973)

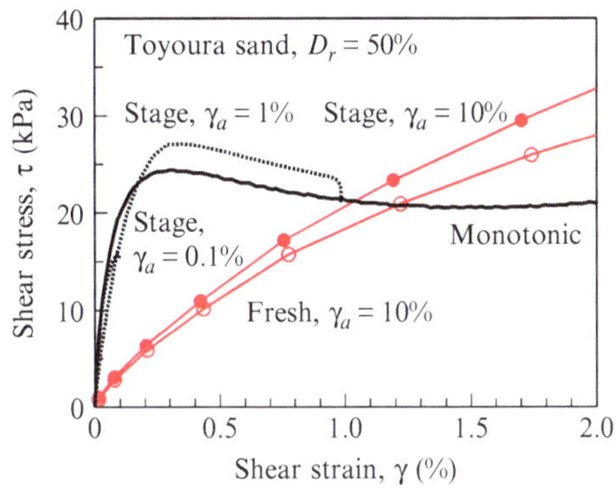

Fig. 6.26 Comparison of monotonic loading behavior under different loading rates

that shear strain amplitude is 0.07 % at the maximum in this test. If shear strain amplitude reaches several percent, which is not extraordinarily large strain in the recent seismic ground response analyses performed in Japan, attained strain level shall be about 100 times larger than that shown in Fig. 6.25; then the strain rate also becomes about 100 times as the tests are carried under constant frequency.

Strain rate-dependent characteristics are commonly observed in the structural elements such as steel and concrete (Iwai et al. 1982). Thus, it is very difficult to believe that soil is an exception, but research available focusing on this aspect is few (Boulanger and Idriss 2006). Therefore, there is no way but to assume that strain rate dependency does not exist in the engineering practice in the current level of understanding.

There are few data available related to the effect of loading rate. Thus, here, the test results in Ref. (Yoshida et al. 2005), shown in Fig. 6.26, are retrieved so that the difference of loading speed can be seen. All tests are carried out in a strain-controlled manner in order to reproduce decrease of shear stress. The monotonic test is carried at a loading speed of 0.1 % strain/min. On the other hand, three cyclic tests in the stage test and one cyclic test in the fresh test were performed at 0.1 Hz frequency under sinusoidal wave based on the JGS standards (JGS 2009). Stress–strain curve at the first cycle of loading and that of the monotonic loaded test are comparable; the only difference is loading speed. The average strain rates are 2.4, 24, and 240 %/min at shear strain amplitude of 0.1 %, 1 %, and 10 %, respectively. The strain rate at the beginning of loading is more than 3,000 times faster in the cyclic loading compared with the one in the monotonic loading. The stiffness degrades as the loading speed increases. Generally, the stiffness increases with increasing loading rate, but test data shows contrary tendency. At the moment, the reason of this discrepancy is not well perceived.

Figure 6.27 shows an example of the strain increments obtained from the strain increment in each step divided by time increment in each step in the effective stress analysis at the Port Island during the 1995 Kobe earthquake (Yoshida 1998). The instantaneous strain rate reaches to more than 10 %/s. The strain rate in the total stress analysis can become almost similar order as can be observed for clays.

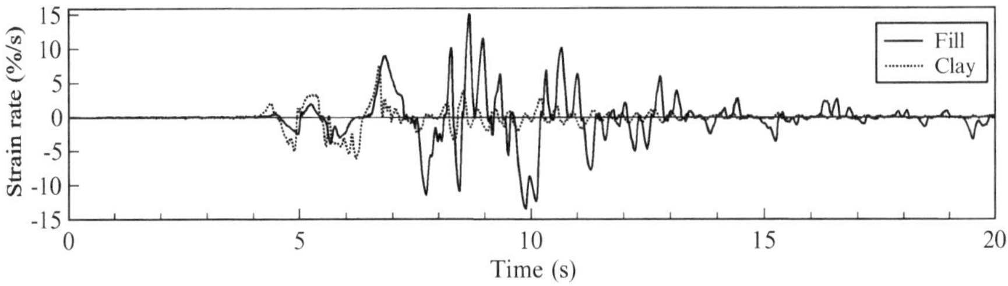

Fig. 6.27 Strain rate in effective stress analysis (Modified from Yoshida 1998)

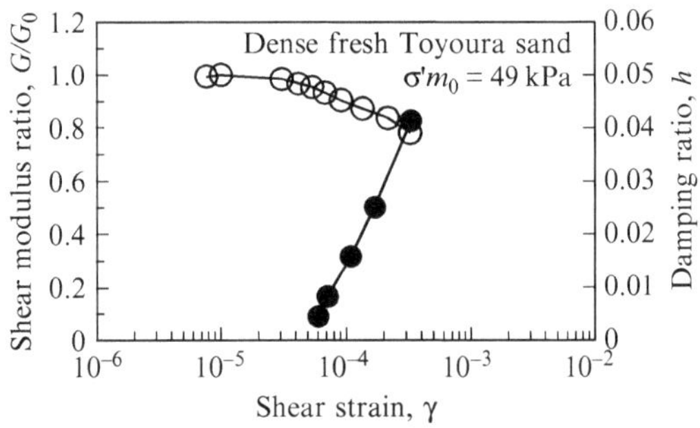

Fig. 6.28 Result of cyclic triaxial test in which damping ratio at small strains is measured accurately by using LDT (Modified from (Sato et al. 1994))

6.7.4 Damping Characteristics

Damping ratios of about 2–4 % are frequently observed even at small strains in cyclic shear deformation characteristic test results. It must be zero if soil behaves in an elastic manner at small strains. Various factors are considered to explain this confliction. Friction in the contact-type displacement transducer and friction between the piston and the cap (top platen) in the cell are considered to be the reason (Kokusho 1982). Bedding error which occurs because of the disturbance at the ends of a test specimen may also be responsible (Yamashita 1992). Many other possible reasons were shown by Sato et al. (1994). The use of local deformation transducer (LDT) (Sato et al. 1994) may be a solution to this problem. A result using the LDT is shown in Fig. 6.28; damping ratio is zero at small strains. It is noted that the LDT is not applicable at large strain, which is the reason why Fig. 6.28 does not include large strain behavior.

Figure 6.29 compares results obtained from cyclic torsional shear test and resonant column test (Tatsuoka et al. 1978). There is a significant difference between the damping ratios. This may occur from the difference of loading cycle; several thousand cycles are loaded in the resonance column test, whereas 10 cycles of loading are used in the cyclic torsional shear test (Iwasaki et al. 1980b), but the effect of the loading cycles is not very clear.

Recently, test results that show approximately zero damping ratios at small strains began to be observed in literature (e.g., see Figs. 6.20 and 7.11). This is a fruit

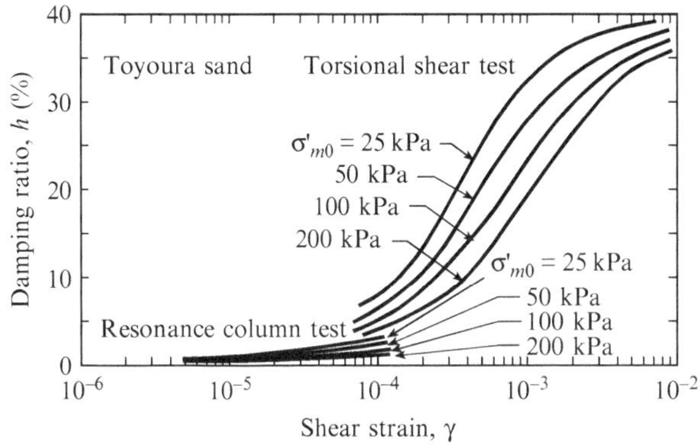

Fig. 6.29 Comparison of damping ratios obtained through resonant column and torsional shear tests (After Tatsuoka et al. 1978)

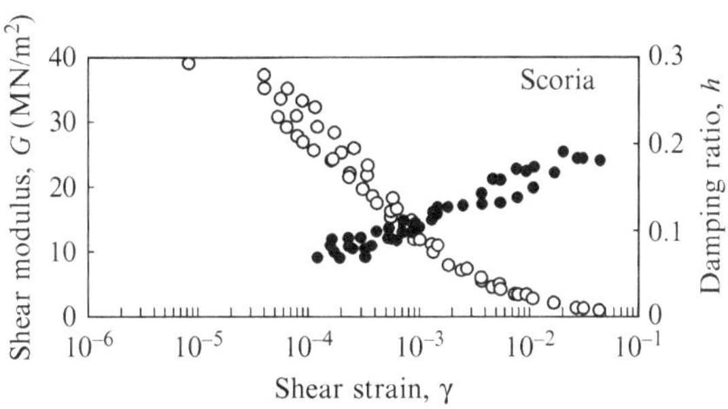

Fig. 6.30 Cyclic shear deformation characteristics of volcanic soil

of efforts to improve the test method such as to set a load cell in the cell and to use a noncontact displacement meter in order to reduce the effect of friction. However, test results that report nonzero damping ratio at small strains are still frequently encountered in practice.

In addition, there is a certain amount of data that shows nonzero damping ratio at small strains, which appear not to be affected by test apparatus. Figure 6.30, for example, shows result on the volcaniclastic material (Satoh et al. 1997). Damping ratios at strains of 10^{-4} reach about 10 %, a very large value that cannot be resulted by the mechanism described above. In the same manner, Fig. 6.31 shows the result of a sedimentary soft rock (clay rock), which shows 3–4 % damping ratios at small strains (Nishi et al. 1986). If hair cracks develop during sampling, initial modulus of the sample becomes very small (Fukumoto et al. 2009) and damping ratio will not be zero at small strains.

Damping ratios at strain 10^{-4} (Kokusho 1982) are compared in Fig. 6.32. Damping ratios become smaller for each soil category as the shear moduli become larger. However, the damping ratios are at the order of several percent in general in all samples. In addition, the damping ratios are greater at low confining stresses although it is not shown in the figure, which is seemingly caused by hair cracks within the samples. This is later discussed in Sect. 7.2.11 in detail.

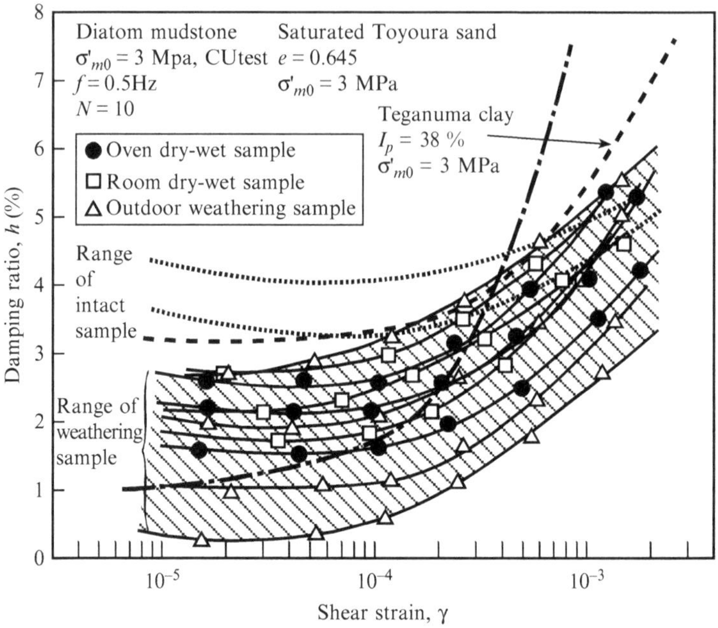

Fig. 6.31 Damping characteristics of mudstone (After Nishi et al. 1986)

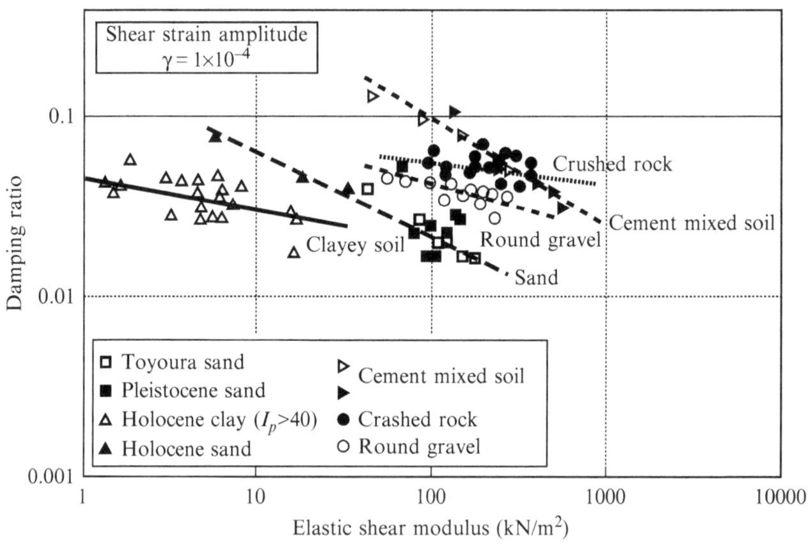

Fig. 6.32 Relationships between elastic shear modulus and damping ratio (After Kokusho 1982b)

This issue is hardly problematic from an engineering point of view because the damping ratio at small strains hardly affects the result of the seismic ground response analysis under a large earthquake motion. It is also pointed out that the damping caused by wave scattering, which cannot be observed in laboratory test and is hardly considered in practical problems, is nearly of the same order (see Sect. 12.3), and it compensates overestimation of the damping ratio at small strains (Yoshida 1994).

A more important issue is that a mechanism similar to that in nonzero damping at small strains also appears at large strains, resulting in larger damping ratios in the whole strain range. Unfortunately, the available data is inadequate for evaluating this issue. In addition, as shown in Sect. 15.7, the damping ratio itself is not a very important mechanical property; therefore, this issue needs not be considered in the practical analysis.

6.7.5 Cyclic Shear Deformation Characteristics and Shear Strength

The shear stress τ is calculated from the secant modulus G in Eq. 4.1, as

$$\tau = G\gamma \tag{4.1}$$

The G–γ relationships are equivalent to the τ–γ relationships, but this simple fact is frequently forgotten in the engineering practice. Let us consider the simple shear deformation in a one-dimensional analysis as an illustrative example. Then the shear strength τ_f is calculated from the Mohr–Coulomb failure condition as

$$\tau_f = c + \sigma'_v \tan\phi \tag{6.28}$$

where c denotes cohesion and ϕ denotes internal friction angle. The internal friction angle ϕ is usually 0 for clay and cohesion c is usually 0 for sand. In the multidimensional state, Eq. 6.28 is rewritten as

$$\tau_f = c + \sigma'_m \sin\phi \tag{6.29}$$

On the other hand, as described in Sect. 6.5, the G–γ relationships obtained in the laboratory test cannot reproduce the in situ relationships because of disturbance effects, and the in situ G is obtained from Eq. 6.6. This indicates that the shear strength is proportional to the elastic shear modulus. However, it is sure that this is not the actual relation.

As described in Sect. 6.6, behavior at large strain causing failure cannot be obtained using conventional laboratory test. Therefore, above observation and the fact that the shear strength is not proportional to the elastic shear modulus may not be competing. The engineer must be, however, careful on this gap between the G–γ relationships and the shear strength.

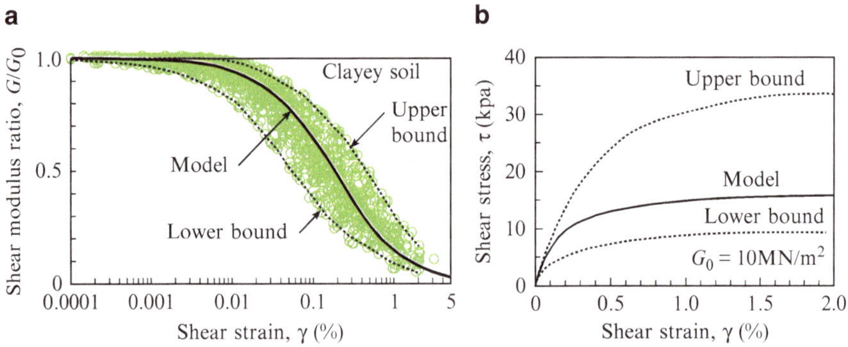

Fig. 6.33 Compilation of G–γ relationships. (**a**) Test result and upper and lower bounds. (**b**) Stress–strain relationships

Empirical equations are generally used when the cyclic shear deformation characteristics test is not carried out. These equations are usually developed as an average value of the test results. However, considering the relationships between the upper bound acceleration and shear strength, using the average value may result in the inaccurate result. An example is shown in Fig. 6.33. This is one of the empirical equations of this intent; original data are shown in Sect. 7.2. The empirical formula is developed so that it passes middle of the test data in Fig. 6.33a. Upper bound and lower bound lines are then added by the author. Stress–strain curves calculated from the proposed model and upper and lower bounds are shown in Fig. 6.33b. Significant difference is observed between the model and upper and lower bound curves, and there may be no engineer who is satisfied by this error, because the difference of a half or double shear strength results in a half and a double difference of peak acceleration when it reaches upper bound acceleration.

As can be seen in this example, the suitability of an empirical equation cannot be measured by its simplicity but by its accuracy. Of course, simplicity is an important factor of an empirical equation, but if accuracy is scarified for simplicity, its usefulness decreases considerably.

When looking at the G/G_0–γ curve, agreement between the empirical equation and the test results appears well suited as the absolute value of G/G_0 is small at large strains. However, when the curve is recalculated into the τ–γ curve, the difference becomes clear, and no engineer wants to use empirical equation that has a half and double error.

6.7.6 *Behavior at Large Strains*

As described in Sect. 6.7.2, the applicability of the conventional cyclic shear deformation characteristics test is up to 0.1 % shear strain or a little more for sands.

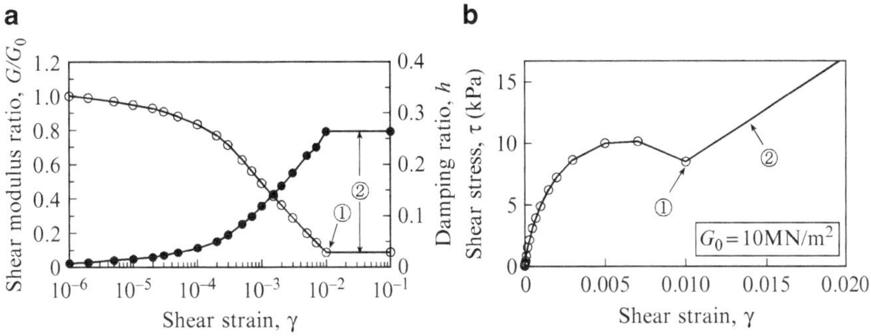

Fig. 6.34 Cyclic shear deformation characteristics and stress–strain curve. (**a**) Cyclic shear deformation characteristics. (**b**) Stress–strain curve

Applicability range is wider for clays, but it cannot become more than several tens percent. Theoretically speaking, one should conduct effective stress analysis so that excess porewater pressure generation can be considered, because limitations of the test arise mainly from the excess porewater pressure generation. However, partly because effective stress analysis is much more difficult than total stress analysis, and partly because effective stress analysis requires more soil data than total stress analysis, many engineers tend to use total stress analysis rather than effective stress analysis. Then, one needs to extrapolate soil data in order to approximate the large strain soil behavior even if the accuracy of the extrapolation may be limited.

Let us show an example for extrapolation. Figure 6.34a shows the cyclic shear deformation characteristics of the Holocene clay illustrated in Sect. 7.2.1 (Iwasaki et al. 1979). Data up to 10^{-2} strain (①) is shown in the figure, whereas Fig. 6.34b shows the stress–strain curve calculated from Fig. 6.34a. The stress decreases at ①, but it is impossible because shear stress amplitude always increases in the cyclic shear deformation characteristics tests. Values of this curve are shown in Table 7.5, in which the last two data are denoted by parentheses. This indicates that they are extrapolated values. Possibly this extrapolation is made in Fig. 6.34a, and the resulting curve seems natural. However, if extrapolation is made by focusing on Fig. 6.34b, the values shall be very much different.

The next line ② in Fig. 6.34a is the region that is not shown in the original data. This extrapolation ② is based on the method employed in SHAKE (Schnabel et al. 1972) which is one of the most frequently used computer programs for the total stress seismic ground response analysis. Constant secant modulus indicates that stress–strain curve is a line that crosses through the origin, and resulting curve cannot be seen to be natural when looking at Fig. 6.34b.

Figure 6.35 shows another example to extrapolation; the stress–strain curve for Toyoura sand is shown (Railway Technical Research 1999). The stress–strain curve also seems aberrant. The author supposes the reason in the following. As described in Sect. 6.7.2, the applicability of the test is 10^{-3} or a little larger strain. Data larger than this strain is, however, requested in practical use. Then the hysteresis

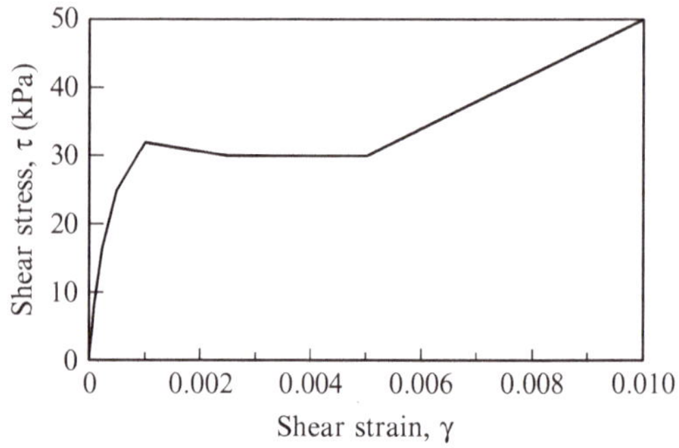

Fig. 6.35 Stress–strain curve (cyclic shear deformation characteristics are shown in Table 7.10)

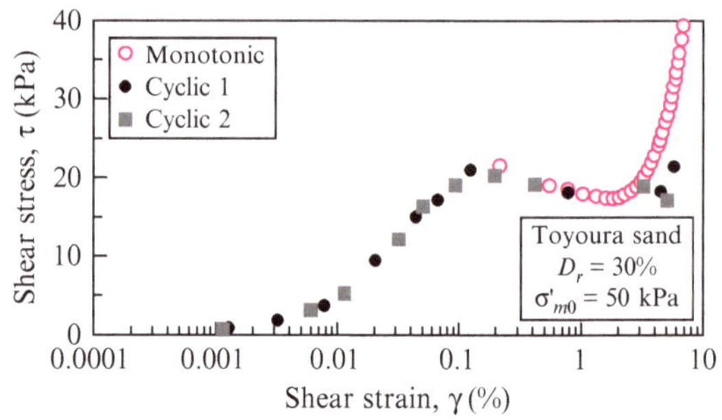

Fig. 6.36 Comparison between monotonic and cyclic loading test (After Kiku and Yoshida 1998)

loop in each cycle of loading, in which shear strain increases as loading cycle, is employed to evaluate the G–γ curve. Subsequently large strain part of the data may be extrapolated one.

Since data at large strains is insufficient, obtaining a good extrapolation is not very likely. A better extrapolation for engineering practice may be, however, possible by considering various behaviors near the shear strength.

Results obtained from the monotonic and the cyclic tests are compared in Fig. 6.36. Agreement looks good. However, the plot of the cyclic loading slips around 10^{-2}, which is obviously caused by the excess porewater pressure generation as explained before. In addition, shear stress measured for cyclic loading is smaller than that for monotonic loading at strains larger than 3 %.

As apparent from the above examples, there are several reasons for this difference between soil behavior under the monotonic loadings and that under the cyclic loadings:

1. Specimen becomes dense as the excess porewater pressure is dissipated between the stages, which results in an increase in shear modulus.
2. Effective mean stress decreases as the excess porewater pressure generates under undrained loading, which results in a decrease of the shear modulus.
3. Cyclic nonlinear behavior may reduce stiffness.

The first factor appears strongly in Fig. 6.19 where the stress under the cyclic loading is larger than that under the monotonic loading. On the other hand, the second feature appears strongly in Fig. 6.36.

As the shear stress of clayey soils increases monotonically as shown in Fig. 7.32, cyclic deformation characteristics test can be performed up to fairly large strains. Then, a shear stress close to the shear strength (say, $0.95\tau_f$) can be taken approximately 5 % shear strain, and a stress–strain curve passing this point may give a good approximation of the stress–strain curve.

On the other hand, as frequently explained, the limitation of the test for sand is 0.1 % or a little larger, which results from the excess porewater pressure generation in sands. It indicates that stress–strain behavior is affected by the past loading history. Therefore, proper extrapolation of sand properties is very difficult or nearly impossible. Knowing this fact, the following procedure may be used.

Onset of the soil liquefaction is identified by either the excess porewater pressure generation or the shear strain amplitude. In the latter case, 5 % double amplitude in cyclic triaxial test is frequently used as liquefaction-onset criterion. This strain corresponds to 3.75 % single amplitude shear strain. It is obviously impossible to trace the post-liquefaction behavior by means of the total stress analysis. In addition, if the excess porewater pressure generation or a decrease in effective stress is not considered, the shear moduli are considered larger than that in the effective stress analysis, which results in an underestimation of shear strains. Considering these situations, a maximum strain of about 2 % is considered to be the upper limit under which the total stress analysis is applied. There is no background on this limit strain but just an imagination by the author; it may change depending on future researches.

In these extrapolations, it should be noted that maximum acceleration may be underestimated if the shear stress is underestimated and vice versa. Many unsolved problem remains on the cyclic shear deformation characteristics at large strain, and they remain for future research (Yoshida and Mikami 2010).

6.7.7 *Effect of Number of Loading Cycles*

Two aspects of the effect of the loading cycles are discussed in this section.

The stiffness and damping ratio are calculated from the hysteresis loop at the 10th cycle of loading. Then it may be interesting to investigate the behavior before the 10th cycle of loading. Figure 6.37 shows measured stiffness and damping ratio at the first, second, and tenth cycles of loading (Yamashita 1992). Decimals on the figure indicate the excess porewater ratio marked at each stage. The excess porewater pressure begins to generate at the strain of 10^{-4}. Then, a change in stiffness and damping ratio in each stage initiates, and shear strain amplitude at the 10th cycle of loading becomes larger than that observed at the first cycle of loading. Increase in shear strain amplitude occurs at the first and second cycles of loading and reaches stable state. The excess porewater pressure ratio reaches 0.35 at the strain of 0.2 %; soil approaches to liquefied state quickly after that.

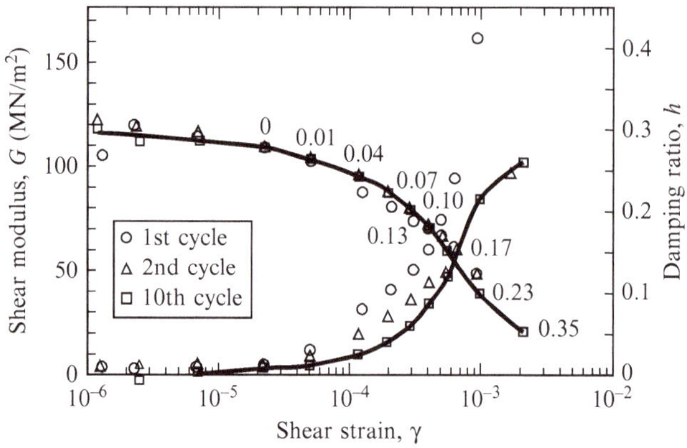

Fig. 6.37 Effect of loading cycles on cyclic shear deformation characteristics (After Yamashita 1992)

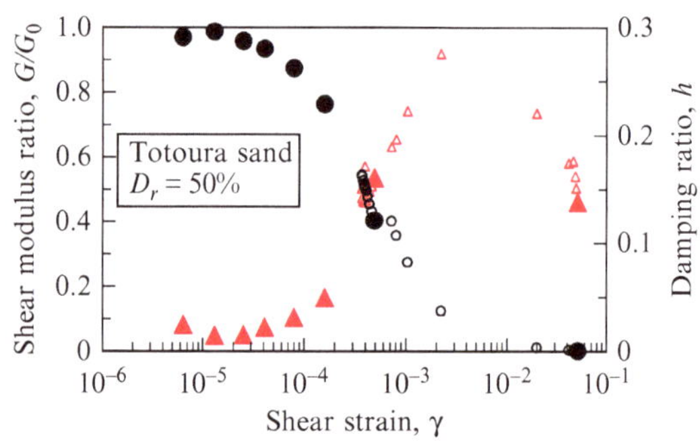

Fig. 6.38 Change in modulus and damping during a stage

The change of shear modulus and damping ratio in each loading stage is shown in Fig. 6.38 (Yoshida et al. 2005). Here large solid symbols represent data at the tenth cycle of loading, and small hollow symbols are data within a stage. It is observed that the shear modulus gradually decreases with each loading cycle even within a stage. The shear modulus at the first cycle of the shear strain amplitude γ_a is about 0.1 greater than that in the last cycle of the previous stage, which clearly results from the densification between the stages. The damping ratio increases in general, but it decreases suddenly at the middle of the last stage, which clearly indicates that significant excess porewater pressure generates.

Results for a cyclic and two monotonic (drained and undrained) loading tests are compared in Fig. 6.39 (Yoshida et al. 2005). The cyclic loading test is a strain-controlled test; load is repeatedly applied in three cycles under constant shear strain amplitude. Excess porewater pressure generated between stages was not allowed to dissipate in order to keep the test specimen at the same density. Shear strain amplitude γ_d is increased 10 times between each stage. The results displayed from larger strain to smaller strain amplitude are shown from the left.

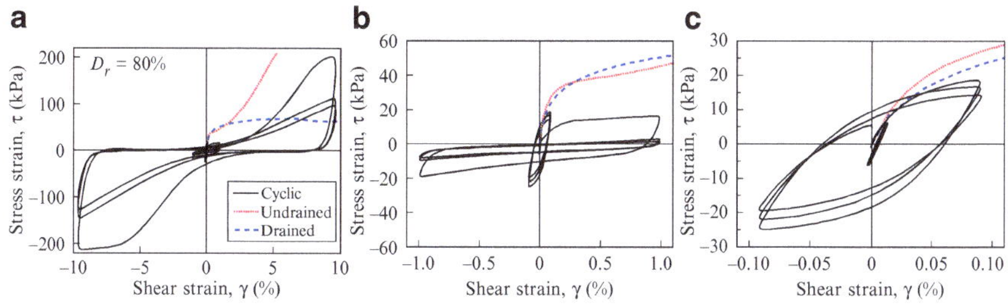

Fig. 6.39 Comparison between cyclic and monotonic loading tests (After Yoshida et al. 2005). (**a**) γ_d=10 %. (**b**) γ_d=1 %. (**c**) γ_d=0.1%

Two monotonic loading test results exhibit very different behaviors at large strains. The shear stress increases to considerably large values because of cyclic mobility in the undrained test, whereas it remains nearly constant because of the confining stress; hence the shear strength is constant in the drained test. The stress–strain curve at the first cycle of loading shows a similar behavior as the undrained test result. As shown in Fig. 6.39a, however, it does not come close to the undrained test result although shear strain increased from 1 to 10 %. The same phenomenon is observed in Fig. 6.39b where strain amplitude increased from 0.1 to 1 %. On the other hand, stress–strain curves at the first loading cycle and two monotonic loading tests show similar behavior in Fig. 6.39c (r_d = 0.1 %) although that of cyclic loading test is somewhat smaller because of the limited excess porewater generation. It is obvious that these differences result from the excess porewater pressure generation.

These test results emphasize a critical issue. The G–γ relationships are assumed implicitly to be a backbone curve or a stress–strain curve under the monotonic loading as explained later in modeling of the stress–strain relationships in Chap. 8. The result in Figs. 6.19 and 6.39 shows that this assumption does not hold in the strain range larger than 0.1 % (or from more small strain). As described several times, behavior at large strains is not made clear in the state of art. Therefore, practical engineers need to use the conventional concept by neglecting this fact.

6.7.8 Initial Stress and Its Effect to Analysis

Stresses that worked in ground before an earthquake are called the initial stresses in the seismic ground response analysis. As discussed in Sect. 10.6, there are several methods to evaluate the initial stresses and each method results in different sets of initial stresses. The effect of the initial stress is discussed in this section without concerning the methodologies used to evaluate them.

Among the initial stresses, the most important one is the initial confining stresses as stiffness and strength of soils are controlled by them. Next, the initial shear is important because apparent shear strength decreases by the existence of the initial shear stress. There are two types of shear stresses (see Sect. 6.4); one is simple shear stress and the other is deviator stress. The simple shear stress exists on the sloped grounds. On the other hand, the deviator stresses exist even in the horizontally layered ground.

The initial deviator stresses can be obtained from the coefficient of earth pressure at rest K_0. The value of K_0 is obtained from in situ measurement such as a borehole lateral loading test. It can also be obtained by the laboratory test on frozen samples (Hatanaka and Suzuki 1995). However, frozen sampling is rarely carried out in practice. Instead, K_0 is decided based on the experience. For example, K_0 is known to be about 0.5 for normally consolidated clays, a little larger than 0.5 for over consolidated clays, and about 0.6 for sandy clays although it increases for dense sand (Glossary dictionary of soil mechanics and Japan Society of Soil Mechanics and Foundation engineering 1985). For sandy soils, Jaky's equation may be used for sandy ground (Glossary dictionary of soil mechanics and Japan Society of Soil Mechanics and Foundation engineering 1985), which is

$$K_0 = 1 - \sin\phi' \tag{6.30}$$

where ϕ' is an internal friction angle based on the effective stress. For instance, Eq. 6.30 yields about 0.5 and 0.36 for sand with $\phi' = 30°$ and $40°$, respectively.

Generally speaking, K_0 is approximately equal to 0.5. Therefore, this value is frequently assumed in the design practice. By the way, Mohr's circle for the case when $K_0 = 0.5$ is drawn in Fig. 6.40 with Mohr–Coulomb failure condition for internal friction angles $\phi = 30°$ and $40°$. The failure line approaches to Mohr's circle for the initial stress, especially in the case of $\phi = 30°$. Therefore, stress–strain curve during earthquake shall be significantly affected by the existence of initial stresses.

The effect of K_0 on cyclic shear deformation characteristics is examined in Fig. 6.41a (Yamashita and Toki 1994). This figure shows that the cyclic shear deformation characteristics are nearly identical at $K_0 = 1$ and 0.5, which seem to conflict with the above discussion. However, it has a trick. The stress–strain curves at a single stage of the tests are compared in Fig. 6.41b. Hysteresis curve is nearly point symmetrical with respect to origin when $K_0 = 1$, whereas it gradually moves toward right when $K_0 = 0.5$. As discussed in Sect. 6.7.2, shear modulus G is evaluated from the shape of hysteresis loop at the 10th cycle of loading. Therefore, Fig. 6.41a just demonstrates that hysteresis loops at the 10th cycle of loading are very similar to

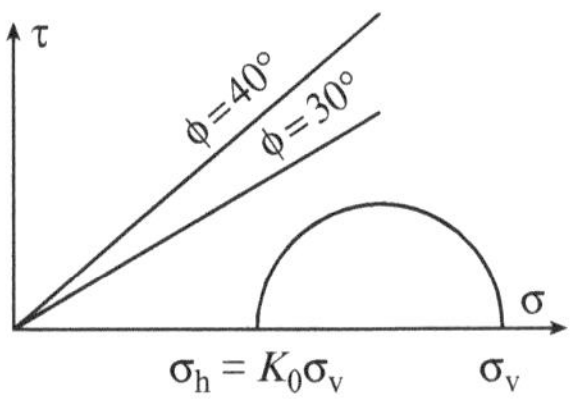

Fig. 6.40 Mohr's circle and failure surface

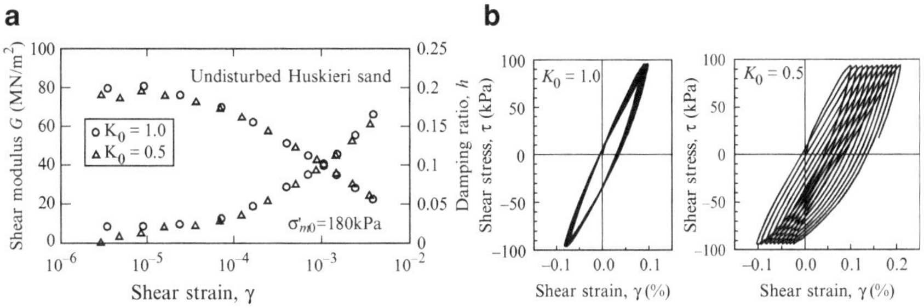

Fig. 6.41 Deformation characteristics under anisotropically consolidated condition. (**a**) Cyclic shear deformation characteristics. (**b**) Stress–strain relationships

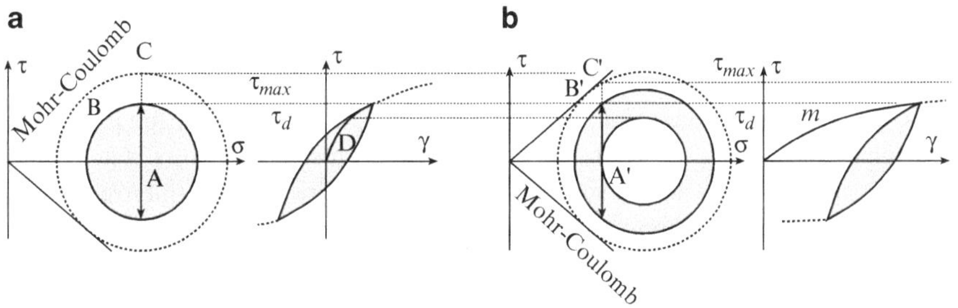

Fig. 6.42 Difference between isotropic and anisotropic initial stresses. (**a**) Isotropic stress state. (**b**) K_0 consolidated stress state

each other. It is noted that the location of hysteresis loop is not counted in the cyclic shear deformation characteristics.

At present, the difference of the effect between the deviator stress and the simple shear stress that causes shear stresses in a separate direction is not well understood. For simplicity, therefore, the effect is discussed by considering that simple shear and deviator stresses work in the same manner. In the horizontally layered ground, only deviator stresses work and shear stresses do not work before the earthquake; stresses are σ_v and $\sigma_h = K_0\sigma_v$ in the actual ground and $\sigma_h = \sigma_v$ in the isotropic stress state. Then the shear stress τ is assumed to work in the horizontal direction under earthquake loading.

Behaviors of an infinitesimally small element for different initial stresses are compared in Fig. 6.42. Let us first consider the case of the isotropic initial stress state. The stress state working in the horizontal plane at the beginning of the earthquake is denoted by point A. This state point shifts back and forth in the vertical direction under the change of the shear stress. Shear stress in the horizontal plane reaches maximum when Mohr's circle reaches the Mohr–Coulomb yield surface, and the shear strength τ_{max} is the same with the radius of Mohr's circle in this state. The shaded circle is the region where Mohr's circle passes under the loading with amplitude τ_d.

Next, let us consider the anisotropic or K_0 initial stress state in Fig. 6.42b. The state point on the horizontal plane at the beginning of loading is point A′. Since Mohr's circle is not a point but a circle, it is initially sheared. Therefore, stiffness at the beginning of loading is smaller than the elastic modulus as shown by a curve *m*. This stress–strain curve is the same with the curve in Fig. 6.42a, but starts from point D where the shear stress equals to the radius of the initial Mohr's circle in Fig. 6.42b. The shear strength τ_{max} in this case occurs at point C' and is obviously smaller than that of isotropic loading case $\left(\overline{A'C'} < \overline{AC}\right)$. The range where Mohr's circle moves is a donut shape (shaded area).

The stiffness at unload is the same with the elastic modulus when unload occurs. Therefore, the hysteresis loop under the constant amplitude loading (shaded area) is located far in the right side from the origin, and it is the same shape with that in Fig. 6.42a. This is the reason why the stiffness and the damping ratio by both test methods are the same.

This fact that the hysteresis loop becomes similar to the one in the isotropic stress state explains many phenomena. For example, the stiffness by the PS logging is the elastic modulus, although the site is usually anisotropic stress state. As the ground is always subjected to small amplitude vibrations such as microtremors or small earthquake, the ground behaves elastically under the strain level used in the PS logging. The behavior is also similar to the ground with isotropic stress state as the natural ground has been subjected to many earthquakes. This may be the reason why the equivalent linear method such as SHAKE has been succeeded although they cannot consider the initial shear. In the numerical analysis, the ground with initial shear behaves similar with the ground with isotropic initial stress state when subjected to several earthquakes with the same magnitude or small earthquakes after subjecting the large earthquake (Yoshida 1996).

On the other hand, there are grounds that do not have hysteresis of the shear loading such as a fill and a reclaimed land. These grounds have suffered significant damage in the past earthquakes, which may be because of initial shear. Therefore it is important to determine the initial stress in these grounds.

References

Boulanger RW, Idriss IM (2006) Liquefaction susceptibility criteria for silts and clays. J Geotech Geoenviron 132(11):1413–1426

Chiba T, Yamashita S, Toki S (1994) Cyclic deformation characteristics of sand by triaxial and cylindrical shear tests. In: Proceedings of the symposium on dynamic deformation characteristics in dynamic problem of ground and soil structures -test and investigation methods and its application, JGS, pp. 181–186 (in Japanese)

Committee on Dynamic Property of Clay (1995) State of art report on dynamic property of clay. In: Proceedings of the symposium on dynamic property of clay, JGS, pp. 98–114 (in Japanese)

Fukumoto S, Yoshida N, Sahara M (2009) Dynamic deformation characteristics of sedimentary soft rock. J JAEE 9(1):46–64 (in Japanese)

Glossary dictionary of soil mechanics, Japan Society of Soil Mechanics and Foundation engineering, 2nd revision, 1985, 648 pp. (in Japanese)

Hara A (1973) Kinematic property of ground and its application. In: Proceedings of the 2nd symposium of earthquake ground motion, AIJ, pp. 33–39
Hardin BO, Drnevich VP (1972) Shear modulus and damping in soils: design equations and curves. Proc Am Soc Civ Eng 98(SM7):667–692
Hatanaka M, Suzuki Y (1995) Two methods for the determination of lateral stress in sand. Soils Found 35(2):77–84
Hatanaka M, Suzuki Y, Kawasaki T, Endo M (1988) Cyclic undrained shear properties of high quality undisturbed Tokyo gravel. Soils Found 28(4):57–68
Ishihara K (1996) Soil behavior in earthquake geotechnics, vol 46, Oxford Engineering Science Series. Oxford Science, Oxford
Iwai S, Nakamura T, Yoshida N, Wakabayashi M (1982) Effects of loading rate on the performance of structural elements – part 1 effects of the strain rate on the stress-strain relationships of concrete and steel. Trans Archit Inst Jpn 314:102–111 (in Japanese)
Iwasaki T, Tokita K, Yoshida S (1979) Dynamic deformation characteristics of Holocene clay, PWRI report, No. 1504, Public Work Research Institute, Tsukuba, Japan
Iwasaki T, Tatsuoka F, Takagi Y (1980a) Experimental study on dynamic deformation characteristics of soils (ii), Report of PWRI, vol. 153
Iwasaki T, Tatsuoka F, Takagi Y (1980b) Experimental study on dynamic deformation characteristics of soils (ii), Report of PWRI, vol. 153 (in Japanese)
Japanese Geotechnical Society (2000) Chapter 6 Triaxial undrained test to determine liquefaction strength of soil, Chapter 7: Cyclic loading test to determine deformation characteristics. In: Standards of Japanese Geotechnical Society for laboratory test, 1st revise, 902 pp (in Japanese)
Japanese Geotechnical Society (2004) Remedial measures against soil liquefaction. Geotechnical engineering practice series 18, 513 pp (in Japanese)
JGS (1995) Japanese standards for geotechnical and geoenvironmental investigation methods. In: Standards and explanations, 889 pp (in Japanese)
JGS (2009) Standards of Japanese Geotechnical Society for laboratory test, pp.730–789 (in Japanese)
Katayama I, Fukui S, Sato M, Makihara Y, Tokimatsu K (1986) Comparison of dynamic soil properties between undisturbed and disturbed dense sand samples. In: Proceedings of the 21st Japan National Conference on geotechnical engineering, pp 583–584
Kiku H, Yoshida N (1998) Dynamic deformation property tests of sand at large strains. In: Proceedings of the 33rd Japan National Conference on geotechnical engineering, pp 869–870 (in Japanese)
Kokusho T (1980) Cyclic triaxial test of dynamic soil properties for wide strain range. Soils Found 20(2):45–60
Kokusho T (1982) Dynamic deformation characteristics of soil and nonlinear response of ground, Report No. 301, Central Research Institute of Electric Power Industry (in Japanese)
Kokusho T (1987) In-situ dynamic soil properties and their evaluation. In: Proceedings of the 8th Asian Regional Conference of SMFE, Kyoto, vol II, pp 215–240
Kondner RL (1963) Hyperbolic stress-strain response; cohesive soils. Proc ASCE SM1:115–143
Momose S, Nakajima T, Tsuchiya H (1996) Evaluation of hydroelastic cyclic loading system for a liquefaction potential test. In: The 31st Japan National Conference of Soil Mechanics and Foundation Engineering, pp 977–978 (in Japanese)
Murono Y (1990) A study on the seismic design methodology of pile foundation considering the non-linear dynamic interaction caused by strong ground motions, RTRI Report, Special No. 32, Railway Technical Research Institute (in Japanese)
Nishi K, Kudo K, Ishiguro K, Nishi T (1986) Dynamic deformation properties of weathered mudstone. In: Proceedings of the 21st Japan National Conference on Geotechnical Engineering, pp. 1019–1022
Proceedings of the symposium on Dynamic deformation characteristics in dynamic problem of ground and soil structures -test and investigation methods and its application, JGS, 1994 (in Japanese)

Railway Technical Research Institute (1999) Design standard and commentary of railway facilities, Maruzen (in Japanese)
Research Committee on Early Effective Use of Wasted Material Fill (1991) Investigation technique. In: Proceedings of the symposium on use of wasted material fill, JGS, pp. 17–20
Sakai K, Kaneko S, Yukawa H (2006) High grade sampling of gravel (GP sampling). Found Eng Equip 34(9):44–47 (in Japanese)
Sato T, Tatsuoka F, Kowata Y (1994) Various problems on accurate measurement in cyclic triaxial test. In: Proceedings of the symposium on dynamic deformation characteristics in dynamic problem of ground and soil structures -test and investigation methods and its application, JGS, pp. 153–156 (in Japanese)
Satoh T, Horike M, Takeuchi Y, Uetake T, Suzuki H (1997) Nonlinear behavior of scoria soil sediments evaluated from borehole record in eastern Shizuoka prefecture Japan. Earthq Eng Struct Dyn 26:781–795
Schnabel PB, Lysmer J, Seed HB (1972) SHAKE A Computer program for earthquake response analysis of horizontally layered sites, Report No. EERC72-12, University of California, Berkeley
Suzuki T, Takenobu M, Sugano T, Nakazawa H (2008) Effect of setting of parameters in FLIP on result of dynamic response analysis. In: Proceedings of the 5th Kanto branch Conference on Geotechnical Engineering, pp. 239–242 (in Japanese)
Tatsuoka F. Shibuya S (1992) Deformation characteristics of soils and rocks from field and laboratory tests. Rep Inst Ind Sci Univ Tokyo 37(1):1–136
Tatsuoka F, Iwasaki T, Tagami Y (1978) Hysteretic damping of sands under cyclic loading and its relation to shear modulus. Soils Found 18(2):25–40
Tokimatsu K (1989) Dynamic property of soil from laboratory test, in-situ rest and earthquake measurement. In: Proceedings of the 2nd symposium on dynamic interaction between soil and structure, pp. 11–16 (in Japanese)
Tokimatsu K, Hosaka Y (1986) Effects of sample disturbance on dynamic properties of sand. Soils Found 26(1):53–64
Tokimatsu K, Oh-hara J (1990) Frozen sampling, Tsuchi-to-Kiso. JGS 38(11):61–68 (in Japanese)
Ueno K, Miura K, Tanise A, Maeda K (1991) Method for measuring the shear band in granular materials. In: The 26th Japan National Conference on Geotechnical Engineering, pp. 533–536 (in Japanese)
Yamashita S (1992) A study on effect of variety of factors that affect cyclic shear deformation characteristics of sand and its application. Thesis in the Department of Civil Engineering presented in partial fulfillment of the requirements for the degree of Doctor of Engineering, Hokkaido University, 258pp (in Japanese)
Yamashita S, Toki S (1994) Effect of initial stress on cyclic deformation characteristics. In: Proceedings of the symposium on deformation characteristics of soils in dynamic problem of soils and soil structures, JSSMFE, pp 163–168 (in Japanese)
Yasuda S, Yamaguchi I (1984) Dynamic shear moduli in the laboratory and the field, Proc. Symposium on evaluation of deformation and strength characteristics of sandy soils and sand deposits, JSSMFE, Tokyo, pp 115–118 (in Japanese)
Yasuda S, Nagase H, Oda S, Kitsuji K (1994) Comparison of dynamic shear modulus and damping ratio between stage-test method and fresh-test method. In: Proceedings of the symposium on dynamic deformation characteristics in dynamic problem of ground and soil structures -test and investigation methods and its application, JGS, pp 127–132 (in Japanese)
Yoshida N (1994) Applicability of conventional computer code SHAKE to nonlinear problem. In: Proceedings of the symposium on amplification of ground shaking in soft ground, JSSMFE, pp 14–31 (in Japanese)
Yoshida N (1996) Initial stress effect on response of level ground. In: Proceedings of the 11th World Conference on Earthquake Engineering, Acapulco, Mexico, Paper No. 1023
Yoshida N (1998) Can effective stress analysis simulates actual phenomena, fundamental concepts of the design of foundation in the liquefiable ground., Committee on Structural Engineering, AIJ, pp. 47–92 (in Japanese)

Yoshida N, Mikami T (2010) Establishment of cyclic shear deformation characteristics test that responds to demand of times. Geotech Eng Mag 58(2):1–5 (in Japanese)
Yoshida N, Tsujino S, Ishihara K (1990) Stress-strain model for nonlinear analysis of horizontally layered deposit. In: Summaries of the technical papers of annual meeting of AIJ, Chugoku, vol B (Structure I), pp. 1639–1640 (in Japanese)
Yoshida N, Mikami T, Sawada S, Kiku H (2005) Laboratory test on cyclic behavior of soil for earthquake response analysis. In: Proceedings of the 40th Japan National Conference on Geotechnical Engineering, pp. 459–460 (in Japanese)
Zen K, Yamazaki H, Umehara Y (1987) Experimental stud on shear modulus and damping ratio of natural deposits for seismic response analysis, Report of the Port and harbour Research Institute, Ministry of Transport, vol. 28, No. 1, pp. 41–113

Chapter 7
Estimation of Mechanical Soil Properties

The best way to obtain the mechanical soil properties for seismic ground response analysis is in situ test for the elastic modulus and laboratory test on undisturbed sample for the nonlinear properties. In the engineering practice, however, this is not always possible. The cyclic deformation characteristics test is a costly test in the practice. The frozen sample requires huge cost to retrieve the highly undisturbed sample, and its applicability is limited to clean sand.

As a means to overcome this issue, from a practical perspective, test data and empirical equations are introduced mainly based on the Japanese researches in this chapter. It is noted that in using an empirical equation, tests on the similar material of concern are to be preferred. In addition, it is essential to understand that test data sometimes scatter very much resulting in a degree of error in analysis. This aspect of the problem is discussed in Sect. 15.1.

7.1 Elastic Properties

A direct in situ measurement of elastic modulus is difficult to perform mainly because accurate measurement of strain is difficult. Therefore, instead, the wave velocities are measured for evaluating the elastic modulus. The elastic modulus can be calculated from the wave velocity by using Eq. (5.8). There are many empirical equations available for evaluating the shear wave velocity from the SPT N-value. Equation (5.8) requires density which is usually obtained during a standard borehole investigation.

N. Yoshida, *Seismic Ground Response Analysis*, Geotechnical, Geological and Earthquake Engineering 36, DOI 10.1007/978-94-017-9460-2_7

7.1.1 Equation by Imai et al.

The pioneering research to correlate shear wave velocity and SPT N-value was performed by Imai et al. (Imai 1977). They gathered V_s and V_p data from various soils and found that there are correlations between the V_s and the SPT N-value. For all soils, the relationship yields

$$V_s = 91.0N^{0.337} \quad (r = 0.889)\,[\mathrm{m/s}] \tag{7.1}$$

where r denotes coefficient of correlation. In addition, the relationships shown in Fig. 7.1 were proposed for soils classified as per soil type and geologic age, expressed as

$$\begin{aligned} V_s &= 102N^{0.292} && \text{Holocene clay (Ac)} \\ V_s &= 80.6N^{0.331} && \text{Holocene sand (As)} \\ V_s &= 114N^{0.294} && \text{Pleistocene clay (Dc)} \\ V_s &= 97.2N^{0.323} && \text{Pleistocene sand (Ds) [m/s]} \end{aligned} \tag{7.2}$$

Dashed lines in Fig. 7.1 are added by the author and they correspond to half and twice of the S-wave velocities evaluated from Eq. (7.2). It can be easily observed that the scatter within the data lies between the indicated boundary lines. In other words, these equations exhibit a variation between the half and the twice of the measured velocities. As the shear modulus is proportional to the square of the S-wave velocity, this yields scatter in the shear modulus ranging from 1/4 to 4 times. Recall that a change in the elastic modulus results in shift in the shear strength (see Sect. 6.5). As explained in Sect. 1.3, in addition, the shear strength is related to the maximum acceleration at the ground surface. This indicates that the maximum acceleration also scatters within 1/4 and 4 times the mean values. This scatter may be outside of allowable range in an engineering practice.

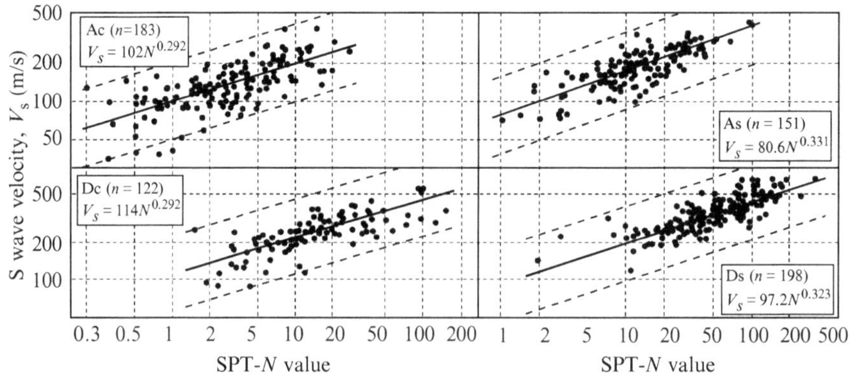

Fig. 7.1 Relationships between SPT N-value and V_s for various soil types (Modified from Imai 1977)

Table 7.1 Relationship between S-wave velocity, elastic shear modulus, and SPT N-value

Soil type	$V_s = AN^m$ [m/s]			$G_0 = BN^n$ [MN/m^2]		
	A	m	r	B	n	r
Clayey fill	98.4	0.248	0.574	15.4	0.557	0.582
Sand or gravel fill	91.7	0.257	0.647	14.2	0.500	0.647
Holocene clay	107	0.274	0.721	17.6	0.607	0.715
Holocene peat	63.6	0.453	0.771	5.37	1.08	0.768
Holocene sand	87.8	0.292	0.690	12.5	0.611	0.671
Holocene gravel	75.4	0.351	0.791	8.25	0.767	0.788
Loam, Shirasu	131	0.153	0.314	22.4	0.383	0.497
Tertiary sand/clay	109	0.319	0.717	20.4	0.668	0.682
Pleistocene clay	128	0.257	0.712	25.1	0.555	0.712
Pleistocene sand	110	0.285	0.714	17.7	0.631	0.729
Pleistocene gravel	136	0.246	0.550	31.9	0.528	0.552

Note: r denotes coefficient of correlation

Subsequent to this report, they gathered more data and proposed empirical equations for the S-wave velocity V_s as well as the elastic shear modulus G_0 through Eq. (5.8) (Imai and Tonouchi 1982a). The results are summarized in Table 7.1. Here coefficients of correlation are similar for V_s and G_0, which indicates that the density can be measured accurately.

S-wave velocities used in this study were obtained through downhole test. As explained in Sect. 5.2, correlation between the S-wave velocity and the SPT N-values may not be appropriate for a detailed analysis because V_s measured by the downhole method is sometimes an average for a certain layer thickness. Therefore, the involved variability might be smaller if S-wave velocity obtained through the suspension logging method is used. In addition, the confining stress dependency of the SPT N-value is not considered in this paper. The confining stress dependency of the SPT N-value is not considered, too, in a majority of the subsequent empirical equations.

7.1.2 *Evaluation by Japan Road Bridge Design Specifications*

As can be seen in Fig. 7.1, range of the SPT N-value depends on the soil type and the geologic age. This signifies that applicability range of an empirical equation need not be infinitely large but can be bounded to a small extent of SPT-N value. Considering this, a more simple equation is provided by the Japan Road Association (1985):

$$\begin{aligned} V_s &= 100N^{1/3} \quad \text{Clayer soil} \\ V_s &= 80N^{1/3} \quad \text{Sandy soil} \end{aligned} \tag{7.3}$$

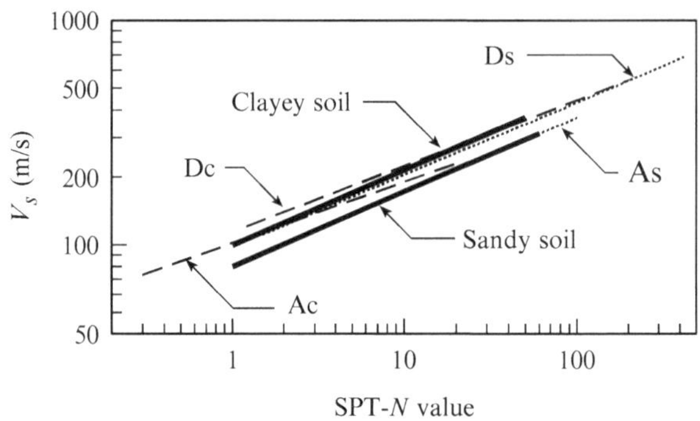

Fig. 7.2 Comparison between Imai et al. and road bridge (After Japan Road Association 2002) (*Thin line* indicates Eq. 7.2, and *thick line* indicates Eq. 7.3)

These equations may be more useful in the practice than the former equations because it is sometimes difficult to distinguish the Holocene and the Pleistocene soils. A comparison between Eqs. (7.2) and (7.3) is shown in Fig. 7.2. Equation (7.3) is similar to the Holocene sand in Eq. (7.2). The applicable ranges are between 1 and 15 SPT *N*-value for clays and between 1 and 50 SPT *N*-value for sands.

7.1.3 Equations Developed for Port Facilities

Various empirical equations that are applicable for the remedial measures against soil liquefaction of fill material are introduced in Coastal Development Institute of Technology (1997). Among them, representative equations are Eq. (7.4) (sandy soil (Imai and Tonouchi 1982b)) and Eq. (7.5) (Clayey soil (Zen et al.1987a)):

$$G_0 = 144N^{0.68} \tag{7.4}$$

$$G_0 = 170q_u \tag{7.5}$$

where q_u denotes unconfined compression strength. A rather good correlation between the elastic modulus and the shear strength is known to exist for clayey soils. In addition, Eq. (7.6) is introduced as confining stress dependency (Zen et al. 1987a; Uwabe et al. 1982):

$$V_s = V_{s0}\left(\frac{\sigma'_v}{\sigma'_{v0}}\right)^B \tag{7.6}$$

where V_{s0} and V_s denote the S-wave velocities before and after the construction, respectively, and σ'_{v0} and σ'_v are effective overburden stresses before and after the construction, respectively. Finally, exponent B is a power of the confining stress dependency and is defined as 0.25 for sandy and 0.5 for clayey soils. Thus, referring

Table 7.2 S-wave velocity of various materials frequently used in port constructions

Material	V_s (m/s)
Riprap of mound	300
Backfill	225
Caisson	2,000

in Eq. (5.8), the elastic modulus of the sand is proportional to the square root of the effective confining stress and that of the clay is proportional with the effective stress itself. Here, equation for sandy soil is relevant for soils with plastic index I_p less than or equal to 30 and that for clayey soils with I_p greater than 30.

Table 7.2 also shows the S-wave velocity of the geomaterial frequently used in the port area.

In addition, Eq. (7.7) is shown in the design specification for port facilities (1989 version) (Ports and Harbours Bureau, Transport Ministry 1999):

$$\begin{aligned}
G_0 &= 98\left(285 - 2I_p\right)\sigma'_m && \left(I_p \geq 30\right)\\
G_0 &= 9.90\frac{\left(1.6I_p + 185\right)(2.973 - e)^2}{1+e}\sigma'^{\,0.5}_m && \left(I_p < 30\right)\\
G_0 &= 6929\frac{(2.17 - e)^2}{1+e}\sigma'^{\,0.5}_m && \text{(Sandy soil with round particle)}\\
G_0 &= 3267\frac{(2.973 - e)^2}{1+e}\sigma'^{\,0.5}_m && \text{(Gravelly soil with angular particle)}
\end{aligned} \tag{7.7}$$

7.1.4 Equations Frequently Used in Buildings Design

Ohta and Goto (1978) classified the indices that affect elastic modulus into four categories when developing an empirical equation. They are SPT N-value, depth, geologic age, and soil type. Here SPT N-value and depth are quantitative indices, whereas geologic age and soil type are quantitative indices which cannot be counted in the ordinary sense. A multivariate analysis is possible to evaluate best parameters if indices are composed of quantitative indices only, but it cannot be used when qualitative indices are involved. They developed an alternative procedure to consider both quantitative and qualitative parameters. Totally 300 data were compiled with V_s ranging from 50 to 620 m/s, SPT N-values between 2 and 200, and depth between 1 and 80 m. The coefficient or correlation depends on the choice of the parameters for an individual analysis set. The highest coefficient of correlation was obtained when considering all four parameters, resulting in

$$V_s = 68.79N^{0.171}H^{0.199}EF \quad (r = 0.856) \tag{7.8}$$

Table 7.3 Values of parameters of Eqs. (7.8) and (7.9)

Equation (7.8)				Equation (7.9)			
Geologic age	*E*	Soil type	*F*	Geologic age	*E*	Soil type	*F*
Holocene	1.000	Clay	1.000	Holocene	1.000	Clay	1.000
Pleistocene	1.303	Fine sand	1.086	Pleistocene	1.448	Fine sand	1.056
		Medium sand	1.066			Medium sand	1.013
		Course sand	1.135			Course sand	1.039
		Sand gravel	1.153			Sand gravel	1.069
		Gravel	1.448			Gravel	1.221

where H denotes layer depth in meters and E and F are influence coefficients for the geologic age and the soil type, respectively, and are tabulated in Table 7.3.

If effect of layer depth is neglected, then the following equation is obtained:

$$V_s = 93.10N^{0.249}EF \quad (r = 0.787) \tag{7.9}$$

Values of E and F are also shown in Table 7.3. Equation (7.8) is frequently referred in publications on the building design such as Building Research Institute, Land, Infrastructure and Transportation Ministry (2001).

A simpler empirical equation derived by using a similar approach was proposed from Japanese Central Disaster Prevention Council (2005) as

$$V_s = 112.73N^{0.256}EF \tag{7.10}$$

where E is 1.000, 1.223, and 1.379 for the Holocene, the Pleistocene, and the Tertiary age soils, respectively, and F is 1.000, 0.855, and 0.90 for clayey, sandy, or gravelly soils, respectively.

7.1.5 Equations by Iwasaki et al.

Iwasaki et al. (1977) derived relationships between the V_s and SPT N-values through least square analysis on the borehole test results in the coastal area in Tokyo, Kawasaki, and Kobe as

$$\begin{aligned}
V_s &= 103N^{0.211} && \text{Holocene sandy soil (As)} && (N = 1\text{–}30)\\
V_s &= 143N^{0.0777} && \text{Holocene clayey soil (Ac)} && (N = 1\text{–}7)\\
V_s &= 205N^{0.125} && \text{Pleistocene sandy soil (Ds)} && (N = 5\text{–}500)\\
V_s &= 172N^{0.183} && \text{Pleistocene clayey soil (Ds)} && (N = 2\text{–}200)
\end{aligned} \tag{7.11}$$

Here, SPT N-values in the parenthesis resemble the region shown in Fig. 7.3. They also reported that strain range in the elastic wave measurement is in the order of

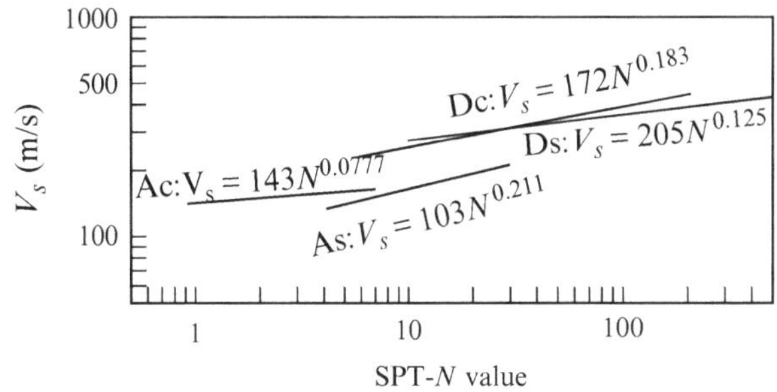

Fig. 7.3 Relationships between SPT N-value and V_s shown by Iwasaki et al.

10^{-5} near the ground surface and order of 10^{-8} at depth greater than 50 m. Note that in this saturation, the power in the SPT N-value is smaller than the prior empirical equations.

7.1.6 Equations Based on Laboratory Tests

As discussed in Sect. 6.5, the elastic moduli obtained through the laboratory test by means of, so-called, undisturbed samples do not have sufficient accuracy. Therefore, the error for reconstituted sample is to be much larger than them. This indicates that the use of laboratory test results instead of the in situ elastic modulus is not suited for practical applications. In addition, it is difficult to obtain the effective mean stress σ'_m because the coefficient of earth pressure at rest K_0 is difficult to measure accurately.

The soils employed in the shaking table or the centrifuge tests are fresh soils. Then, empirical equations based on laboratory test are applicable. In the same manner, they may be applicable to the filled soil as they are also freshly reconstituted soils.

Many empirical equations have been proposed in the past, which are summarized in Table 7.4. Generally speaking, each empirical equation is expressed as functions of the void ratio e and the effective confining stress σ'_m as

$$G_0 = Af(e)\sigma'^n_m \tag{7.12}$$

where A denotes equation coefficient and $f(e)$ denotes the function of the void ratio e. Effects resulting from the presence of the pores can also express by a relative density D_r, but the expression by means of the void ratio e is considered to be good for the clean sands (Adachi and Tatsuoka 1986). The term "2.17−e" appears frequently. This term appeared in the first study of this kind and the following studies seem to use the same term. As pointed out in Iida (1939), S-wave does not propagate in the sand when e is larger than 2, and this term seems to be consistent with it.

Table 7.4 Empirical equations on elastic shear modulus obtained through laboratory test

	References	First author	A	$f(e)$	n	Sample	Test	Comments
Sand	Hardin and Richart (1963)	Hardin	7,000	$(2.17-e)^2/(1+e)$	0.5	Round Ottawa sand	R	$\gamma \leq 10^{-4}$
			3,300	$(2.97-e)^2/(1+e)$	0.5	Angular quartzes sand		
	Shibata and Soelarno (1975)	Shibata	42,000	$0.67-e/(1+e)$	0.5	Three clean sands	P	
	Iwasaki et al. (1978)	Iwasaki	9,000	$(2.17-e)^2/(1+e)$	0.38	11 clean sands	R	$\gamma = 10^{-6}$
	Kokusho (1980b)	Kokusho	8,400	$(2.17-e)^2/(1+e)$	0.5	Toyoura sand	T	
	Yu and Richart (1984)	Yu et al.	7,000	$(2.17-e)^2/(1+e)$	0.5	Three clean sands	R	
	Lo Presti et al. (1997)	Lo Presti	9,014	$e^{-1.3}$	0.45	Toyoura sand	H, R	
	Numata et al. (2000)	Numata	29,718	$(0.79-e)/(1+e)$	0.55	Three clean sands	T	Frozen sample
	Asonuma et al. (2002)	Asonuma	10,276	$e^{-2.46}$	0.52	Undisturbed Shirasu	T	
	Saxena and Reddy (1989)	Saxena	3062	$1/(0.3+0.7e^2)$	0.574	Monterey sand	R	$\gamma = 10^{-5}$
	Hardin and Black (1968)	Hardin	31.5	$\frac{(32.17-14.8e)^2}{1+e}$	0.5	Ottawa sand, $\sigma'_m > 95.75$	R	$\gamma = 2.5 \times 10^{-4}$
			23.15	$\frac{(22.52-10.6e)^2}{1+e}$	0.5	Ottawa sand, $\sigma'_m > 95.75$		
Clay	Hardin and Black (1968)	Hardin	3,300	$(2.97-e)^2/(1+e)$	0.5	Kaolinite et al.	R	Over consolidation
	Marcuson and Wahls (1972)	Marcuson	4,500	$(2.97-e)^2/(1+e)$	0.5	Kaolinite, $I_p = 35$	R	Over consolidation and time effect
			450	$(4.4-e)^2/(1+e)$	0.5	Bentonite, $I_p = 60$		
	Zen et al. (1978)	Zen	2,000–4,000	$(2.97-e)^2/(1+e)$	0.5	Disturbed clay, $I_p = 0 \;\; 50$	R	A depends on I_p
	Kokusho et al. (1982a)	Kokusho	141	$(7.32-e)^2/(1+e)$	0.6	Undisturbed clay, $I_p = 40 \;\; 85$	R	Over consolidation

Organic	Ishihara et al. (2003)	Ishihara	0.236	$(91.5 - e)^2/(1 + e)$	0.65	Undisturbed organic soil	R	Block sample
Gravelly soil	Prange (1981)	Prange	7,230	$(2.97 - e)^2/(1 + e)$	0.38	Ballast, $D_{50} = 40$ mm, $U_c = 3.0$	R	
	Kokusho and Esashi (1981)	Kokusho	13,000	$(2.17 - e)^2/(1 + e)$	0.55	Quarry, $D_{50} = 30$ mm, $U_c = 10$	T	
			8,400	$(2.17 - e)^2/(1 + e)$	0.6	Round gravel, $D_{50} = 10$ mm, $U_c = 2.0$		
	Tanaka et al. (1987)	Tanaka	3,080	$(2.17 - e)^2/(1 + e)$	0.6	Gravel, $D_{50} = 10$ mm, $U_c = 2.0$	T	
	Goto et al. (1987)	Goto	1,200	$(2.17 - e)^2/(1 + e)$	0.85	Gravel, $D_{50} = 10.7$ mm, $U_c = 13.8$	T	
	Nishio et al. (1985)	Nishio	9,360	$(2.17 - e)^2/(1 + e)$	0.44	Undisturbed gravel, $D_{50} = 10.7$ mm, $U_c = 13.8$	T	
	Asonuma et al. (2002)	Asonuma	2,488	$e^{-0.56}$	0.68	Volcanic ash soil, $D_{50} = 6.6$ mm, $U_c = 6.0$	T	
	Tanaka et al. (1985)	Tanaka	2,056	$(2.17 - e)^2/(1 + e)$	0.62	Gravel contents = 25 %	T	Reconstituted sample
			3,079	$(2.17 - e)^2/(1 + e)$	0.6	Gravel contents = 50 %		
	Nishi et al. (1983)	Nishi	2,393	$(2.17 - e)^2/(1 + e)$	0.66			

Test methods: *R* resonance column, *T* triaxial, *P* pulse, and *H* torsional shear tests

When using the empirical equations, one needs to take into consideration the test method and the effective strain range for which the interested relationship is pertinent. According to Shibata and Soelarno (1975), elastic modulus obtained using the simple shear method is a half of that determined by the ultrasonic pulse test. It was very difficult to measure behaviors at small strains accurately in the old days by static test. Therefore, resonant column tests were frequently used. However, at present, small strains with 10^{-6} level can be measured according to the development of measurement devices. As described in Sect. 6.7.4, on the other hand, many engineers feel that the damping ratio becomes more meaningful when shear strains are larger than 10^{-5} or 5×10^{-5} (JGS 1994). This may indicate that the accuracy may be taken less at small strain for nonacademic purposes.

Figure 7.4 shows variation of the exponent n of the confining stress dependency (Kokusho 1980a). The value of n is less than 0.5 at the strain of 10^{-6}. It gradually reaches 0.5 at strains 10^{-5}–10^{-4} and becomes larger with strains. Therefore, values of the coefficient A as well as exponent n in Eq. (7.12) depend on the effective strain level of the test procedure. The grain size also affects the coefficient A for gravelly soils.

On the other hand, $n = 0.5$ is frequently used in the engineering practice. As shown in Fig. 7.4, however, this n value corresponds to 10^{-5}–10^{-4} shear strain and does not resemble the elastic state.

There is another method to evaluate appropriate value for the exponent n. As V_s, from Eq. (7.3), is proportional to $N^{1/3}$, elastic shear modulus is proportional to the effective confining stress with exponent 2/3 from Eq. (5.8). Note that confining stress dependency is not considered in Eq. (7.3). SPT N-value is proportional to the power of a half of the effective overburden stress when corrected SPT N-value N_1 is identical to that in Eq. (5.6). In conclusion, the elastic shear modulus appears to be proportional to the effective confining stress with exponent of 1/3. This exponent seems to agree with Fig. 7.4 at very small strain.

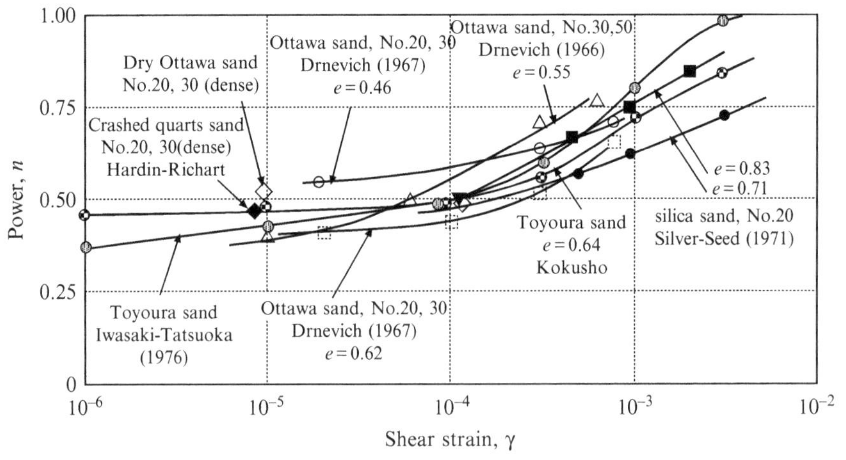

Fig. 7.4 Confining stress dependency of shear modulus exponent (After Kokusho 1980a)

7.2 Nonlinear Properties

As mentioned, nonlinear soil properties are obtained through the cyclic shear deformation characteristics test and the obtained results are usually expressed in terms of the G–γ and the h–γ relationships. Equations used in the practical design and test results on various materials are introduced in this section. Notes on how to employ them were already discussed in Sect. 6.7.

General feature of nonlinear soil properties is introduced first. A schematic figure of a typical cyclic shear deformation characteristics test result is shown in Fig. 7.5. Two sets of test data, denoted as L and M, are shown in the figure. As explained in Sect. 6.5, left ordinate displays the shear modulus ratio for which the maximum value that can be attained is 1.0. In practice, however, the maximum value of test results may become slightly smaller than 1.0 when the maximum shear modulus is evaluated using, for example, the Hardin–Drnevich model, and it may be a little larger than 1.0 when, for example, the shear modulus at the smallest shear strain is used as denominator of the shear modulus ratio. On the other hand, the damping ratio is written in the right ordinate. It is expressed as ratio or in percent.

The larger shear modulus ratio performance (L in the figure) can be interpreted as the nonlinear behavior appears at larger strain range than the smaller shear modulus ratio data such that for M. Therefore, the damping ratio of data L is generally smaller than that of data M.

Shear strain at $G/G_0 = 0.5$ is often denoted as $\gamma_{0.5}$ and is frequently used to be a representative value for the G/G_0–γ curve. It is used in many mathematical models such as Hardin–Drnevich model (Sect. 6.6), hyperbolic model (Sect. 8.2.3), and Ramberg–Osgood model (Sect. 8.2.4), in which it is termed the reference strain and is designated as γ_r.

The damping ratio at small strains is sometimes close to 0, while in some cases finite value 2–4 % (Sect. 6.7.4). As the maximum damping ratio scatters from less than about 0.1 (10 %) to nearly 0.4 (40 %) depending on soil type, the maximum value of the left ordinate may significantly differ from figure to figure. It may be fixed to 1.0 (100 %), 0.6 (60 %), 0.4 (40 %), 0.3 (30 %), or 0.2 (20 %) for various

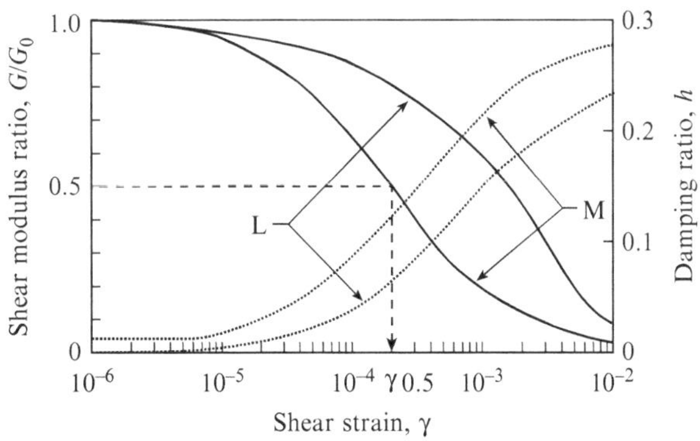

Fig. 7.5 Example of cyclic shear deformation characteristics

purposes. Coordinate axis becomes the same in both left and right ordinates when 1.0 is used. As a maximum value of 0.6 is close to theoretical maximum value ($2/\pi$), almost all data can be drawn when coordinate axis is set to 0–0.6, whereas data is better outlined when the maximum of 0.2–0.4 is chosen.

7.2.1 *Equations by PWRI*

Public Work Research Institute, Land, Infrastructure, and Transportation Ministry of Japan carried out series of test on various soils and reported for separate soil types: Holocene clay (Iwasaki et al. 1979, 1980a), Pleistocene clay (Yokota and Tatsuoka 1982), and sand (Iwasaki et al. 1980b); overall results are compiled in PWRI (1982).

7.2.1.1 Holocene Clayey Soils

Undisturbed clay soil samples at the Kawasaki City and the Nagoya City were consolidated by an in situ overburden stress and were tested by means of cyclic triaxial test. The shear modulus reduction curve is shown in Eq. (7.13):

$$\begin{aligned} \frac{G}{G_0} &= \left[A \cdot {\sigma'}_m^B\right]_{\gamma=\gamma_i} & & \left(10^{-6} \le \gamma \le 5 \times 10^{-4}\right) \\ \frac{G}{G_0} &= \left[A \cdot {\sigma'}_m^B\right]_{\gamma=5\times10^{-4}} \cdot [K]_{\gamma=\gamma_i} & & \left(5 \times 10^{-4} \le \gamma \le 2 \times 10^{-2}\right) \end{aligned} \quad (7.13)$$

where σ'_m is the mean effective stress, and the values for coefficients A, B, and K are tabulated in Table 7.5a. On the other hand, only one averaged curve, shown in Fig. 7.6, is shown for the damping ratio because of the limited number of data. It is read as in Table 7.5b (PWRI 1982). Example of cyclic shear deformation characteristics for various mean effective stresses is shown in Fig. 7.6.

7.2.1.2 Pleistocene Clayey Soils

Soils sampled at the Nagoya City were tested and compiled. Samples are composed of low to medium plastic clay with the SPT N-value ranging between 15 and 35 and $V_s \approx 300$ m/s. Both resonant column test and cyclic torsional shear test (low-frequency test) are performed. Initial axial stress was set same with the in situ overburden stress, and lateral stress was set half of the overburden stress to reproduce the in situ stress state. Test results are summarized in Fig. 7.7. Data of the Toyoura sand is also shown in the figure.

Shear modulus reduction curves for clay and Toyoura sand with high initial stress (196 kPa) are similar to each other. There are slight differences between the resonant column test and the cyclic torsional shear test results; they do not lie in one line.

Table 7.5 Coefficients for cyclic shear deformation characteristics for Holocene clay

(a) G–γ relationships

γ	A	B	γ	K
2×10^{-6}	0.979	0.00258	5×10^{-4}	1.000
5×10^{-6}	0.896	0.0160	10^{-3}	0.831
10^{-5}	0.826	0.0275	2×10^{-3}	0.655
2×10^{-5}	0.740	0.0443	5×10^{-3}	0.431
5×10^{-5}	0.617	0.0727	10^{-2}	0.282
10^{-4}	0.515	0.101	2×10^{-2}	0.170
2×10^{-4}	0.431	0.129	5×10^{-2}	(0.06)
5×10^{-4}	0.301	0.185	10^{-1}	(0.03)

(b) h–γ relationships

γ	h	γ	h
10^{-6}	(0.02)	5×10^{-4}	0.073
2×10^{-6}	(0.023)	10^{-3}	0.092
5×10^{-6}	(0.028)	2×10^{-3}	0.110
10^{-5}	(0.032)	5×10^{-3}	0.140
2×10^{-5}	(0.036)	10^{-2}	0.161
5×10^{-5}	0.044	2×10^{-2}	0.176
10^{-4}	0.051	5×10^{-2}	0.192
2×10^{-4}	0.057	10^{-1}	0.200

Modified from Iwasaki et al. (1979)
Note 1: Values expressed in parenthesis denote estimated data points
Note 2: Some B values were expressed as negative in the original paper but are replaced to positive values partly because figures in the original paper are drawn based on positive value and partly because inverse proportional nature does not resemble the natural soil behavior

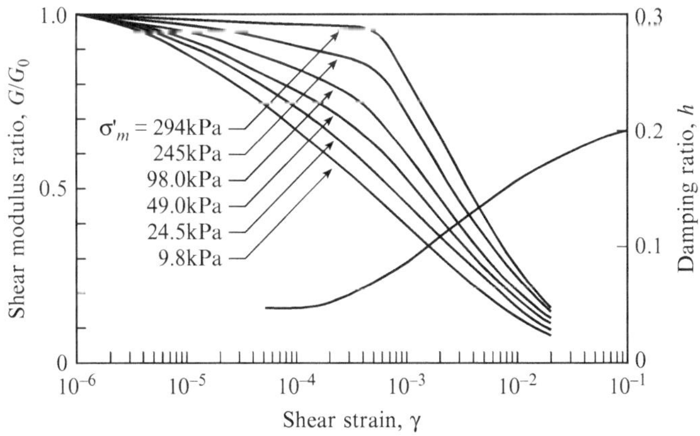

Fig. 7.6 Cyclic shear deformation characteristics of Holocene clay

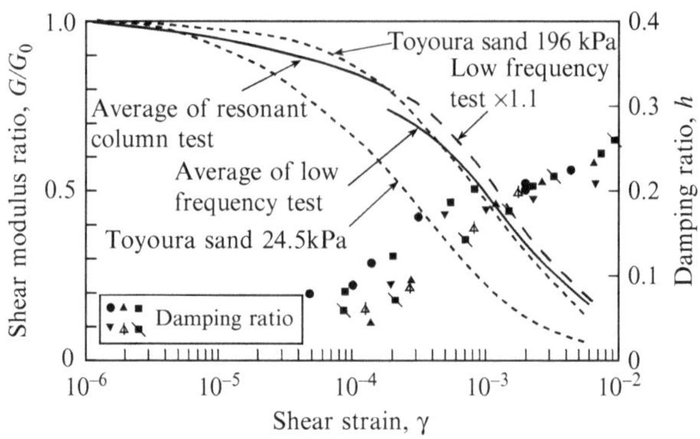

Fig. 7.7 Cyclic shear deformation characteristics of Pleistocene clay

Table 7.6 Cyclic shear deformation characteristics of Pleistocene clay

γ	G/G_0	h (%)
10^{-6}	1.000	0.7
2×10^{-6}	0.990	0.9
5×10^{-6}	0.970	1.3
10^{-5}	0.950	1.6
2×10^{-5}	0.928	1.9
3×10^{-5}	0.908	2.3
5×10^{-5}	0.880	2.8
10^{-4}	0.834	3.7
2×10^{-4}	0.769	5.0
3×10^{-4}	0.715	6.3
$5 \times 10-4$	0.627	8.4
$7 \times 10-4$	0.563	9.9
10^{-3}	0.491	11.9
$1.5 \times 10-3$	0.415	14.2
$2 \times 10-3$	0.362	15.8
$3 \times 10-3$	0.288	18.3
$5 \times 10-3$	0.200	21.7
$7 \times 10-3$	0.145	23.3
10^{-2}	0.085	26.4

However, if the shear modulus ratio obtained through the torsional shear test result is multiplied by 1.1, resultant curve seems continuous from resonant column test result; hence, they suggest using this curve for shear modulus reduction curves for Pleistocene clay. Average damping ratio is employed the same as that for Holocene clay. Representative values are read from this figure and summarized in Table 7.6 (PWRI 1982).

7.2.1.3 Holocene Sands

The cyclic torsional shear test results for Toyoura sand were compiled. They concluded that the effect of the void ratio is small but the confining stress dependency is significant, and they indicated that

$$\frac{G}{G_{\gamma=10^{-6}}} = \left(\frac{G}{G_{\gamma=10^{-6}}}\right)_{p=98\text{ kPa}} \left(\frac{\sigma'_m}{98}\right)^{m(\gamma)-m\left(10^{-6}\right)} \tag{7.14}$$

where σ'_m denotes the effective confining stress. In order to use Eq. (7.14), it is necessary to define the shear modulus reduction curve at the reference stress (98 kPa) and exponent of the confining stress $m(\gamma)-m(10^{-6})$. Among them, the values for $m(\gamma)$ and $m(10^{-6})$ can be extracted from Fig. 7.4. Obtained $m(\gamma)-m(10^{-6})$ values are tabulated in Table 7.7 (PWRI 1982) with the shear modulus ratio at the reference stress.

Test results for various sands, including Toyoura sand, are shown in Fig. 7.8. Results on sands with small amount of fines are included in the figure. All the sands show almost identical nature. Therefore, the shear modulus reduction curve for the Holocene sand can be expressed by (7.14).

Subsequently, damping characteristics can be expressed as (PWRI 1982):

$$h = 0.3\left(1 - \frac{G}{G_0}\right) \tag{7.15}$$

Table 7.7 Cyclic shear deformation characteristics of Holocene sands

γ	G/G_0	$m(\gamma)-m(\gamma=10^{-6})$
10^{-6}	1.000	0.000
2×10^{-6}	0.989	0.018
5×10^{-6}	0.978	0.028
10^{-5}	0.959	0.040
2×10^{-5}	0.928	0.058
3×10^{-5}	0.905	0.064
5×10^{-5}	0.867	0.080
10^{-4}	0.789	0.116
2×10^{-4}	0.689	0.156
3×10^{-4}	0.606	0.190
5×10^{-4}	0.500	0.260
10^{-3}	0.356	0.350
2×10^{-3}	0.228	0.422
3×10^{-3}	0.170	0.448
5×10^{-3}	0.108	0.476
10^{-2}	0.058	0.480

G/G_0 values are defined at $\sigma'_m = 98$ kPa

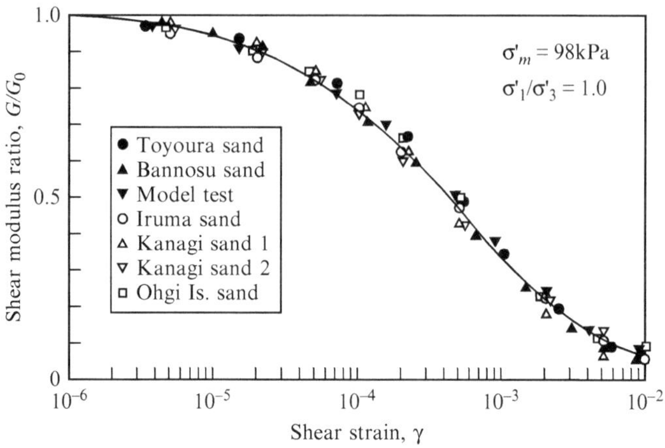

Fig. 7.8 Strain-dependent shear modulus reduction curves for various sands (After (Iwasaki et al. 1980b))

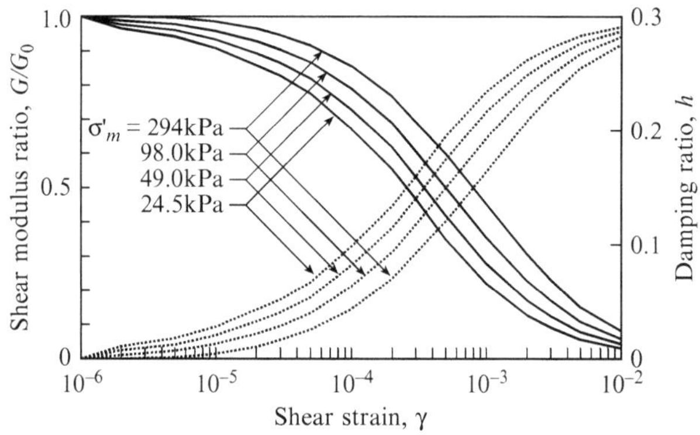

Fig. 7.9 Example of cyclic shear deformation characteristics of sandy soil

Relationships demonstrated in by Eqs. (7.14) and (7.15) are exhibited in Fig. 7.9, for several typical confining pressures. They have tendencies that G/G_0 becomes small and h becomes large as σ'_m decreases with decreasing σ'_m. This indicates that sandy soil becomes more easily exposed to nonlinear behavior when σ'_m is small.

It is noted that this test was carried out under the drained condition, whereas almost all other data shown in this book were carried out under the undrained condition. Difference between the results under each condition was already discussed in Sect. 6.7.2. The difference between shear moduli under two conditions is insignificant, but considerable difference is observed in damping ratios, as shown in Fig. 6.15; the damping ratio under the undrained condition is small, which is also observed in Fig. 7.30.

7.2.1.4 Pleistocene Sandy Soils and Gravelly Soils

It is recommended to use the cyclic shear deformation characteristics of Holocene sands into that of the Pleistocene sands (PWRI 1982) because test data on the Pleistocene sand by Kokusho and Sasaki (1980) are similar to the one of Toyoura sand. On the other hand, data on gravelly soils was missing at 1982 when report by Kokusho and Sasaki (1980) was published. Importance to make the cyclic shear deformation characteristics test by using samples taken at the site is also pointed out.

7.2.2 Equations Involved in Technical Standards for Port and Harbor Facilities

Design specification for port facilities (1989 version) (The Overseas Coastal Area Development Institute of Japan 1989) expressed shear modulus reduction curve as a function of the shear strain amplitude γ and the plasticity index I_p:

$$\frac{G}{G_{\max}} = \overline{A}\left(I_p, \gamma\right)\left(\frac{\sigma'_m}{98}\right)^{n(I_p, \gamma)} \tag{7.16}$$

where $\overline{A}\left(I_p, \gamma\right)$ and $n(I_p, \gamma)$ are given in Table 7.8. These equations were developed by Zen et al. (1987b).

On the other hand, the damping characteristics are tabulated in Table 7.9; the confining stress dependency is not considered as the effect is not clear.

This equation was eliminated in the 1999 version of design specification. Instead the same set of relationships was introduced as a simplified procedure in *Handbook*

Table 7.8 $\overline{A}\left(I_p, \gamma\right)$ and $n(I_p, \gamma)$

	Plasticity index, I_p					
	NP less than 9.4		9.4 less than 30		More than or equal to 30	
Shear strain amplitude γ	$\overline{A}\left(I_p, \gamma\right)$	$n(I_p, \gamma)$	$\overline{A}\left(I_p, \gamma\right)$	$n(I_p, \gamma)$	$\overline{A}\left(I_p, \gamma\right)$	$n(I_p, \gamma)$
10^{-6}	1.	0.	1.	0.	1.	0.
10^{-5}	0.93	0.01	0.96	0.	0.97	0.
5×10^{-5}	0.83	0.03	0.91	0.01	0.93	0.
10^{-4}	0.75	0.05	0.84	0.02	0.89	0.
2.5×10^{-4}	0.56	0.10	0.74	0.05	0.82	0.
5×10^{-4}	0.43	0.16	0.59	0.09	0.70	0.
10^{-3}	0.30	0.22	0.45	0.16	0.58	0.
2.5×10^{-3}	0.15	0.30	0.26	0.22	0.40	0.
5×10^{-3}	–	–	0.12	0.26	0.25	0.
10^{-2}	–	–	–	–	0.18	0.

Table 7.9 Value of damping ratio

Shear strain amplitude, γ	Plasticity index $I_p < 30$			Plasticity index $I_p \geq 30$		
	Average	Maximum	Minimum	Average	Maximum	Minimum
10^{-6}	0.026	0.040	0.015	0.025	0.050	0.010
10^{-5}	0.030	0.040	0.018	0.030	0.054	0.010
5×10^{-5}	0.033	0.042	0.020	0.034	0.062	0.014
10^{-4}	0.037	0.048	0.026	0.038	0.070	0.018
2.5×10^{-4}	0.055	0.068	0.040	0.050	0.088	0.030
5×10^{-4}	0.080	0.098	0.060	0.066	0.108	0.042
10^{-3}	0.120	0.145	0.092	0.086	0.133	0.056
2.5×10^{-3}	0.174	0.200	0.148	0.118	0.174	0.080
5×10^{-3}	0.200	0.222	0.178	0.144	0.208	0.100
10^{-2}	0.220	0.240	0.200	0.175	–	0.125

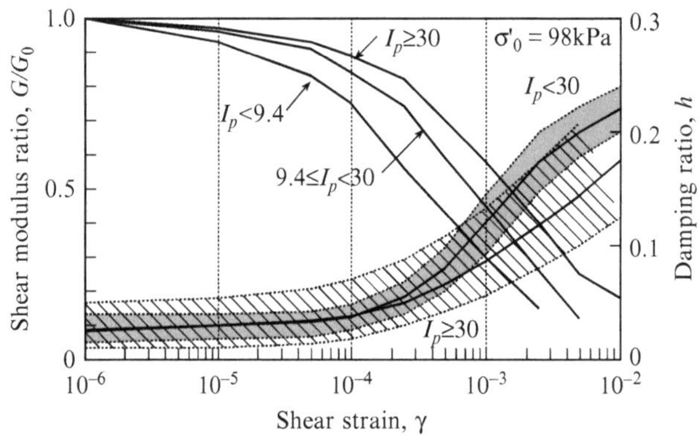

Fig. 7.10 Cyclic shear deformation characteristics described by technical standards for port and harbor facilities

on Liquefaction Remediation of Reclaimed Land (Coastal Development Institute of Technology 1997). Data on $I_p < 30$ is relevant when the equations are adopted for sands. Note that $I_p = 30$ is used as a boundary between sandy and clayey soils in Sect. 7.1.3 as well. However, in reality, the relation between sandy soils and I_p is not unique. For example, sands with clay are classified as sandy soil, but I_p for this soil type is usually large. According to the author's experience, in practice, it is likely that I_p for sands can exceed 30.

The cyclic shear deformation characteristics at $\sigma'_m = 98$ kPa are shown in Fig. 7.10. In Table 7.8, $n(I_p, \gamma) = 0$ indicates that the confining stress dependency is not considered or observed. Generally speaking, confining stress dependency becomes small when I_p is larger than 30–40.

Compared with the results by PWRI, the shear modulus reduction curve shows similar shape. Nonetheless, there exists a significant difference between the damping ratios at large strains induced by the difference in the drainage conditions.

In addition, some scattering is seen on the damping ratio, whereas unique value is given by the PWRI data as there is no significant difference between soils.

7.2.3 Equations Involved in Standards for Railway Structures

Several cyclic shear deformation characteristics test results obtained through triaxial test are shown by Railway Technical Research Institute (1999) for use when test is not performed. Data for crushed stones used for the mechanical stabilization, Toyoura sand, Inagi sand, and Iwate loam are tabulated in Table 7.10 and are displayed in Fig. 7.11.

Compared with the previously described test data, these results have several distinct features. The damping ratio of the Toyoura sand at small strains is small to be approximately 0.005 and at large strains as large as 0.3. Damping ratio begins

Table 7.10 Cyclic shear deformation characteristics test of various soils suggested by the Railway Technical Research Institute

	Crushed stone for mechanical stabilization		Toyoura sand		Inagi sand		Iwate loam	
Strain	G/G_0	h	G/G_0	h	G/G_0	h	G/G_0	h
1×10^{-6}	1.00	0.02	1.00	0.005	1.00	0.03	1.00	0.02
1×10^{-5}	0.99	0.024	0.99	0.007	0.98	0.032	0.98	0.02
5×10^{-5}	0.90	0.045	0.89	0.036	0.90	0.041	0.85	0.02
1×10^{-4}	0.82	0.058	0.80	0.064	0.80	0.051	0.77	0.023
2.5×10^{-4}	0.64	0.077	0.66	0.114	0.59	0.069	0.65	0.025
5×10^{-4}	0.48	0.093	0.50	0.174	0.40	0.09	0.55	0.04
1×10^{-3}	0.32	0.111	0.32	0.235	0.27	0.115	0.46	0.071
2.5×10^{-3}	0.15	0.126	0.12	0.284	0.16	0.151	0.32	0.12
5×10^{-3}	0.05	0.124	0.06	0.29	0.10	0.172	0.20	0.151
1×10^{-2}	(0.06)	(0.122)	0.05	0.282	0.07	0.18	0.12	0.145

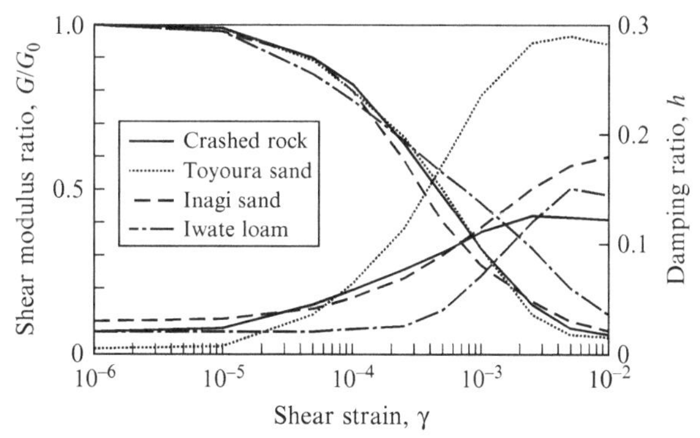

Fig. 7.11 Cyclic shear deformation characteristics presented in Table 7.10

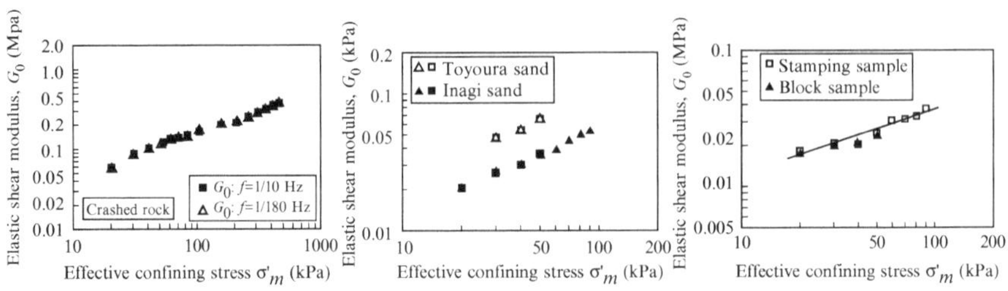

Fig. 7.12 Confining stress dependency on elastic modulus of various soils

to decrease at strain 0.005, which is a typical behavior of the undrained test results (see Fig. 6.20 as well). Furthermore, only one set of test data is shown for each soil, and σ'_m dependence is not considered.

The confining stress dependence of the elastic shear modulus is shown in Fig. 7.12. The value of the exponent of σ'_m is 0.52 for crushed stones, 0.56 for Toyoura sand and Inagi sand, and 0.46 for Iwate loam. These values are not denoted explicitly, but can be easily extracted from the figure.

7.2.4 Equation in Building Standard Law

The cyclic shear deformation characteristics for the clayey and the sandy soils were provided in the Ministerial Notification (2000) listed in Table 7.11. This data was most probably developed using the Ramberg–Osgood model (explained in Sect. 8. 2.4) because they agree by using the model parameters shown in Table 7.12. This notification discarded in 2006, however, appears in the older technical papers, which is the reason why this equation is discussed here.

As shown in the previous and following sections, cyclic shear deformation characteristics are affected by various factors such as confining stress, plasticity index, etc. Therefore, without doubt, adopting a single characteristic curve for each soil is unacceptable. It may be an easy way for engineers because they need to consider on nonlinear characteristics, but it may lead to significant error as already discussed in Sect. 6.7.5.

7.2.5 Equation by Central Research Institute of Electric Research Industry

Cyclic shear deformation characteristics of various soils have been investigated by the Central Research Institute of Electric Research Industry. Among them, some typical test results are discussed in this subsection.

Table 7.11 Cyclic shear deformation characteristics of sandy and clayey soils according to Ministerial Notification

Sand						Clay					
γ	G/G_0	h	γ	G/G_0	h	γ	G/G_0	h	γ	G/G_0	h
0.00001	1.000	0.02	0.002	0.23	0.218	0.00001	1.000	0.02	0.002	0.377	0.163
0.00002	0.979	0.02	0.003	0.183	0.231	0.00002	0.99	0.02	0.003	0.311	0.181
0.00003	0.962	0.02	0.004	0.155	0.239	0.00003	0.983	0.02	0.004	0.269	0.192
0.00004	0.942	0.02	0.005	0.137	0.244	0.00004	0.975	0.02	0.005	0.24	0.199
0.00005	0.922	0.022	0.006	0.123	0.248	0.00005	0.966	0.02	0.006	0.219	0.205
0.00006	0.901	0.028	0.007	0.112	0.251	0.00006	0.957	0.02	0.007	0.202	0.209
0.00007	0.881	0.034	0.008	0.104	0.254	0.00007	0.948	0.02	0.008	0.188	0.213
0.00008	0.861	0.039	0.009	0.097	0.256	0.00008	0.939	0.02	0.009	0.176	0.216
0.00009	0.842	0.045	0.01	0.091	0.257	0.00009	0.93	0.02	0.01	0.167	0.218
0.0001	0.823	0.05	0.02	0.06	0.266	0.0001	0.92	0.021	0.02	0.114	0.232
0.0002	0.678	0.091	0.03	0.047	0.27	0.0002	0.834	0.044	0.03	0.091	0.238
0.0003	0.583	0.118	0.04	0.04	0.272	0.0003	0.763	0.062	0.04	0.077	0.242
0.0004	0.517	0.137	0.05	0.035	0.273	0.0004	0.706	0.077	0.05	0.068	0.244
0.0005	0.468	0.151	0.06	0.031	0.274	0.0005	0.659	0.089	0.06	0.062	0.246
0.0006	0.429	0.162	0.07	0.028	0.275	0.0006	0.62	0.1	0.07	0.056	0.247
0.0007	0.398	0.17	0.08	0.026	0.276	0.0007	0.587	0.108	0.08	0.052	0.248
0.0008	0.373	0.177	0.09	0.024	0.276	0.0008	0.558	0.116	0.09	0.049	0.249
0.0009	0.351	0.184	0.1	0.023	0.277	0.0009	0.533	0.122	0.1	0.023	0.250
0.001	0.333	0.189					0.511	0.128			

Table 7.12 Parameters for Ramberg–Osgood model used for obtaining values in Table 7.11

	α	β	γ_r	h_{max}
Sand	2.64	2.62	0.000432	0.26213
Clay	3.07	2.4	0.00106	0.285

7.2.5.1 Holocene Clays

Test data for undisturbed clayey soil (Teganuma Holocene clayey soil) obtained by means of the thin-walled tube sampler was evaluated (Kokusho et al. 1982a, b). The shear modulus was observed to have confining stress dependency, but this disappears when I_p becomes larger than 40 and effect of the plasticity index governs nonlinear soil characteristics. Further, it was shown that stress history such as an over consolidation or a long-term consolidation does not affect the shear modulus in a significant manner. An empirical equation was not proposed in their study, but the relationships between strain γ and $\log I_p$ were demonstrated by a bilinear line with several G/G_0 values. These two lines can be defined by specifying three corner points, provided in Table 7.13. Strains at $G/G_0 = 0.5$ in the table is a reference strain, which is larger than that observed by Zen et al. (1978), but the shape of the two lines is similar.

Table 7.13 Coordinates of corner points for the bilinear relationships

	First point		Second point		Third point	
G/G_0	I_p	γ	I_p	γ	I_p	γ
0.125	0	0.00526	50	0.0301	110	0.0664
0.25	0	0.00181	54	0.0122	110	0.0209
0.5	0	0.00068	45	0.00322	110	0.00491
0.75	0	0.00021	40	0.00087	110	0.00093

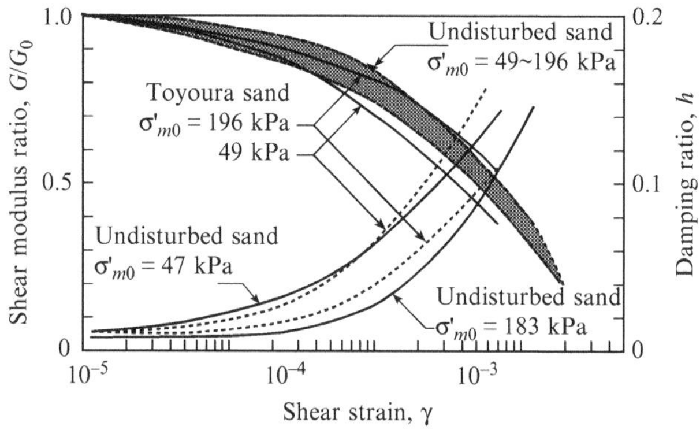

Fig. 7.13 Cyclic shear deformation characteristics of sand (After Nishi et al. 1983)

7.2.5.2 Sands and Gravels

Soils sampled near the JPDR nuclear power plant at the Tokai village, Ibaraki Prefecture, were tested (Nishi et al. 1983). The undisturbed sand samples were retrieved by means of the triple tube sampler, whereas the gravels are obtained in a disturbed fashion.

The cyclic shear deformation characteristics of sand are shown in Fig. 7.13. Obtained results are similar to those of the Toyoura sand in general. Nonlinear soil behaviors of gravel are shown in Fig. 7.14. Gravel begins to exhibit nonlinear behavior at smaller strains, while the damping ratio at large strains is smaller than that of the Toyoura sand.

7.2.5.3 Nevada Sand

Nevada sand is the sand in the USA and was used in VELACS project (Arulanandan and Scott 1993). Disturbed dense ($D_r = 90–95$ %) and loose ($D_r = 50–60$ %) samples are tested (Kanatani et al. 1999). Test results were shown in Fig. 7.15. Effect of the density on the cyclic deformation characteristics is shown to be small, whereas the damping ratio is very large compared with other test data on sands. This may be a consequence of tests being carried out under drained condition.

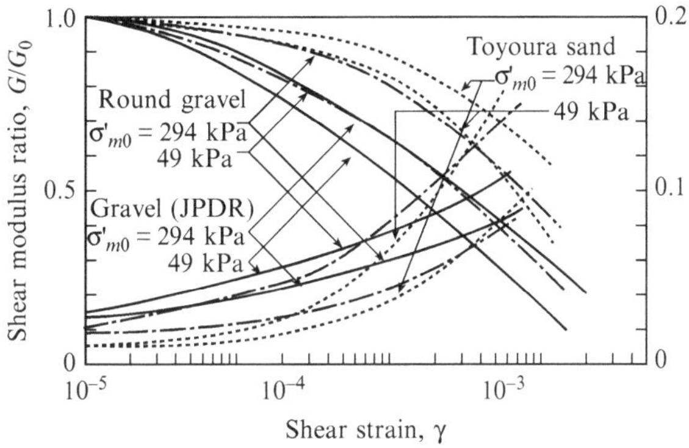

Fig. 7.14 Cyclic shear deformation characteristics of gravel (After Nishi et al. 1983)

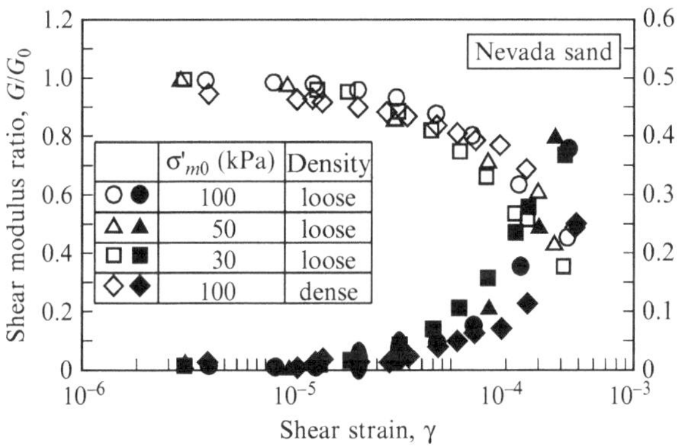

Fig. 7.15 Cyclic shear deformation characteristics of Nevada sand

7.2.6 Study Compiled by Seed and Idriss

Seed and Idriss gathered and compiled various test data (Seed and Idriss 1970). Range and mean of the reviewed data are shown in Fig. 7.16. The effect of the confining stress, the internal friction angle, saturation ratio, and the coefficient of earth pressure at rest are discussed for sand. Among them, the effect of confining stress dependency shows very similar with previously described stress dependency behavior. The shear modulus ratio increases and the damping ratio decreases with an increase in internal friction angle. The effect of K_0 is minor, but strains exhibiting nonlinear behavior are seen to be largest when $K_0 = 1$, which may result from the effect of initial shear discussed in Sect. 6.7.8.

They also showed that evaluation of accurate cyclic shear deformation characteristics is difficult for clayey soil partly because sample disturbance has a very considerable effect on the test results and partly because there is no available in situ technique to measure the deformation characteristics at large strain. This difficulty

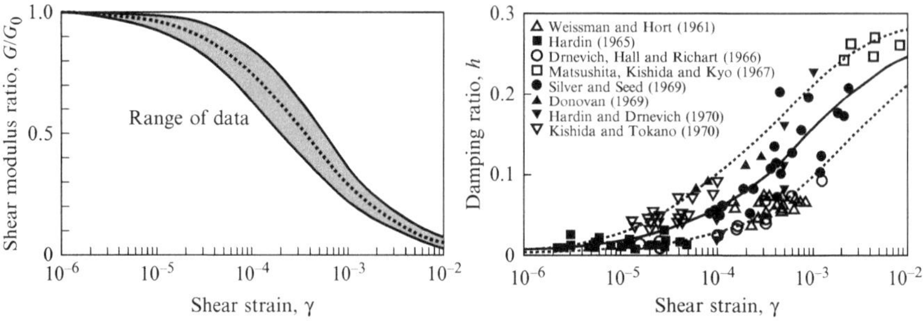

Fig. 7.16 Cyclic shear deformation characteristics of sand

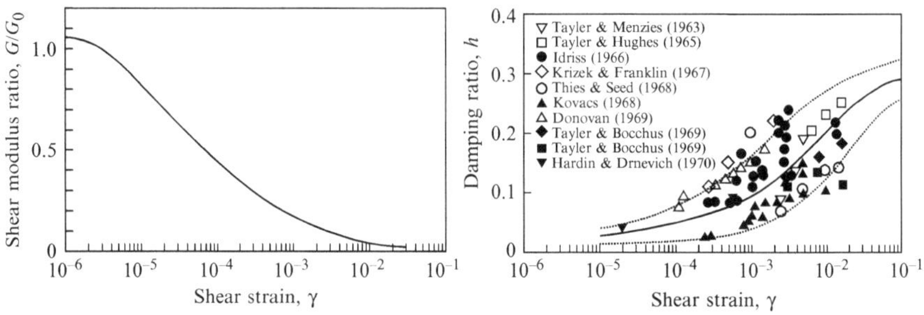

Fig. 7.17 Cyclic shear deformation characteristics of saturated clay

appears as a disagreement of in situ and laboratory elastic moduli already discussed in Sect. 6.5.

Data on the San Francisco Bay mud, the Union Bay clay, etc. are reviewed. The mean values of the modulus reduction curve are shown in Fig. 7.17a; individual test data and existing range are shown in Fig. 7.17b. Here, the shear modulus is normalized by the shear modulus at 3×10^{-6} strain; thus, the shear moduli at strains smaller than 3×10^{-6} are larger than unity. As compared with the data described in this book as well as other data, shear modulus ratio in Fig. 7.17a is very small, which indicates that the nonlinear behavior begins to develop at smaller strains. Therefore, use of this data should be performed with care.

Average deformation characteristics read from these figures are tabulated in Table 7.14.

7.2.7 Equation by Yasuda et al.

Yasuda and Yamaguchi (1985) proposed Eq. (7.17) based on the cyclic shear deformation characteristics test result on 103 undisturbed samples retrieved at 13 sites from Hokkaido to Kyushu, Japan:

Table 7.14 Average deformation characteristics read (From Figs. 7.16 and 7.17)

γ	Sand		Clay	
	G/G_0	h	$G/G_{\gamma=3\times10^{-6}}$	h
1.0×10^{-6}	1.00	0.008	1.056	–
3.16×10^{-6}	0.988	0.011	0.991	–
1.0×10^{-5}	0.962	0.018	0.823	0.028
3.16×10^{-5}	0.890	0.036	0.628	0.037
1.0×10^{-4}	0.747	0.059	0.445	0.050
3.16×10^{-4}	0.524	0.099	0.290	0.068
1.0×10^{-3}	0.290	0.158	0.172	0.094
3.16×10^{-3}	0.137	0.210	0.092	0.141
1.0×10^{-2}	0.053	0.247	0.037	0.203
3.16×10^{-2}	–	–	0.021	0.261
1.0×10^{-1}	–	–	–	0.291

Table 7.15 Coefficients for Eq. (7.17)

γ	A_1	A_2	B_1	B_2
10^{-4}	0.827	−0.044	0.056	0.026
3×10^{-4}	0.670	−0.068	0.184	0.086
10^{-3}	0.387	−0.099	0.277	0.130
3×10^{-3}	0.189	−0.089	0.315	0.147
10^{-2}	0.061	−0.054	0.365	0.167
3×10^{-2}	0.041	−0.019	0.403	0.183
γ	C_1	C_2	D_1	D_2
10^{-4}	0.035	0.005	−0.559	−0.258
10^{-3}	0.136	0.036	−0.375	−0.173
10^{-2}	0.234	0.037	0.0	0.

$B_1 = B_2 = D_1 = D_2 = 0$ when $D_{50} \le 0.007$ mm

$$\frac{G}{G_{\max}} = (A_1 + A_2 \log D_{50}) \left(\frac{\sigma'_m}{98}\right)^{(B_1 + B_2 \log D_{50})}$$

$$h = (C_1 + C_2 \log D_{50}) \left(\frac{\sigma'_m}{98}\right)^{(D_1 + D_2 \log D_{50})} \tag{7.17}$$

Values of parameters A_1 to D_2 are shown in Table 7.15 for typical shear strains. Equation (7.17) is applicable under $19.6 \le \sigma'_m \le 294$ kPa and $0.002 \le D_{50} \le 1$ mm.

As behaviors for wide range of soils from clay to coarse gravel are expressed by one equation, this relationship is very convenient for use in the engineering practice. The minimum strain in the table is 10^{-4}, and G/G_0 and h at small strains can be interpolated by setting 1.0 and 0 or 0.02 (2 %), respectively, at $\gamma = 10^{-6}$. Calculated cyclic shear deformation characteristics for the typical soils are shown Fig. 7.18.

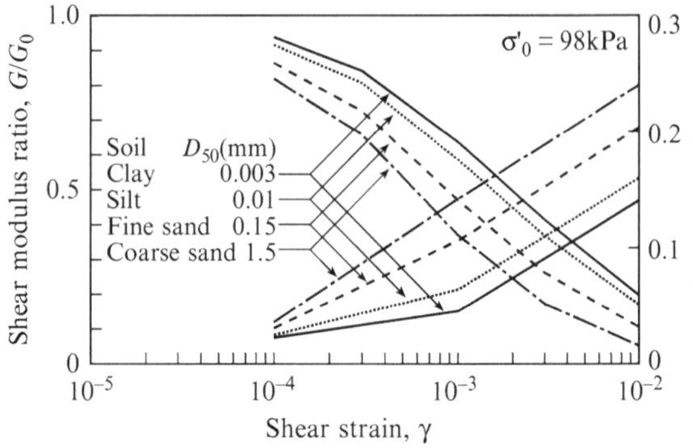

Fig. 7.18 Cyclic shear deformation characteristics of various soils calculated based on the proposal by Yasuda et al.

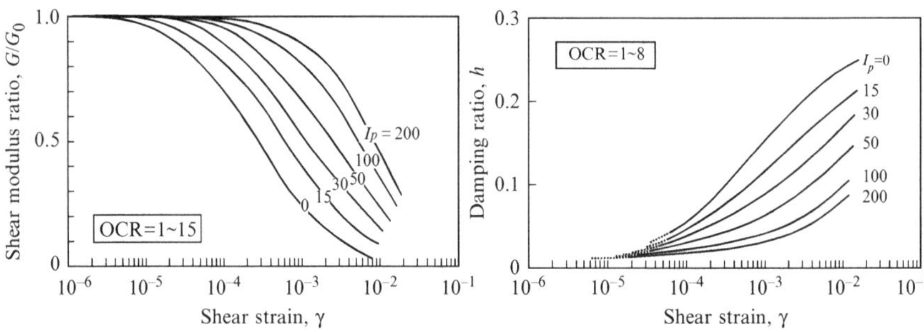

Fig. 7.19 Cyclic shear deformation characteristics with I_p (After Vucetic and Dobry 1991)

7.2.8 Study by Vucetic and Dobry

Vucetic and Dobry (1991) gathered 16 technical papers on the cyclic shear deformation characteristics for various soils and concluded that the plasticity index is the most predominant factor that affects the cyclic deformation characteristics. They compiled the gathered data as a function of I_p, as shown in Fig. 7.19. Digitized data read from this figure is shown in Table 7.16. These deformation characteristics can be used for wide range of soils from sand to clay. However, they sometimes do not agree well with those in Japan. For example, soils with $I_p = 0$–15 are clearly classified as sand, but confining stress dependency is not considered. Shear modulus ratio around strain 10^{-2} is very small compared with other data.

7.2.9 Study by Oyamada et al.

Koyamada et al. (2003) gathered triaxial and torsional cyclic shear test data for 167 undisturbed soils at 45 sites in Japan. In these tests, initial isotropic stresses were

Table 7.16 Digitized cyclic deformation characteristics (From Fig. 7.19)

I_p	200		100		50		30		15		0	
γ	G/G_0	h	G/G_0	h	G/G_0	h	G/G_0	h	G/G_0	h	G/G_0	h
1×10^{-6}	1.000		1.000		1.000		1.000		1.000		1.000	
3.16×10^{-6}	1.000		1.000		1.000		0.999		0.998		0.996	
1×10^{-5}	0.999	0.012	0.997	0.015	0.995		0.980		0.964		0.991	
3.16×10^{-5}	0.997	0.014	0.998	0.017	0.991	0.020	0.981	0.023	0.932	0.026	0.871	0.031
1×10^{-4}	0.991	0.018	0.979	0.023	0.955	0.031	0.892	0.038	0.815	0.047	0.703	0.056
3.16×10^{-4}	0.965	0.023	0.922	0.030	0.840	0.044	0.742	0.061	0.639	0.078	0.475	0.100
1×10^{-3}	0.893	0.032	0.811	0.042	0.666	0.063	0.522	0.089	0.400	0.116	0.241	0.153
3.16×10^{-3}	0.743	0.049	0.619	0.063	0.453	0.094	0.327	0.125	0.207	0.159	0.105	0.202
1×10^{-2}	0.457	0.081	0.355	0.099	0.239	0.134	0.153	0.170	0.088	0.199	0.018	0.240
1.65×10^{-2}	0.323	0.097	0.240	0.117		0.154		0.190		0.215		0.251

set the same as the in situ effective overburden stress. They evaluated cyclic shear deformation characteristics by using the Hardin–Drnevich model (see Sect. 6.6 for details) as

$$\frac{G}{G_0} = \frac{1}{1 + \gamma/\gamma_{0.5}}, \quad h = h_{\max}\left(1 - \frac{G}{G_0}\right) \tag{7.18}$$

where $\gamma_{0.5}$ denotes the reference strain and $h_{\max}$ denotes maximum damping ratio.

They gave average feature of several parameters such as void ratio, plasticity index, etc. for each soil category and evaluated the model parameters by a curve fitting as shown in Table 7.17. Among them, models based on overall sandy or clayey soils were recommended as the proposed relationship, which is shown in Fig. 7.20. The line BSL in the figure represents the equation in Sect. 7.2.4 (Ministerial Notification).

7.2.10 *Study by Imazu and Fukutake*

Imazu and Fukutake (1986a) collected cyclic deformation characteristics test results from 23 technical papers. Range of the data is shown in Fig. 7.21. They also proposed the following empirical formulae (Imazu and Fukutake 1986b):

$$\frac{G}{G_{\max}} = \frac{1}{1 + a(\gamma)^b}, \quad h = c(\gamma)^d \tag{7.19}$$

where *a*–*d* are parameters provided in Table 7.18 (Imazu and Fukutake 1986a) and μ and σ denote the mean and the standard deviation, respectively.

In the developed relationships, the damping ratio of sands appears to be too large compared with other data as it is greater than 0.3 at 10^{-2} strain. In reality, the rate

Table 7.17 Values of parameters for Hardin–Drnevich model determined by Oyamada et al.

Prefecture	Layer	Soil type	σ'_0 (kPa)	e	I_p (%)	F_c (%)	V_s (m/s)	$\gamma_{0.5}$	h_{max}
Tokyo	HP (Yurakucho layer)	Clayey	92	1.77	30.6	82.1	150	0.0019	0.16
		Sandy	74	0.89	–	17.2	190	0.0009	0.21
	PP (Tokyo gravel layer)	Loam	48	3.71	44.7	96.1	190	0.0019	0.14
		Clayey	174	1.45	37.4	91.5	270	0.0020	0.13
		Sandy	167	0.91	–	10.4	370	0.0009	0.21
Kanagawa	HP	Clayey	150	1.54	33.8	90.1	180	0.0019	0.15
		Sandy	122	0.88	–	20.1	200	0.0008	0.21
	PP (Kazusa formulation)	Clayey	265	1.45	54.2	77.2	280	0.0021	0.21
		Sandy	311	0.95	–	18.1	330	0.0013	0.23
Osaka	HP	Clayey	166	1.39	33.5	94.9	160	0.0014	0.19
		Sandy	119	0.50	–	13.7	200	0.0008	0.19
	PP (Osaka formulation)	Clayey	379	1.30	55.7	98.3	260	0.0016	0.17
		Sandy	332	0.60	–	15.7	340	0.0016	0.21
Average of all data		Clayey	205	1.66	39.8	30.2	210	0.0018	0.17
		Sandy	193	0.86	–	15.6	290	0.0010	0.21

HP Holocene period, *PP* Pleistocene period, σ'_0 initial confining stress, e void ratio, I_p plasticity index, F_c fines contents. S-wave velocity V_s is an average value

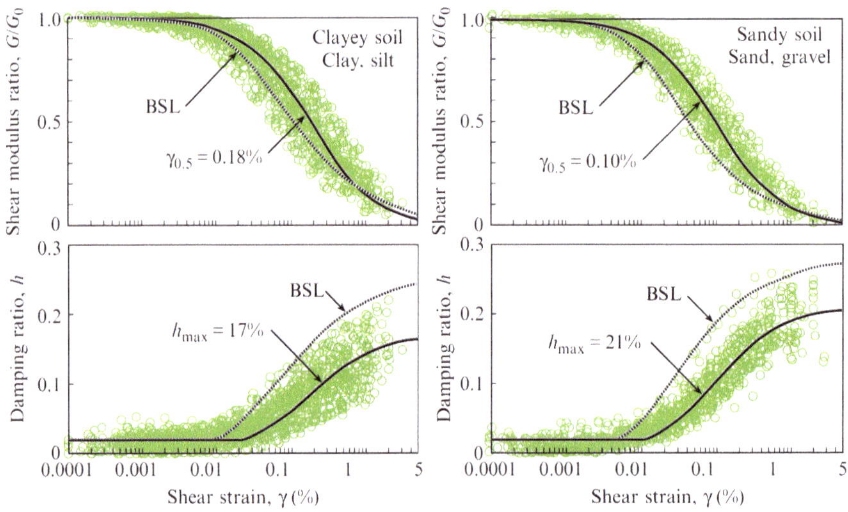

Fig. 7.20 Cyclic shear deformation characteristics test results and proposed curve (Modified from Koyamada et al. 2003)

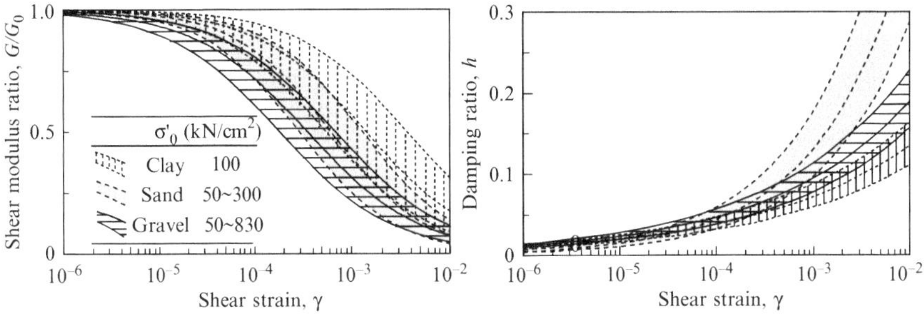

Fig. 7.21 Cyclic shear deformation characteristics of various soils (After Imazu and Fukutake 1986a)

Table 7.18 Coefficients for Eq. (7.19)

		a	*b*	*c*	*d*
Sand	$\mu+\sigma$	1,419	0.89	456.7	0.47
	μ	729.7		338.0	
	$\mu-\sigma$	375.4		250.1	
Clayey soil	$\mu+\sigma$	381.3	0.79	56.97	0.27
	μ	179.1		46.84	
	$\mu-\sigma$	84.02		38.49	
Gravel	$\mu+\sigma$	640.0	0.75	90.97	0.30
	μ	392.8		75.36	
	$\mu-\sigma$	241.3		62.38	

Fig. 7.22 Damping characteristics of various soils (Modified from Fig. 7.21 (After Kokusho 1980b))

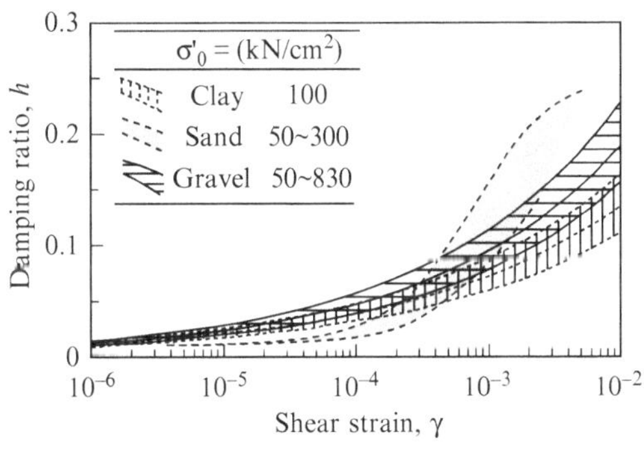

of increase of damping ratio becomes small and sometimes damping ratio decreases by the excess porewater pressure generation. Nonetheless, this behavior cannot be expressed by Eq. (7.19) as it is a monotonically increasing function of shear strain, which is the reason of the disagreement described above. Then Fig. 7.21 is modified as Fig. 7.22 (JGS 1992) by referring to Kokusho (1980b).

It is noted that not all of the existing soil types are referred in Figs. 7.21 and 7.22; they are just generic types of soil. Comparison with other data is discussed in Sect. 7.2.12.

According to Fig. 7.21, the modulus reduction curve for gravel is located beneath the ones for sand and clay. This indicates that nonlinear behavior appears at the smallest strain in gravel and then sand and clay. On the other hand, damping ratio is largest for sand and then gravel and clay especially at large strains. These features are observed in the other researches and can be considered as a general feature of soil.

7.2.11 Study by Fukumoto et al.

Recent study indicates (Yoshida 2001; Fujikawa et al. 2000) nonlinear behavior of soft rock, which is considered to consist engineering seismic base layer or engineering bedrock under the very strong ground motions such as level 2 ground motion. Here, in many design specifications, two types of the earthquake motion are termed as level 1 and level 2 ground motions. Level 1 ground motion is the ground motion that the structure will experience in its life with high possibility, whereas level 2 ground motion is a very rare but very strong ground motion. Unfortunately, test data on the cyclic shear deformation characteristics of the soft rock is very few (e.g., Nishi et al. (1985)) and the knowledge on their mechanical properties is limited because nonlinear behavior of soft soil has been not expected.

Fukumoto et al. (2009) collected test data on the sedimentary soft rocks and investigated their fundamental mechanical properties.

Results of the monotonic loaded tests on the consolidated silt samples retrieved at Yokohama City, Japan, are shown in Fig. 7.23. Elastic moduli are measured from the slope of the small unload–reload loading cycles. Figure 7.23a shows the effect of strain measurement. LDT is a local displacement transducer that is designed to accurately measure small strains. On the other hand, EDS is an external

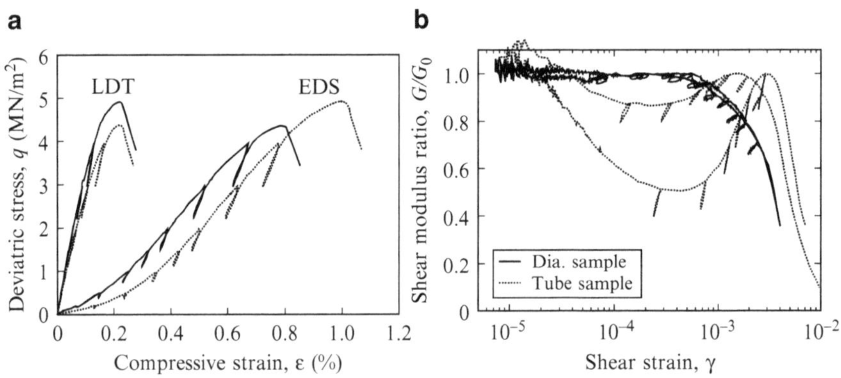

Fig. 7.23 Effect of measurement procedure and fine cracks in samples. (**a**) Stress–strain curves (**b**) Secant modulus expression

displacement sensor that is usually used in the triaxial tests. There are significant differences between the measured strains by two instruments. Bedding error and existence of small cracks are believed to be responsible of this discrepancy. In addition, the stress–strain curve by EDS shows concave shape because stiffness increases as cracks close.

The quality of samples is discussed through Fig. 7.23b. The ordinate is the shear modulus ratio (secant modulus/elastic modulus), and the abscissa is the shear strain measured by means of LDT. Usually, soft rocks are sampled by means of diamond core cutter or triple tube sampler to take undisturbed samples, which are referred as dia-sample or tube sample, respectively, in the following. Data in Fig. 7.23a are only tube sampled data. The shear modulus ratios for the tube samples decrease once, then increase again, and drop significantly. This curious behavior can be recognized that fine cracks close. It is noted that not all dia-sample shows behavior similar with Fig. 7.23b; existence of the samples with cracks is reported in the original paper even for the dia-sample. It is, however, not known whether these cracks exists in situ or created at sampling.

Next, they collected 71 test data on the soft rocks that were investigated statistically. It is found that the cyclic shear deformation characteristics are affected by the plasticity index, but the effect of confining stress is rather small. Hence, they classified test data according to the plasticity index and proposed empirical equation.

Fine cracks are demonstrated to affect the deformation characteristics of soft rock under monotonically loading conditions in the preceding. Then it is natural to expect that the cyclic deformation characteristics are also affected by cracks. For instance, the shear modulus reduction curves for samples with I_p 20–40 are shown in Fig. 7.24. It is easily noticed that there exists a large scatter. Test result for the monotonically loaded samples shown in Fig. 7.23b is also provided in the figure. It can be easily observed that monotonically loaded test result and the upper range of the test data (■ in the figure) are similar to each other. Thus, this sample was considered as less affected by the presence of cracks and employed to develop the model.

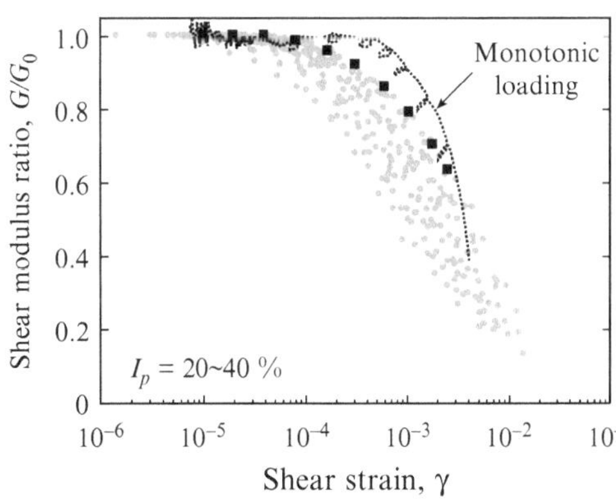

Fig. 7.24 Example of shear modulus reduction curve

Shear modulus reduction curve is expressed by using the Ramberg–Osgood model (see Sect. 8.2.4) as

$$1 = \frac{G}{G_0}\left[1 + \alpha\left(\frac{G}{G_0}\cdot\frac{\gamma}{\gamma_r}\right)^{\beta-1}\right] \tag{7.20}$$

where α, β, and γ_r are model parameters. Next, damping characteristics are expressed by the equation proposed by Hardin and Drnevich (Kokusho 1980a) (Sect. 6.6.1) as

$$h = h_{\max}\left(1 - \frac{G}{G_0}\right) + h_0 \tag{7.21}$$

where $h_{\max}$ and h_0 denote the maximum and minimum damping ratios, respectively. It is noted that Eqs. (7.20) and (7.21) are entirely independent. Application of Masing's rule (see Sect. 8.2.2) to Eq. (7.20) results in an equation that has the same form with (7.21). In this case, the maximum damping ratio $h_{\max}$ can be expressed by means of β, but $h_{\max}$ in Eq. (7.21) is an independent parameter. Table 7.19 shows values of parameters on all categories, and the cyclic deformation characteristics are shown in Fig. 7.25.

Even if a good agreement between the monotonically load test results and the cyclic shear deformation characteristics test results (■) is said to agree in the original paper, it is important to emphasize certain differences between them. The

Table 7.19 Model parameters for Eqs. (7.20) and (7.21)

		Plasticity index I_p (%)			
Equations	Parameters	0–20	20–40	40–60	60–
G/G_0–γ Eq. (7.20)	α	2.1905	2.2262	2.4508	2.2505
	β	2.1313	2.1546	2.2933	2.1703
	γ_r	0.00410	0.00485	0.00645	0.0145
h–γ Eq. (7.21)	h_0	0.014	0.009	0.010	0.015
	$h_{\max}$	0.130	0.122	0.100	0.121

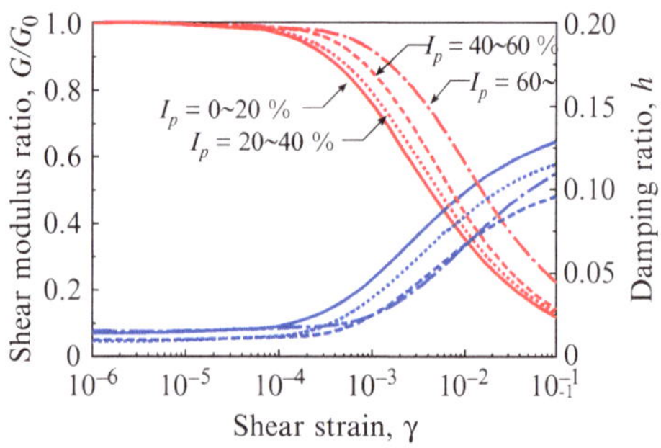

Fig. 7.25 Cyclic shear deformation characteristics calculated through Eqs. (7.20) and (7.21)

stiffness obtained through cyclic load test is smaller at strains around 10^{-4}–10^{-3} and is larger at strains larger than 10^{-2} compared to the monotonic load test results. This difference may be explained by considering the existence of crack. Unfortunately, it is not identified whether this crack exists prior the test or appears during the test. If it existed prior the test, behavior similar to tube sample in Fig. 7.23b is expected to appear. Therefore, the author believes that cracks are generated as a result of loading.

7.2.12 Study by Wakamatsu et al.

Wakamatsu et al. (2010) collected more than 400 cyclic shear deformation characteristics test data in the Kanto district, Japan. They classified soils in 26 categories based on the geologic age and the depositional conditions as shown in Table 7.20. Average cyclic shear deformation characteristics for each category are given in Fig. 7.26. Data on clay scatter more than that in sand. Mean results for fills (Bc and Bs), Holocene soils (Ac and As) and Pleistocene soils (Dc and Ds) are shown in Fig. 7.27. Younger soil formations develop nonlinear behavior at lower strains than older soil formations; aging effect is clearly visible. However, scatter in Fig. 7.27 is much smaller than that in Fig. 7.26. This indicates importance of a detailed classification such as in Table 7.20. Figure 7.28 shows the effect of confining stress and plasticity index. When I_p becomes greater than 30, effect of the confining stress becomes small for clayey soils, whereas effect of the confining stress becomes large for the clayey soil with $I_p < 30$ and sandy soil. Figure 7.29 compares average cyclic shear deformation characteristics and range within the standard deviation with data by Imazu and Fukutake (Kokusho 1980b); both data agree well. Comparisons with data by PWRI (Iwasaki et al. 1980b) are shown in Fig. 7.30; the shear modulus ratios agree but the damping ratio by PWRI is larger because of the difference of the drainage conditions. Finally, Fig. 7.31 shows the comparisons with data by Zen et al. (1987b); shear modulus ratio is larger than that by Zen et al. Soft soil data may be prone in the data by Zen et al. because samples are taken in port areas.

These comparisons indicate that a more accurate empirical equation can be obtained by using the soil classification based on the geologic age and the depositional conditions.

7.2.13 Remaining Literatures

Test data of the cyclic shear deformation characteristics have been reported by various researchers in the past in addition to the study in the preceding subsections. Woods (1991) compiled test methods on the cyclic shear test and introduced technical papers by these test methods. Richart (1977) expressed test data by the

Table 7.20 Soil classification based on geologic age and depositional environment

Geologic age		Depositional environment	Soil type	Geologic category	Number of data
Man-made		Fill	Clayey soil	1-Bc	10
			Sandy soil	2-Bs	14
Holocene	Upper	Aeolian	Sandy soil	3-As	2
		Marine	Sandy soil	4-As	15
			Gravel	5-Ag	0
		Brackish water	Clayey soil	6-Ac	5
			Sandy soil	7-As	20
		Fluvial	Peat	8-Ap	5
			Clayey soil	9-Ac	29
			Sandy soil	10-As	27
			Gravel	11-Ag	0
	Lower	Marine	Clayey soil	12-Ac	107
			Sandy soil	13-As	32
		Fluvial	Gravel	14-Ag	0
Pleistocene	Upper	Marine, brackish water	Clayey soil	15-Ac	18
			Sandy soil	16-As	4
	Upper	Volcanic ash fall	Loam	17-Lm	14
			Clayey soil	18-Dc	7
			Sandy soil	19-Ds	0
			Gravel	20-Dg	0
	Middle	Volcanic ash fall	Loam	21-Lm	0
			Clayey soil	22-Dc	75
			Sandy soil	23-Ds	93
			Gravel	24-Dg	1
Pliocene–Pleistocene	Lower Pleistocene, Upper Pliocene	Marine	Mudstone	25-Dc	4
			Sandy–gravel rock	26-Dsg	0

Hardin–Drnevich model and the Ramberg–Osgood model. Kokusho (1987) and Ishihara (1982) also compiled test data and showed general feature. JGS (2007) also provides a comparison of test data.

7.3 Behavior Under Large Strain: Shear Strength

Cyclic deformation characteristics shown in the previous section may not be adequate to describe soil behavior under very large earthquake motions. Shear stress may reach the shear strength under the large ground motions, but, as discussed in

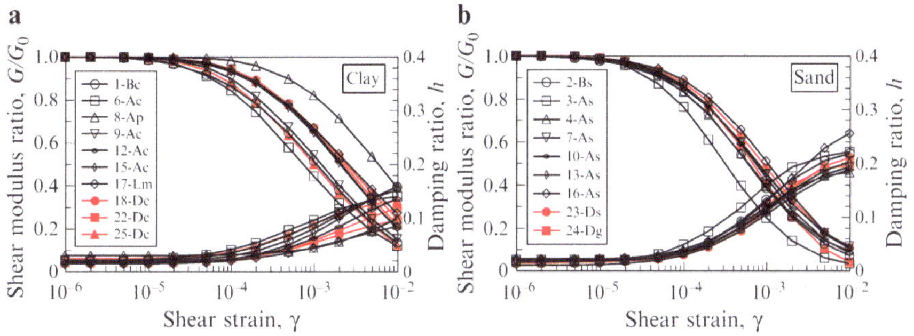

Fig. 7.26 Mean cyclic shear deformation characteristics in each category. (**a**) Clayey soil (**b**) Sandy soil

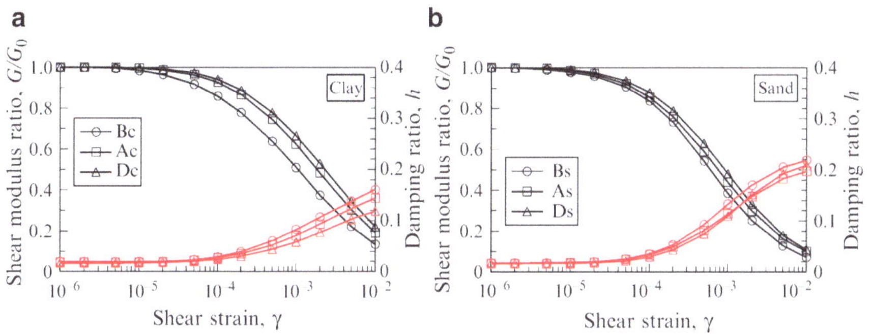

Fig. 7.27 Effect of depositional age. (**a**) Clayey soil (**b**) Sandy soil

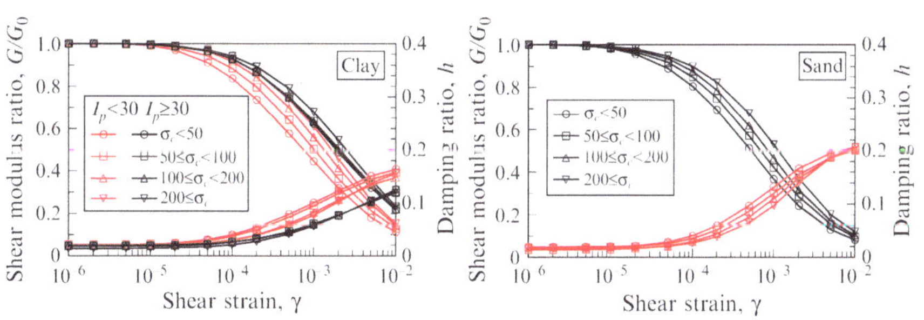

Fig. 7.28 Effect of plasticity index and confining stress

Sect. 6.7, behavior up to the shear strength cannot be obtained in the conventional test method. Shear strength is very important because, as explained in Sect. 1. 3, it controls the upper bound acceleration. Here, it is noted that available data are very limited on the relationships between shear strength expressed as the Mohr–Coulomb criteria and maximum shear stress obtained by the cyclic shear

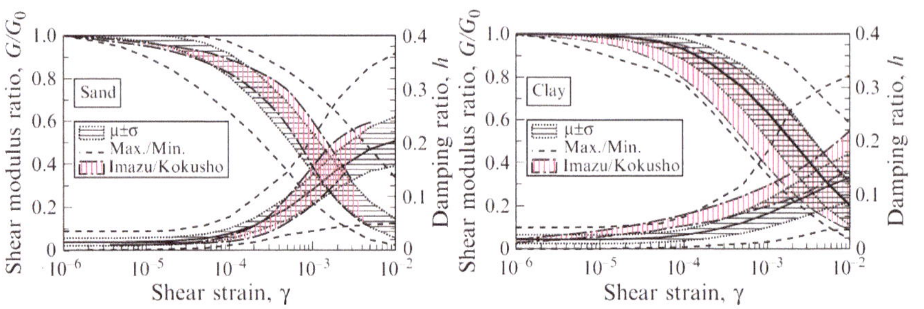

Fig. 7.29 Comparison with data by Imazu and Fukutake

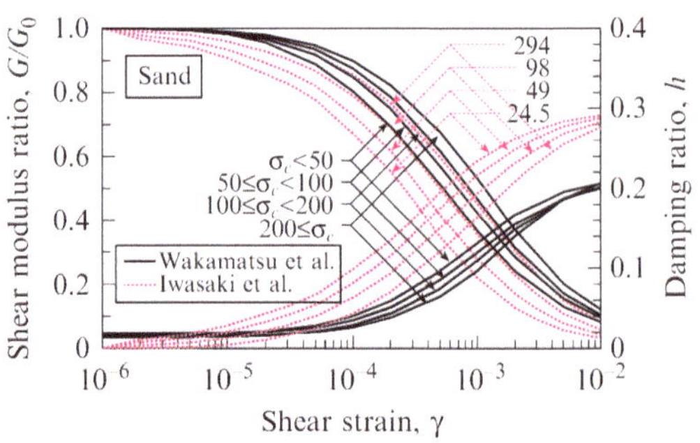

Fig. 7.30 Comparison with PWRI data

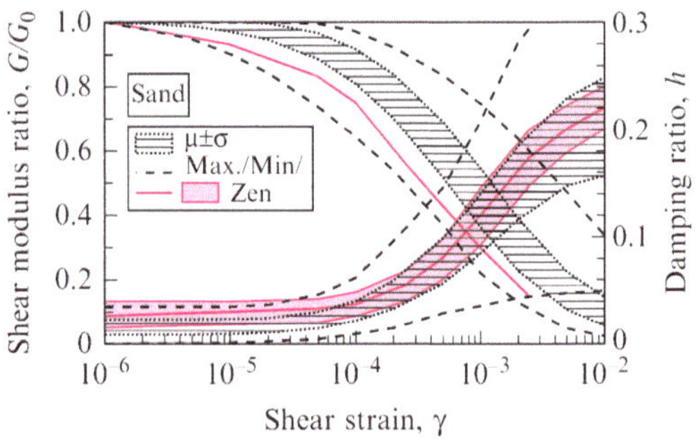

Fig. 7.31 Comparison with data in port areas in Japan (After Zen et al.)

deformation characteristics test, but, in this section, they are treated as identical, and strength parameter such as internal friction angle and cohesion is described.

Figures 7.32 and 7.33 show typical stress–strain curves for clay and sand up to large strains. Stress increases monotonically reaching the shear strength for clay; the shape is very simple. On the other hand, that of sand is complicated. The stress path for sand reaches the Mohr–Coulomb yield surface gradually, but it is difficult to express the behavior in one-dimensional stress–strain curve.

Conventional cyclic shear deformation characteristics tests are insufficient for determination of shear strength of soil because they do not mention on failure strain. The test result for monotonic loading tests can hardly be adopted for evaluating the maximum stress for a one-dimensional model under the cyclic loading such as an

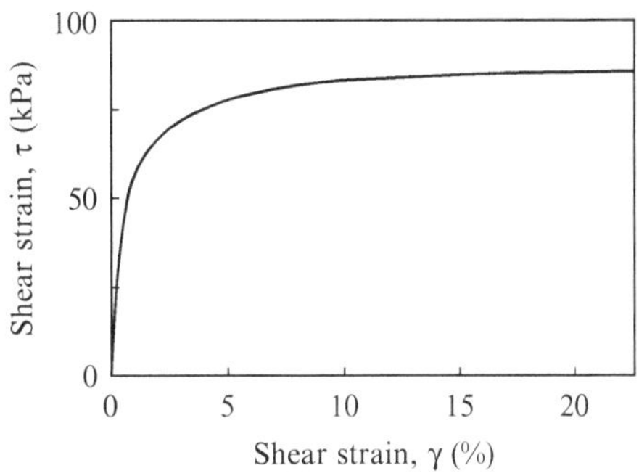

Fig. 7.32 Stress–strain curve for clay (Modified from JGS 1995)

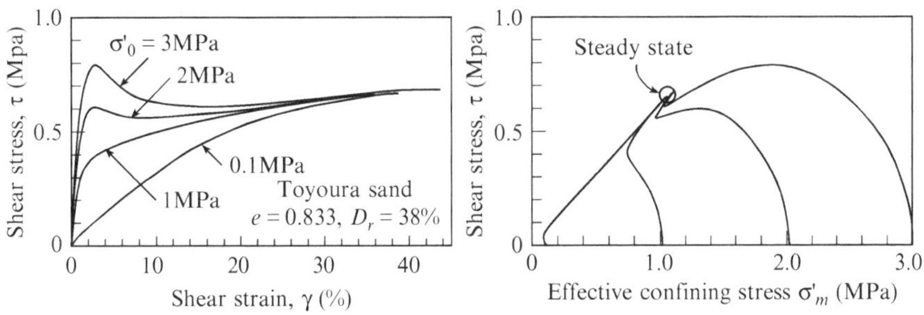

Fig. 7.33 Stress–strain curves and stress paths for sand (Modified from Ishihara 1996)

earthquake. This was discussed in detail in Sect. 6.7.6. Therefore, only empirical equations on shear strength are introduced in this section.

7.3.1 Shear Strength of Sand

Empirical equation proposed by Ohsaki (Kitazawa et al. 1959) has been frequently used to evaluate the internal friction angle of sand in Japan:

$$\phi = 15 + \sqrt{20N} \quad \text{[degree]} \tag{7.22}$$

This equation was derived from the box shear test result from the soils in the Holocene layer, the terrace, and the Tokyo formation. A linear equation was derived from the test data in Fig. 7.34. In addition, relationships between the relative density and range of internal friction angle and SPT N-value are summarized in Table 7.21. Here, relative density is the same as that defined by Terzaghi (Terzaghi and Peck 1948).

Similar relationships have been proposed in the past. For example, Japan Road Association proposed that (1996)

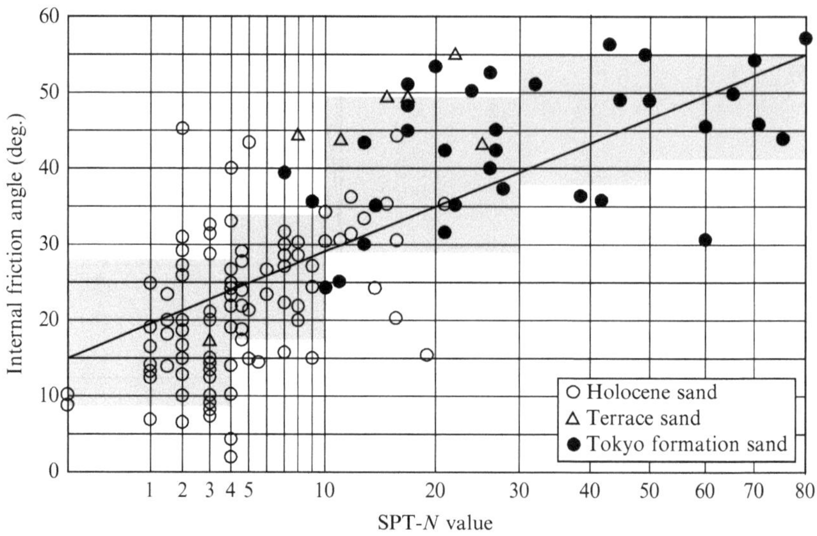

Fig. 7.34 Relationships between internal friction angle and SPT N-value (After Kitazawa et al. 1959)

Table 7.21 Range of relative density and internal friction angle under various SPT N-values

SPT N-value		0–4	4–10	10–30	30–50	50–
Relative density		Very loose	Loose	Medium	Dense	Very dense
Internal friction angle in degree	Average	18.3	25.9	39.0	46.4	47.9
	Region	9–28	18–34	29–49	38–55	41–55

$$\phi = 15 + \sqrt{15N} \quad [\text{degree}] \tag{7.23}$$

It is, however, noted that Eq. (7.23) is not an average value but a lower limit for conservative design. As typically shown in this example, empirical formulae for strength parameters sometimes define the lower bound because they are frequently developed for the stability analysis in which the lower strength results in a conservative design. Therefore, use of these equations may result in an underestimation of the acceleration in the seismic ground response analysis.

There was skeptical outlook, often expressed by some researcher, on evaluating the internal friction angle from the SPT N-value. A typical example is seen in Fig. 7.35 (Kutsuzawa and Morita 1991) which is obtained from Tanaka (1990) and Nishigaki (1980). Internal friction angles scatter between 35° and 45° and there is no correlation with the SPT N-value. In addition, Ohsaki's equation (Eq. 7.22) significantly underestimates internal friction angle at small SPT N-values. Tanaka (1990) suggested that this underestimation of the internal friction angle is caused by omitting of the effect of the confining stress. Further, Nishigaki (1980) suggested that the internal friction angle correlates with effective overburden stress as well as stress history of soil.

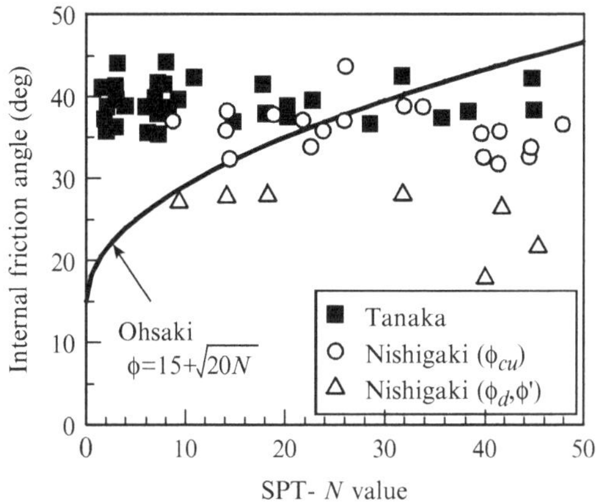

Fig. 7.35 Relationships between SPT N-value and internal friction angle (Modified from Kutsuzawa and Morita 1991)

Ohsaki's formula (Kitazawa et al. 1959) has been frequently used in the engineering practice. One interesting point regarding the data adopted by Ohsaki is that numerous results with an internal friction angle less than 20° exist, which is hardly observed through current tests. Several potential reasons can be imagined. For instance, the effects of dilatancy and confining stress dependency are not considered. Unfortunately, the author cannot decide on the actual basis of this disparity. It is, however, to be noted that this equation is less reliable for evaluating the internal friction angle for the seismic ground response analysis.

Equation (7.23) was removed from the subsequent design specification (2002 version (Japan Road Association 2002)) and the following equation was provided:

$$\phi = 4.8 \ln N_1 + 21 \quad (N > 5) \quad [\text{degree}] \tag{7.24}$$

where N_1 is the corrected SPT N-value considering the overburden stress $\sigma_v^{'}$,

$$N_1 = \frac{170N}{\sigma_v' + 70} \tag{7.25}$$

Here, Eq. (7.25) is the same as Eq. (5.4), but the acceleration of gravity is set to 10 m/s^2 in this equation. Importance to consider the effective stress dependency is emphasized, especially when N_1 is less than 20 because the internal friction angle changes significantly at small N_1.

Equation (7.26) was introduced by Railway Technical Research Institute (1997), which again considers confining stress dependency:

$$\phi = \min\left\{1.85\left(\frac{N}{\sigma_v'/98 + 0.7}\right)^{0.6} + 26, \quad 0.5N + 24\right\} \tag{7.26}$$

Various different equations were also provided by Ishihara (1997) and JGS (1976, 1988, 1995).

As mentioned, empirical equations in the design specifications for obtaining the internal friction angle are frequently proposed for the stability analysis but not for the seismic ground response analysis. Smaller shear strengths result in conservative design in stability analysis, whereas it may result in a critical side design in the seismic ground response analysis because, as described in 1.3, the maximum accelerations and the ground motion intensity are underestimated. Importance to consider the confining stress dependency of the SPT N-value is a rather well-acknowledged understanding. Even then, it is not considered in many past empirical equations. Recently, however, equations that consider confining stress dependency of SPT N-value began to appear such as Eqs. (7.24) and (7.26). For this intent, empirical equation proposed by Hatanaka and Uchida (1996) is listed in this book, as the frozen samples were used for deriving this equation:

$$\phi = 20 + \sqrt{20N_1} \pm 3 \tag{7.27}$$

where N_1 is the corrected SPT N-value by Eq. 5.6. A plot of this equation along with the original data is shown in Fig. 7.36. It is noted that the minimum internal friction angle is a little less than 30 and there is not any data with an internal friction angle less than 25. Further, internal friction angle of Shirasu (volcanic soil) seems to have different characteristics than sand although they are classified as the same.

7.3.2 *Shear Strength of Clay*

An unconfined compression test is usually used to obtain the shear strength of clay. The unconfined compression strength q_u is obtained through this test. Shear strength (cohesion) c is obtained as

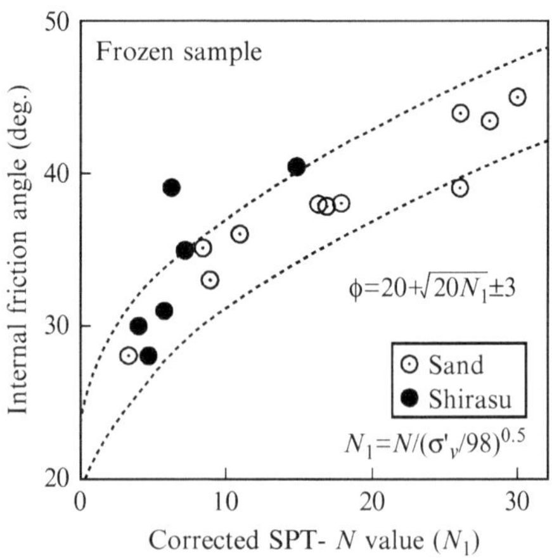

Fig. 7.36 Relationships between SPT N-value and internal friction angle (After Hatanaka and Uchida 1996)

$$c = \frac{q_u}{2} \tag{7.28}$$

Although empirical equations are frequently proposed to evaluate q_u, it is not converted to c in the following because conversion is easy. It is noted that use of laboratory test based on the undisturbed samples is suggested in many references because the relationships between shear strengths and SPT N-values scatter significantly, which indicates that evaluated shear strength through empirical equations may have significant error.

One of the most famous empirical equations is the one proposed by Terzaghi and Peck (1948), that is,

$$q_u = 12.5N \quad \text{[kPa]} \tag{7.29}$$

Ohsaki (Kitazawa et al. 1959) proposed Eq. (7.30) based on the Japanese data as

$$q_u = 40 + 5N \quad \text{[kPa]} \tag{7.30}$$

There are many other similar equations. For instance, design specification of road bridges (Japan Road Association 1996), in Japan, introduced Eq. (7.31)

$$c = (6 - 10)\,N \quad \text{[kPa]} \tag{7.31}$$

This equation is omitted in the subsequent version (2002 edition) (Japan Road Association 2002).

Takenaka and Nishigaki (1974) reported based on the tests on the Tenma clay, Osaka, that scattering of the SPT N-value becomes small if the automatic falling apparatus is used to measure the blow count, which indicates that scattering of the relationships between shear strengths and SPT N-values is caused by the scattering of SPT N-value. They also reported that the shear strength by triaxial test has less affected by the disturbance because it is larger than that by the unconfined compression test. In addition, q_u obtained by well-undisturbed samples is fairly larger than that by Eq. (7.29). Okumura (1982) pointed out that past empirical equations have a tendency to underestimate the shear strength compared with the test results in the port areas.

Shear strengths are compared with the above empirical equations in Fig. 7.37 (JGS 1995). It is seen that shear strengths obtained by Eqs. (7.29) and (7.30) correspond to the lower bound of the shear strengths. Shear strengths obtained by Takenaka and Nishigaki (1974) and by Okumura (1982) can be expressed as (JGS (1995))

$$q_u = (25 - 50)\,N \quad \text{[kPa]} \tag{7.32}$$

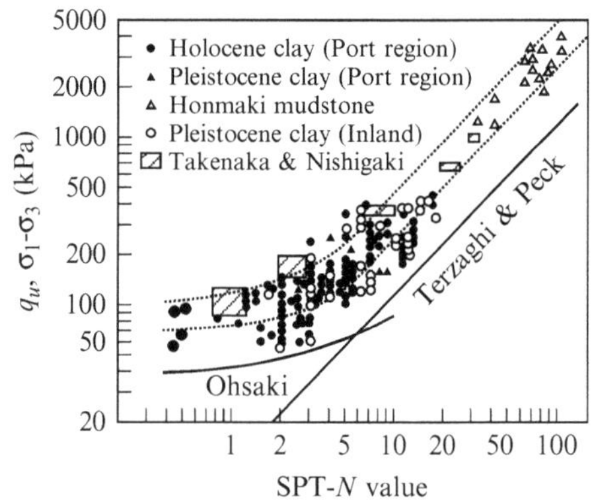

Fig. 7.37 Relationships between SPT *N*-value and q_u (Modified from JGS 1995)

Table 7.22 Soil classification and representative soil parameters

Soil classification	γ_{t2} (kN/m^3)	γ_{t1} (kN/m^3)	D_{50} (mm)	F_c (%)
Surface soil	17	15	0.02	≈100
Clay	15	14		100
Silt	17.5	15.5	0.025	90
Sandy silt	18	16	0.04	70
Silty fine sand	18	16	0.07	≈50
Very fine sand or silty sand	18.5	16.5	0.1	20
Fine sand	19.5	17.5	0.15	10
Medium sand	20	18	0.35	5
Corse sand	20	18	0.6	0
Gravel	21	19	2.0	0

Note: γ_{t2} unit weight under water table, γ_{t1} unit weight above water table, D_{50} average grain size, F_c fines contents

7.4 Other Parameters

Density or unit weight is necessary in a seismic ground response analysis. Unit weight is measured through the density log. When it is not measured in situ in the practice, empirical equations become necessary. Two data are shown in this book.

Japanese Geotechnical Society (2004) summarized representative soil parameters from Japan Road Association (1996) and Yasuda (1988) as in Table 7.22. Average grain size and fines contents are also shown in the table. In many researches, it is suggested to measure fines contents in situ because fines contents scatter very much (Yoshida and Wakamatsu 2010). It is noted that γ_{t1} is not dry unit weight as soil is not dry even above the water table.

Unit weights are shown in Table 7.23 (Railway Technical Research 1997) with range of SPT *N*-values. Here, buoyancy unit weight is obtained by subtracting the unit weight of the water from the one of the saturated sand. In the actual situation, however, soil above the water table is not dry and that below the water table is not perfectly saturated but unsaturated. Considering these situations, design

Table 7.23 Unit weight of soil

SPT N-value		Unit weight	
		Saturated	Buoyancy
Sandy soil	50–	20	10
	30–50	19	9
	10–30	18	8
	–10	17	7
Clayey soil	30–	19	9
	20–30	17	7
	10–20	15–17	5–7
	–10	14–16	4–6

specification of road bridge suggests to subtract 9 when the buoyancy unit weight is used for the earth pressure calculation, and design specification for port facilities suggests to subtract 5–8 in the backfill ground (Kutsuzawa and Morita 1991). However, 10 is frequently used in the practice of the seismic ground response analysis same as in Table 7.23.

References

Adachi N, Tatsuoka F (1986) Soil mechanics III – consolidation, shear and dynamic analysis-. Gihodo, Tokyo, 339 pp

Arulanandan K, Scott RF (eds) (1993) Proc. verification of numerical procedures for the analysis of soil liquefaction problems. Balkema, Davis

Asonuma T, Miura S, Yagi K, Tanaka H (2002) Dynamic deformation characteristics of volcanic soils and their evaluation methods. J Geotech Eng, Proc JSCE 708, III-59:161–173 (in Japanese)

Building Research Institute, Land, Infrastructure and Transportation Ministry (2001) Technical background on structural aspect of revision of the Building Standard Law. Gyousei, Tokyo (in Japanese)

Central Disaster Prevention Council (2005) Report of the 10th meeting of Task force on countermeasures against Tokai earthquake (in Japanese)

Coastal Development Institute of Technology (1997) Handbook on liquefaction remediation of reclaimed land, Revised version (in Japanese)

Committee on Technology Promotion of Soil Mechanics (1976) How to treat SPT N-value, c and f, JGS, Tokyo, 110 pp

Committee on Technology Promotion of Soil Mechanics (1992) Easy-to-understanding principle of soil mechanics, First revision, JGS, Tokyo, 243 pp (in Japanese)

Committee on Dynamic Properties of Clayey Soil (ed) (1995) Proc., Symposium on dynamic properties of clayey soil, JGS, Tokyo

Editorial Committee on Investigation of Ground (1995) Investigation approach of ground. JGS, Tokyo, 648 pp

Editorial Committee on Dynamic Analysis of Ground – from Foundation to Application (2007) Dynamic analysis of ground -from fundamental to application, JGS, Tokyo, 152 pp

Fujikawa T, Hayashi Y, Fukutake K (2000) Study for the effect of the nonlinearity of engineering bedrock on the seismic response of surface ground, Summaries of the technical papers of annual meeting of AIJ (Tohoku), pp 297–298 (in Japanese)

Fukumoto S, Yoshida N, Sahara M (2009) Dynamic deformation characteristics of sedimentary soft rock. J JAEE 9(1):46–64 (in Japanese)
Goto S, Shamoto Y Tamaoki S (1987) Dynamic properties of undisturbed gravel samples obtained by the in situ freezing method. In: Proceedings of the 8th Asian regional conference on soil mechanics and foundation engineering, Kyoto, pp 233–236
Hardin BO, Black WL (1968) Vibration modulus of normally consolidated clay. J Soil Mech Found Div ASCE 94(SM2):353–369
Hardin BO, Richart FE (1963) Elastic wave velocities in granular soils. J Soil Mech Found Div ASCE 89:33–65, No.SM1
Hatanaka M, Uchida A (1996) Empirical correlation between penetration resistance and internal friction angle of sandy soils. Soils Found 36(4):1–9
Iida K (1939) Velocity of elastic waves in a granular substance. Bull Earthq Res Inst 17:189–199
Imai T (1977) P- and S-wave velocities of the ground in Japan. In: Proceedings, 9th ISSMFE, Tokyo, vol 2, pp 257–260
Imai T, Tonouchi K (1982a) Correlation of N value with S-wave velocity. In: Proceedings of the 2nd European symposium on penetration testing, Amsterdam, pp 67–72
Imai T, Tonouchi K (1982b) Correlation of N value with S wave velocity and shear modulus. Proc., 2nd ESOPT
Imazu M, Fukutake K (1986a) Dynamic shear modulus and damping of gravel materials. In: Proceedings of the 21st Japan national conference of soil mechanics and foundation engineering, pp 509–512 (in Japanese)
Imazu M, Fukutake K (1986b) A consideration on data transaction of dynamic shear modulus and damping. In: Proceedings 21st Japan national conference of soil mechanics and foundation engineering, pp 533–536 (in Japanese)
Ishihara K (1982) Evaluation of soil properties for use in earthquake response analysis. In: Proceedings of the international symposium on numerical models in geomechanics, Zurich, Balkema, pp 237–259
Ishihara K (1996) Soil behavior in earthquake geotechnics, Oxford engineering science series 46. Oxford Science Publications, Oxford, 350 pp
Ishihara K (1997) Soil constants such as N value, c and ϕ. The Foundation Engineering and Equipment, Monthly, Sogo Doboku Institute 25(12):31–38 (in Japanese)
Ishihara K, Kokusho T, Tsutsumi C, Ishida H (2003) Study on dynamic properties of highly organic soil. In: Proceedings of the 58th annual conference of the Japan Society of civil engineering, JSCE, Tokyo, vol 3, pp 167–168 (in Japanese)
Iwasaki T, Tatsuoka F, Saheki M (1977) An example of correlation between SPT N-value and strain dependent S wave velocity. In: Proceedings of the 12th Japan national conference on geotechnical engineering, JGS, Tokyo, pp 477–780 (in Japanese)
Iwasaki T, Tatsuoka F, Takagi Y (1978) Shear moduli of sands under cyclic torsional shear loading. Soils Found 18(1):39–56
Iwasaki T, Tokita K, Yoshida S (1979) Dynamic deformation characteristics of Holocene clay, PWRI report, No. 1504. Public Work Research Institute
Iwasaki T, Tokida K, Yoshida S (1980a) On the dynamic deformation and strength characteristics of Alluvial clay. In: Proceedings of the 15th Japan national conference of soil mechanics and foundation engineering, JGS, Tokyo, pp 625–628 (in Japanese)
Iwasaki T, Tatsuoka F, Takagi Y (1980b) Experimental study on dynamic deformation characteristics of soils (ii), Report of PWRI, vol 153. Public Work Research Institute
Japan Read Association (1985) Specifications for highway bridges and commentary, Part V, Earthquake Resistant Design, Maruzen (in Japanese)
Japan Road Association (1996) Specifications for highway bridges, Part V, Seismic design, Maruzen (in Japanese)
Japan Road Association (2002a) Specifications for highway bridges, Part V, Seismic design, Maruzen (in Japanese)
Japan Road Association (2002b) Specifications for highway bridges, Maruzen (in Japanese)

Japanese Geotechnical Society (2004) Remedial measures against soil liquefaction. Geotechnical engineering practice series 18, 513 p (in Japanese)
JGS (1988) Strength constants -*c*, ϕ, *N* value-, JGS, Tokyo, 279 pp (in Japanese)
JGS (1994) Proceeding symposium on dynamic deformation characteristics in dynamic problem of ground and soil structures -test and investigation methods and its application, JGS, Tokyo, (in Japanese)
Kanatani M, Tochiki H, Kawai T (1999) Study on seismic stability of seawall in man-made island (part 4) -Deformation analysis of seawall during earthquake by DEM-FEM coupled analysis method (SEAWALL-2D)-, Abiko Research Laboratory Report No. U98025. Central Research Institute of Electric Power Industry, Tokyo (in Japanese)
Kitazawa G, Kitayama K, Suzuki K, Ohkawa H, Ohsaki Y (1959) Tokyo ground map. Gihodo, Tokyo, 23 pp
Kokusho T (1980a) Triaxial test of dynamic soil properties for wide strain range. Soils Found 20(2):45–60
Kokusho T (1980b) Cyclic triaxial test of dynamic soil properties for wide strain range. Soils Found 20(2):45–60
Kokusho T (1987) In-situ dynamic soil properties and their evaluations. In: Proceedings of the 8th Asian regional conference on soil mechanics and foundation engineering, vol 2, pp 215–240
Kokusho T, Esashi Y (1981) Cyclic triaxial test on sands and coarse materials. In: Proceedings of the 10th international conference on soil mechanics and foundation engineering, Stockholm, vol 1, pp 673–676
Kokusho T, Sasaki M (1980) Dynamic deformation characteristics test of undisturbed Pleistocene sand by cyclic triaxial test. In: Proceedings of the 15th Japan national conference of soil mechanics and foundation engineering, pp 537–540 (in Japanese)
Kokusho T, Yoshida Y, Esashi Y (1982a) Dynamic properties of soft clay for wide strain range. Soils Found 22(4):1–18
Kokusho T, Yoshida Y, Nishi K, Isashi Y (1982b) Dynamic properties of alluvial clayey soils for wide strain range, Central Electric Research Center, Report No. 381027 (in Japanese)
Koyamada K, Miyamoto Y, Miura K (2003) Nonlinear property for surface strata from natural soil samples. In: Proceedings of the 38th Japan national conference on geotechnical engineering, pp 2077–2078 (in Japanese)
Kutsuzawa S, Morita Y (1991) Design specification and soil constant. Tsuchi-to-Kiso JGS 39(12):63–67
Lo Presti DC, Jamiolkowski M, Pallara O, Cavallaro A, Pedroni S (1997) Shear modulus and damping of soils. Geotechnique 47(3):603–617
Marcuson WF, Wahls HE (1972) Time effects on dynamic shear modulus of clays. J Soil Mech Found Div ASCE 98(SM12):1359–1373
Ministerial Notification 1457 on Building Standard Law (2000) Appendix table
Nishi Y (1980) SPT N-value in soil ground. In: Proceedings of symposium on sounding, JGS, pp 109–114
Nishi K, Yoshida Y, Sawada Y, Iwadate I, Kokusho T Hanada K (1983) Soil-structure interaction of JPDR reactor (2) static and dynamic properties of sand and sandy gravel by laboratory tests and proposal of model of ground, Central Electric Research Center, Report No. 383002, 44 pp (in Japanese)
Nishi K, Esashi Y, Kokusho T (1985) Dynamic strength-deformation properties of soft rocks during cyclic loading, Central Electric Research Center, Report No. 383050 (in Japanese)
Nishio S, Tamaki K Machida Y (1985) Dynamic deformation characteristic of crushed rock by large scale triaxial test apparatus. In: Proceedings of the 20th Japan national conference on geotechnical engineering, pp 603–604 (in Japanese)
Numata A, Suda Y, Kokusho T, Yoshida A, Suwa Y, Sato M (2000) Consideration of estimate expression on initial shear modulus for Masado. In: Proceedings of 55th annual conference of the Japan Society of civil engineering. vol 3, pp 144–145 (in Japanese)
Ohta Y, Goto N (1978) Empirical shear wave velocity equations in terms of characteristics soil indexes. Earthq Eng Struct Dyn 6(2):167–187

Okumura J (1982) Frame of SPT N-value and usage in design of port facilities. The Foundation Engineering and Equipment, Monthly, Sogo Doboku Institute 10(6):57–62 (in Japanese)
Ports and Harbours Bureau, Transport Ministry (ed) (1999) Technical standards and commentary of port and Harbour facilities in Japan. The Ports & Harbours Association of Japan, Tokyo
Prange B (1981) Resonant column testing of railroad ballast. In: Proceedings of the 10th international conference on soil mechanics and foundation engineering, Stockholm, vol 1
PWRI (1982) Numerical analysis of seismic response of ground, Report of PWRI, No. 1778, pp 1–47 (in Japanese)
Railway Technical Research Institute (1997) Design standard of railway facilities and commentary, Maruzen, Tokyo, 557 pp (in Japanese)
Railway Technical Research Institute (1999) Design standard and commentary of railway facilities, Maruzen, Tokyo, 469 pp (in Japanese)
Richart FE Jr (1977) Dynamic stress-strain relationships for soils, S-O-A paper. In: Proceedings of 9th ICSMFE, Tokyo, vol 3, pp 605–612
Saxena AK, Reddy KR (1989) Dynamic moduli and damping ratios for Monterey No.0 sand by resonant column tests. Soils Found 29(2):37–51
Seed HB, Idriss IM (1970) Soil moduli and damping factors for dynamic response analyses, Report No. EERC70-10. Earthquake Engineering Research Center, University of California, Berkeley, 40 pp
Shibata T, Soelarno DS (1975) Stress-strain characteristics of sands under cyclic loading. In: Proceedings of JSCE, No. 239, pp 57–65
Takenaka J, Nishigaki Y (1974) Fundamental study on standard penetration test III. In: Proceedings of the 9th Japan national conference on geotechnical engineering, JGS, Tokyo, pp 14–15 (in Japanese)
Tanaka H (1990) Frame of SPT N-value and usage in design of port facilities. The Foundation Engineering and Equipment, Monthly, Sogo Doboku Institute, vol 18, No. 3, pp 77–82 (in Japanese)
Tanaka Y, Kudo K, Kokusho T, Kato M Kato A (1985) Dynamic shear modulus and damping of gravelly soils. In: Proceedings of the 20th Japan national conference on geotechnical engineering, pp 599–602 (in Japanese)
Tanaka Y, Kudo K, Yoshida Y Ikemi M (1987) A study on mechanical property of gravel -Dynamic strength and deformation characteristics of reconstituted sample-, Report No. U87019, Central Research Institute of Electric Power Industry, pp 1–47
The Overseas Coastal Area Development Institute of Japan (1999) Technical standards and commentary of port and Harbour facilities in Japan
Terzaghi K, Peck RB (1948) Soil mechanics in engineering practice. Wiley, New York, 566 pp
Uwabe T, Tsuchida H, Kurata E (1982) Coupled hydrodynamic response analysis based on strong motion earthquake records of fill type breakwater in deep sea, Report of the Port and harbour Research Institute, Ministry of Transport, vol 22, No. 2, pp 286–326 (in Japanese)
Vucetic M, Dobry R (1991) Effect of soil plasticity on cyclic response. J Geotech Eng ASCE 117(1):89–107
Wakamatsu K, Yoshida N, Mikami T (2010) Controlling factors and effect of geologic age and depositional environment on cyclic shear deformation characteristics of soils. Japan Geotech J 5(3):463–478
Woods RD (1991) Field and laboratory determination of soil properties at low and high strains, SOA paper. In: Proceedings of second international conference on recent advances in geotechnical earthquake engineering and soil dynamics, St. Louis, pp 1727–1741
Yasuda S (1988) From investigation to remedial measures against liquefaction, Kajima Institute Publishing, Tokyo, 243 pp (in Japanese)
Yasuda S, Yamaguchi I (1985) Dynamic soil properties of undisturbed samples. In: Proceedings of 20th annual meeting of JSSMFE, Tokyo, pp 539–542 (in Japanese)
Yokota K, Tatsuoka F (1982) On the shear deformation coefficient of undisturbed Pleistocene clay. In: Proceedings of 32nd annual conference of the Japan Society of civil engineering, JSCE, Tokyo, vol 3, pp 257–258 (in Japanese)

Yoshida N (2001) Effect of nonlinear behavior of seismic base layer on evaluation of level 2 ground motion. In: Proceedings of first annual meeting of JAEE, IAEE, Tokyo, p 37 (in Japanese)
Yoshida N, Wakamatsu K (2010) Statistical characteristics of sand for liquefaction potential evaluation. In: Proceedings of the 45th Japan national conference on geotechnical engineering, JGS, Tokyo, pp 337–338
Yu P, Richart FE (1984) Stress ratio effects on shear modulus of dry sands. J Geotech Eng Div ASCE 110:331–345, GT3
Zen K, Umehara Y, Hamada K (1978) Laboratory tests and in-situ seismic survey on vibratory shear modulus of clayey soils with various plasticities. In: Proceedings of 6th Japan symposium of earthquake engineering, pp 721–728 (in Japanese)
Zen K, Yamazaki H, Umehara Y (1987a) Experimental stud on shear modulus and damping ratio of natural deposits for seismic response analysis. Report of the Port and harbour Research Institute, Ministry of Transport 28(1):41–113
Zen K, Yamazaki H, Umehara Y (1987b) Experimental stud on shear modulus and damping ratio of natural deposits for seismic response analysis, Report of the Port and harbour Research Institute, Ministry of Transport, 28(1):41–113 (in Japanese)

Chapter 8
Modeling of Mechanical Soil Properties

Methods for modeling the mechanical properties are described in this chapter. Many data are required in the detailed seismic ground response analysis. Therefore, engineers need to know their mechanical meaning.

8.1 Elastic Modulus

Elastic modulus of the soil depends on an effective confining stress. In the total stress analysis, however, elastic modulus is frequently set constant because change in confining stresses is small, which is a big difference from the effective stress analysis.

8.1.1 Elastic Shear Modulus

Based on the laboratory tests described in Sect. 7.1.6, elastic shear modulus is expressed as functions of void ratio and confining stress as

$$G_0 = Af(e)\sigma'^{n}_{m} \tag{7.12}$$

Values of coefficients A, function $f(e)$, and power n are shown in Table 7.4 and Fig. 7.4. As these parameters are obtained from the laboratory test, it may be different from the in situ value partly because of the disturbance of the soil samples taken in situ or partly because they are made in the laboratory. If the in situ elastic modulus is used without considering the confining stress dependency, Eq. (7.12) is not necessary to use.

N. Yoshida, *Seismic Ground Response Analysis*, Geotechnical, Geological and Earthquake Engineering 36, DOI 10.1007/978-94-017-9460-2_8

On the other hand, when considering confining stress dependency either in the effective stress analysis or in the total stress analysis, evaluation of A and $f(e)$ becomes important or necessary. In addition, value of power n is also important because, as will be explained in Sect. 10.3, it affects the acceleration distribution near the ground surface. In addition, it is required in the input data for constitutive models. Some explanation may be necessary on the latter case.

In the analysis of the ground, $Af(e)$ and n are set the same in the same layer or in some extent. The value of n is usually around 0.5 based on the laboratory test result (Sect. 7.1). When looking at the in situ elastic modulus, however, quite different feature is sometimes seen.

Figure 8.1 shows the soil profiles in the fill area in the Tokyo Bay area (Sato et al. 1998). The same feature can be seen in Figs. 1.12 and 5.3, too. Here, S-wave velocities obtained by the PS logging are nearly constant. In addition, the SPT N-values also keep constant, although the S-wave velocity and the SPT N-value must increase with depths if confining stress dependency exists. The only exception is the sand layers in the type E ground where both the SPT N-value and the S-wave velocity increase as depths. In addition, power on depth H in Eq. (7.8) is 0.199, which is very small compared with the conventional value of the confining stress dependency, 0.5.

If the power on the confining stress is evaluated from Fig. 8.1, it becomes nearly zero and is quite different from the power in the laboratory test result. The reason

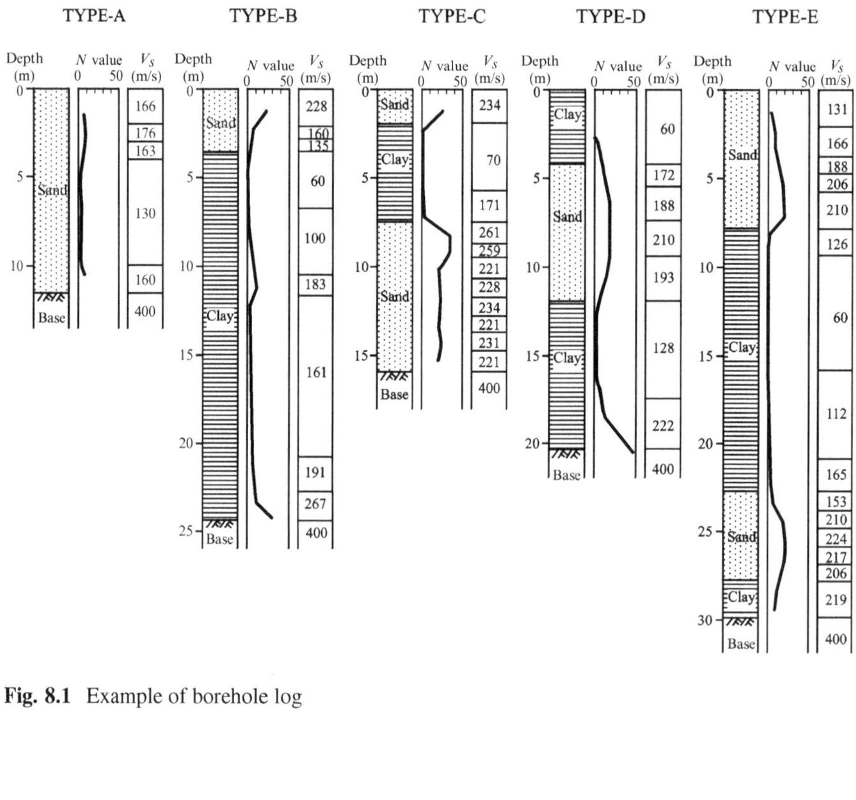

Fig. 8.1 Example of borehole log

of this inconsistency is not clear although several reasons can be imagined. For example, average S-wave velocities are output within certain thickness in the downhole method. Aging effect is larger for the soil at deep depths. However, different opinion can be made; the result of the suspension method also shows similar behavior, and the elastic modulus is to be larger at deep depths because of the aging effect (Schmertmann 1991). As discussed, the reason is not clear at present, but engineers must recognize that change of the elastic moduli is sometimes very small in the vertical direction.

These discussions may not be a big issue in the one-dimensional seismic ground response analysis because elastic moduli are specified layer by layer. On the other hand, elements are grouped and the same $Af(e)$ and n are used in a same group in multidimensional analyses. Resultant elastic modulus increases with depth in the analysis to evaluate initial stress by, for example, a switch-on-gravity analysis, which may be different from the in situ value. Change of the wave velocity is large near the ground surface. Therefore, it is suggested to make a simple analysis such as one-dimensional analysis to grasp the effect of the confining stress dependency in the practice.

8.1.2 Bulk Modulus and Poisson's Ratio

Description in this subsection may not be necessary in the one-dimensional analysis because it deals with only shear deformation, but is necessary for the multidimensional analysis.

Many constitutive models use incremental formulation of Eq. (4.2) for the volume change characteristics, which is expressed as

$$d\sigma'_m = K_0 {\sigma'}_m^n \cdot d\varepsilon_{vd} \tag{8.1}$$

This formulation is called an incremental elasticity since this is not a perfect elastic behavior (Tobita and Yoshida 2004).

The bulk modulus is also set constant in the total stress analysis same with the elastic shear modulus. Method to set the elastic modulus is explained in this book by two standpoints.

As frequently described, soil is composed of soil skeleton and pore material. If the elastic moduli are evaluated based on the in situ measurement such as the PS logging, there will be no difficult problem. Poisson's ratio ν is first evaluated from Eq. (5.9) and the bulk modulus is computed based on Table 4.1. However, when the in situ measurement is not made, V_s can be evaluated, for example, from the SPT N-value, but empirical equation that evaluates other elastic moduli is few.

The bulk modulus hardly affects the result of the seismic ground response analysis because shear deformation is prone during an earthquake. The bulk modulus affects behavior under P-wave propagation, but P-wave propagation does not affect the whole behavior. The bulk modulus is important in the effective stress analysis because it is directly related to the excess porewater development, but the error in the bulk modulus can be canceled by adjusting the dilatancy parameter so as to explain the excess porewater pressure development. Reflecting this situation, it has been determined roughly compared with the shear modulus in the past practices.

The bulk modulus is frequently computed from the elastic shear modulus and Poisson's ratio. In the saturated soil, it is sometimes necessary to separate the behavior into the soil skeleton and the pore material. Here, Poisson's ratio ν is sometimes assumed to be 0.5 for the saturated soil under the undrained condition in the static analysis such as consolidation analysis. However, it cannot be set 0.5 in the dynamic analysis because the P-wave velocity is infinite at $\nu = 0.5$. Therefore, relevant Poisson's ratio is to be set.

Poisson's ratio of 1/3 is frequently used in the practical calculation because of the following reason. Coefficient of earth pressure at rest K_0 is nearly 0.5 in the soft soil ground. When the soil is uniaxially compressed keeping the lateral strain zero, K_0 becomes 0.5 under $\nu = 1/3$. However, there is a big misunderstanding in this issue. If it is a linear material, $K_0 = 0.5$ when $\nu = 1/3$. However, deviator stress appears under this loading, which will cause nonlinear behavior of the shear deformation. It indicates that Poisson's ratio must be smaller than 1/3 to make $K_0 = 0.5$ when considering the nonlinear behavior.

Figure 8.2 shows Poisson's ratios from several references (Lade and Nelson 1987; Kokusho 1980; Nakagawa et al. 1997) by variety of test methods.

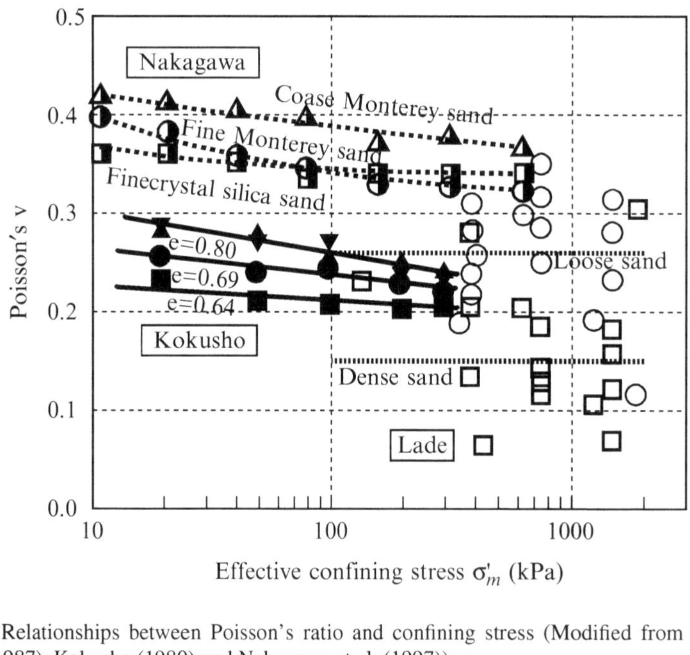

Fig. 8.2 Relationships between Poisson's ratio and confining stress (Modified from Lade and Nelson (1987), Kokusho (1980) and Nakagawa et al. (1997))

Lade et al. (1987) made a triaxial test by using the Santa Monica Beach sand ($D_{50} = 0.27$ mm). Both loose ($D_r = 20$ %) and dense ($D_r = 90$ %) sands are tested. Test was carried out keeping the confining pressure constant, and Poisson's ratio was calculated from the unloaded points. They reported that the confining stress dependency is not observed although data scatter, and Poisson's ratio ν is expressed as

$$\begin{aligned} \nu &= 0.26 \pm 0.06 \text{ Loose sand} \\ \nu &= 0.15 \pm 0.08 \text{ Dense sand} \end{aligned} \tag{8.2}$$

Kokusho (1980) used an improved triaxial test apparatus that can measure strain from 10^{-6} by using a high-sensitivity gap sensor and by installing a load cell into the cell. Poisson's ratio of the Toyoura sand is evaluated from the shear modulus G obtained from the undrained test and Young's modulus E obtained from the drained test. He reported that Poisson's ratio lies between 0.2 and 0.3, and it depends on the confining stress as well as the void ratio.

Nakagawa et al. (1997) measured P- and S-wave velocities of the dry and the saturated sands by the ultrasonic pulse method. They found that the P-wave velocity is constant, whereas S-wave velocity depends on the confining stress. Therefore, Poisson's ratio has confining pressure-dependent nature.

From these data, it can be concluded that the confining stress dependency is observed under small confining stresses such as less than 100 kPa, but confining stress dependency becomes small at large confining stresses. Data by Lade et al. and Kokusho seems to be consistent but that by Nakagawa et al. shows much larger than others. The reason is not clear by the author although difference of strain level by these tests can be pointed out.

Shear modulus is calculated from the cyclic shear deformation characteristics tests. Therefore, it may be consistent to use similar test result to evaluate bulk moduli and then Poisson's ratio. In this meaning based on Kokusho or Lade, Poisson's ratio less than 0.3 is consistent. This is also reasonable to explain the fact that K_0 is nearly 0.5 as discussed above.

There is another problem when the bulk modulus is determined based on the effective stress; power on the confining stress in Eq. (8.1) for the bulk modulus and in Eq. (7.12) for elastic shear modulus may not be the same. If they are the same value, Poisson's ratio becomes constant and the confining stress dependency does not appear, but it does not agree with Fig. 8.2. Many constitutive models use one of the following two methods:

1. Same power for the bulk modulus and shear modulus.
2. Power of Eq. (8.1) is 1.0.

Here, the mathematical treatment is easy in the former method. The bulk modulus can be calculated by assuming a relevant Poisson's ratio. On the other hand, the latter method is consistent with the linear $e - \log \sigma'_m$ relationships which are used in the consolidation analysis (see Eq. 4.19). Poisson's ratio depends on the confining stress in this case.

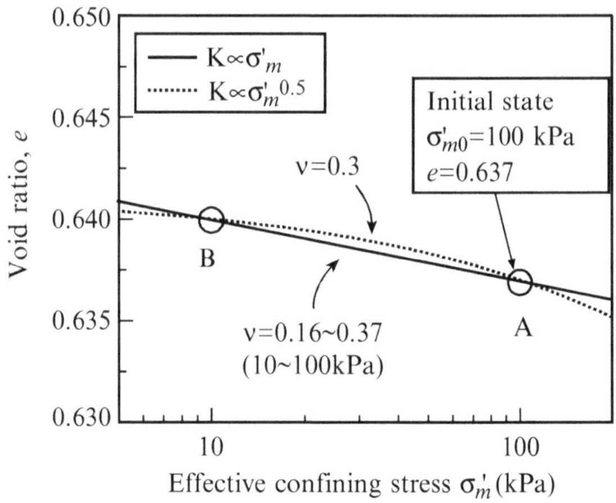

Fig. 8.3 Change of Poisson's ratio

Both methods seem to be quite different, but it is not true as shown in Fig. 8.3 as an example, in which two $e - \log \sigma'_m$ curves are drawn. They are determined to pass point A ($\sigma'_m = 100\,\text{kPa}$ and $e = 0.637$). At first, $e - \log \sigma'_m$ curve is evaluated by keeping Poisson's ratio constant (0.3), which is shown by a dotted line in the figure. Then a straight line is drawn to pass point B which lies on the dotted line and $\sigma'_m = 10\,\text{kPa}$. Poisson's ratio changes from 0.17 to 0.37 between $\sigma'_m = 10 - 100\,\text{kPa}$. However, there is no significant difference between the solid and dotted lines. Since change of the confining stress is not so large in the total stress analysis, this agreement seems sufficient in the engineering sense. It indicates that either method described above can be used if the coefficient is set relevantly.

When the bulk modulus of the soil skeleton is obtained, apparent bulk modulus of the soil (mixture of the soil skeleton and the porewater) K_t is calculated as

$$K_t = \frac{K + K_w}{n} \tag{8.3}$$

where K_w denotes a bulk modulus of water, and $n = e/(1 + e)$ is porosity. Therefore, two more parameters are to be evaluated in the saturated sand. The bulk modulus of water is $K_w = 2.22 \times 10^6\ \text{kN/m}^2$, but actual value is less than it as soil is not usually perfectly saturated. The porosity n is to be measured from the laboratory test.

If the P-wave velocity is not obtained by the PS logging, actual bulk modulus is difficult to evaluate. As mentioned, however, the bulk modulus does not affect whole response of ground under horizontal ground motion. Therefore, a set of representative values may be sufficient to set these values. In situ P-wave velocities frequently lie between 1,300 and 1,400 m/s, and apparent second term of Eq. (8.3) is evaluated as

$$K_w = n\left(\rho V_p^2 - K - \frac{4G}{3}\right) \tag{8.4}$$

8.2 Nonlinear Model for One-Dimensional Analysis

Modeling of the shear deformation characteristics is described in this section. Shear stress τ_{vh} versus shear strain γ_{vh} relationships are dealt with in the one-dimensional total stress analysis. The effective confining stress dependency is frequently replaced by the effective overburden stress dependency as the lateral stress is out of consideration. The shear deformation characteristics are frequently written as τ–γ relationships without adding subscript vh. The same expression is used throughout this book.

There are many constitutive models for the one-dimensional problem, among which models frequently used in the Japanese practice is explained in this book. If the fundamental concept can be understood through these explanations, it is not difficult to use other models. At first, relationships between the cyclic shear deformation characteristics and the constitutive models are explained in Sect. 8.2.1. Then the hysteresis rule under the cyclic loading is explained in Sect. 8.2.2. Finally, explanation of the individual constitutive model is explained.

8.2.1 Relation Between Cyclic Shear Deformation Characteristics and Mathematical Models

Figure 8.4a shows result of a cyclic shear deformation characteristics test. Test was carried out in nine stages in total, among which hysteresis loops in the first seven stages are shown in Fig. 8.4b. Unlike the ordinary method to calculate G and h, hysteresis curves are drawn using the cumulated shear strain from the beginning of the test. The hysteresis loop gradually moves toward right. However, this shift is not considered in evaluating the shear modulus because secant modulus is calculated as a slope angle connecting the unload points. This procedure is equivalent to rewrite the stress–strain curve so that centers of the hysteresis loops locate at the origin as shown in Fig. 8.4c.

Strains have tendency to shift toward elongation side in the triaxial test as explained in Sect. 6.7. On the other hand, equivalent linear analysis such as SHAKE (Schnabel et al. 1972) and FLUSH (Lysmer et al. 1975), which will be explained later, cannot consider stress–strain curve such as in Fig. 8.4b. Therefore, modeling into Fig. 8.4c has been accepted. Applicability of these computer codes may be reevaluated in the future because of the reason that they cannot consider drift of the hysteresis loops, in addition to the fact that effect of the initial stress (see Sect. 6.7.8) cannot be considered.

Many stress–strain models employ a hysteresis rule under the cyclic loading shown in Fig. 8.5a. At first, stress–strain curves are classified into the behavior under the monotonic loading and those after unloading. The former curve is frequently

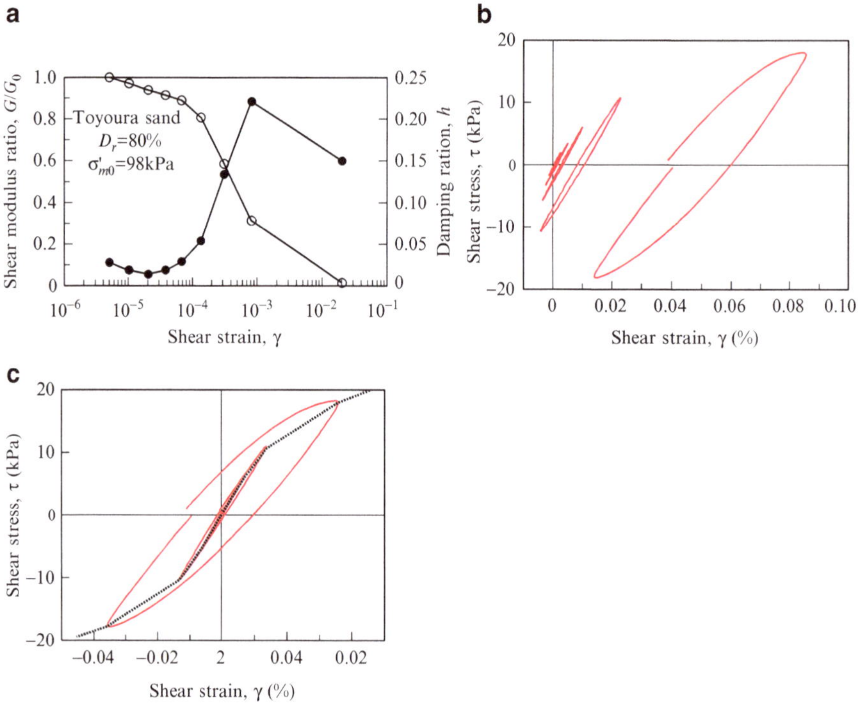

Fig. 8.4 Modeling of stress–strain curve for analysis. (**a**) Cyclic shear deformation characteristics. (**b**) Hysteresis loop at the tenth cycle of loading. (**c**) Assumption for stress–strain relationship modeling

called a skeleton curve or a backbone curve. The state point, therefore, moves along the skeleton curve under the first loading. When direction of loading or strain is reversed, the state point moves apart from the skeleton curve. This state is called unloading, and the state when direction of loading again changes is called reloading. However, it is difficult and impossible to distinguish unloading and reloading in the seismic ground response analysis. Therefore, they are called a hysteretic behavior, and the stress–strain curve that corresponds to the hysteretic behavior is called a hysteresis curve. In summary, stress–strain curves are composed with the skeleton curve and the hysteresis curve. The term "hysteresis loop" is also used when a stress–strain curve closes under the cyclic loading.

The skeleton curve is point symmetry so that behaviors in the positive and negative directions are the same. In addition, hysteresis curve unloaded at point A is frequently modeled to be headed toward A′ which is located point symmetrically of point A against the origin.

Comparison between Figs. 8.4c and 8.5a will give an idea that the G–γ curve corresponds to the skeleton curve and the hysteresis loop at the tenth cycle of loading corresponds to the hysteresis curve. As described in Subsect. 6.7.6, behaviors in

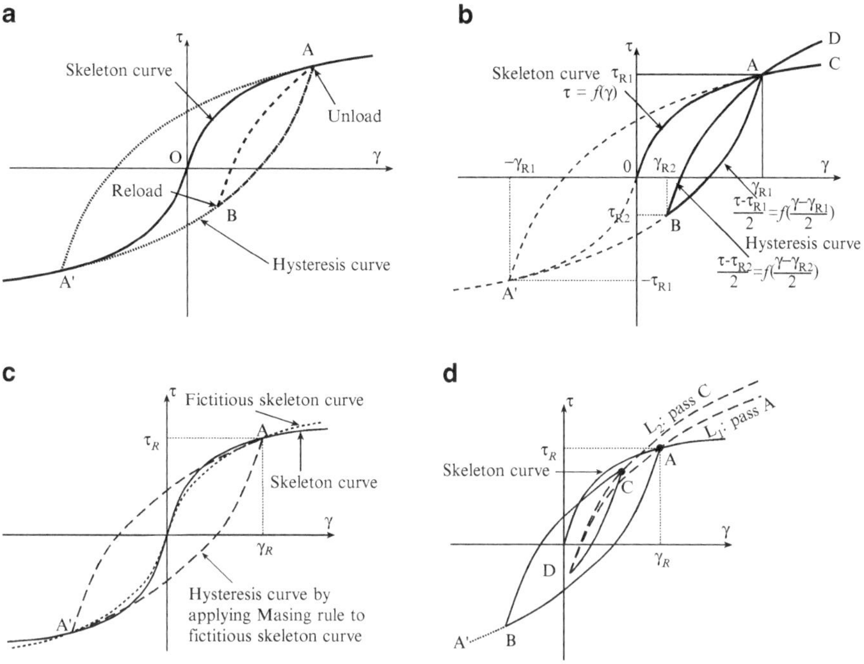

Fig. 8.5 Modeling of stress–strain curve under cyclic loading. (**a**) Modeling of cyclic behavior. (**b**) Masing's rule and hysteresis curve. (**c**) Improved hysteresis rule. (**d**) Hysteresis rule to save memory

the monotonic loaded test and the cyclic loaded test are not completely identical. However, they are conventionally considered to be identical in the engineering practice.

8.2.2 Hysteresis Rules

As the G–γ and h–γ relationships are defined separately, it is natural to define the skeleton curve and the hysteresis curve separately. Here, modeling of the skeleton curve is not a difficult problem as it requires just to find an equation that fit the skeleton curve. On the other hand, as there are waves with different amplitude, determination of the mathematical equation is not so simple under the earthquake loading. The hysteresis rule is a rule to determine the hysteresis curve under the cyclic loading. Both the skeleton curve and the hysteresis curve are required to be defined in the constitutive model for the seismic ground response analysis, but the hysteresis rule is sometimes not sufficient to describe the whole paths when a constitutive model is developed by considering only a simple path such as constant amplitude loading.

8.2.2.1 Masing's Rule

Many constitutive models do not define two behaviors separately, but employ Masing's rule by which hysteresis curves are developed from the skeleton curve. It is schematically shown in Fig. 8.5b.

A stress–strain curve during the path O → A → B → A which includes unload and reload behaviors is considered. When the skeleton curve is expressed as a function of the strain as

$$\tau = f(\gamma), \tag{8.5}$$

then, according to Masing's rule, the hysteresis curve after unload at (τ_R, γ_R) is expressed as

$$\frac{\tau - \tau_R}{2} = f\left(\frac{\gamma - \gamma_R}{2}\right) \tag{8.6}$$

It means that both the skeleton curve and the hysteresis curve are similar shape with a homothetic ratio of 2 and the starting point moved to the unload point. The homothetic ratio need not be 2 in the original definition of Masing's rule, but 2 is used in many constitutive models because the hysteresis curve unloaded at point A is headed to the point symmetrical point A′ and goes back to the skeleton curve at this point. The slope angle of the hysteresis curve and that of the skeleton curve are the same at this point. Therefore, the state point moves smoothly from the hysteresis curve to the skeleton curve at point A′. Next, when reloading occurs at point B on the unloaded path, the state point is headed to the previously unloaded point A. The hysteresis curve can be developed with ease by this rule.

Masing's rule is sometimes explained to be composed of two rules: stiffness at unload is the same with the initial stiffness and the hysteresis curve is two times larger than the skeleton curve (Adachi and Tatsuoka 1986). If the latter rule holds, however, the first condition is automatically satisfied. Therefore, separation of the rule is not necessary.

One problem occurs when the state point comes from the unload or the reload path to a point (e.g., point A in Fig. 8.5b) because no rule is mentioned above. There are two possibilities: the one is a simple elongation of B→A such as B → A → D, and the other is a switch back to the original skeleton curve such as B → A → C in Fig. 8.5b. The same problem occurs on the hysteresis curve.

The latter rule (B → A → C) seems natural because B → A → D shows, for example, a curious curve when infinitesimal small unloading–reloading cycle is considered. Actually the latter rule is employed in many one-dimensional constitutive models. However, in the multidimensional problem, stress point usually does not come back to the unload stress point after an unloading–reloading cycle; this rule is employed only when coming back to the skeleton curve and the former rule is used when unload occurs on the hysteresis curve.

Definition of unload–reload should be made based on the stress in the original meaning in the multidimensional analysis, whereas it is usually defined based on the strain in the one-dimensional analysis partly because direction of the stress increment also changes when the direction of the strain increment changes and partly because it is easy to handle especially in the softening model.

The original definition of Masing's rule is to make a hysteresis curve that has similar shape to the skeleton curve. Additional rules that the homothetic ratio is 2 and that the state point comes back to the original curve when previous unloading occurred are also jointed as the name "Masing's rule" in practice.

Stresses and strains of the unloaded points are to be memorized in Masing's rule when unload occurs at smaller strain. Computer programs sometimes require large number of memorized unload points. According to the author's experience, it is less than 30 in many cases and 40 is a safe side number in almost all analyses. Many computer codes prepare for the case that the number of unloading point increases more than the specified value by, for example, erasing some unload points. Shear strain amplitude is very small when this problem is expected to occur and then the effect is usually negligibly small. The extreme case is a damped free vibration, in which huge number of unloading point is to be memorized if Masing's rule is rigidly applied. However, only recent unload point is used in this case; erasing previous unloading points does not cause problem.

Masing's rule is very convenient because all the behavior can be calculated by setting a skeleton curve. This is, however, a mathematical assumption, and applicability of this rule in the actual behavior is a quite different issue.

Figure 8.6 shows the cyclic deformation characteristics of the Toyoura sand (Yamashita 1992). The damping characteristics obtained from the G–γ curve by applying Masing's rule are also drawn as solid line in the figure. Two damping characteristics agree well at small strains, but difference becomes large at large strains. Many case studies of this kind show similar tendency. Sometimes differences appear at small strains because, as described in Sect. 6.7.4, damping ratio at small strains is sometimes not zero in the test where the damping ratio is nearly zero when it is calculated based on Masing's rule.

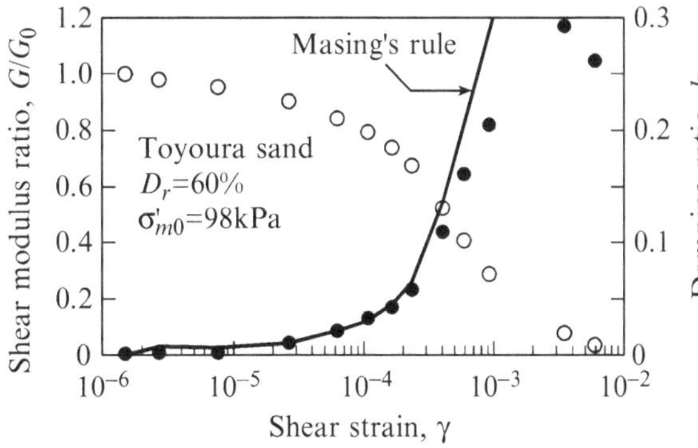

Fig. 8.6 Applicability of Masing's rule

8.2.2.2 Improved Hysteresis Rules

Several hysteresis rules have been proposed to overcome the limitations of Masing's rule shown above.

The first improvement is made to solve the limitation that hysteresis loops obtained by applying Masing's rule cannot explain the damping characteristics well (Ishihara et al. 1985). Figure 8.5c shows the fundamental concept of the improved hysteresis rule. Instead of applying Masing's rule to the skeleton curve, a fictitious skeleton curve is considered and Masing's rule is applied to this fictitious skeleton curve. This fictitious skeleton curve must satisfy at least the following two conditions:

1. It passes unload point A as hysteresis curve must pass the point A′ (point symmetric point of A).
2. Damping ratio coincides with that by test or that specified by the user such as the Hardin–Drnevich model.

As there are two conditions, mathematical models that have more than two parameters can be used as the fictitious skeleton curve. Both the hyperbolic model and the Ramberg–Osgood model (see Sect. 8.2.4) are introduced in the original paper. The hyperbolic model is very useful since the maximum possible damping is the same with theoretically maximum value $2/\pi$.

The hyperbolic model has only two parameters which have mechanical meanings such as elastic modulus and shear strength. Therefore, if values of the two parameters are determined from the above two conditions, stiffness at unload is not the same with the elastic modulus. In addition, continuity of the tangential slope when the state point comes back to the previous unload point is lost. The latter is not a big issue because discontinuity of slope always occurs when unload occurs from the hysteresis loop and the analysis can be carried out without any problem in this situation. However, considering the fact that many constitutive models use elastic modulus for the modulus at unload, the former issue may not be relevant. The third condition is added in this case.

3. Stiffness at unload is the elastic modulus.

Then the number of conditions becomes three; models that have more than three parameters are required. The Ramberg–Osgood model is used in this case.

The value of the parameters is evaluated every time when unload occurs from the skeleton curve.

As discussed at the beginning of this subsection, unload points are to be memorized when unload occurs at smaller strain amplitude than before in Masing's rule, but it may require large amount of memory. Then the second improvement was made to reduce this limitation. Figure 8.5d shows an idea schematically. Here, direction of loading reverses at point D after the path A → B → C → D. The hysteresis curve obtained by applying the previous hysteresis rule gives in line L_2 that passes previous unloaded point C. On the other hand, hysteresis curve is set to pass the unloaded point A where unload occurs from the skeleton curve by applying a new hysteresis rule. A number of the memorized unload points decrease significantly because only one point, the point where unload occurs from the skeleton curve, is to be memorized.

Shapes of hysteresis curves by different hysteresis rules are not the same. However, there is no data by which accuracy of the hysteresis curve can be evaluated. Therefore, it is determined only from the mathematical reasons.

8.2.3 Hyperbolic Model

The hyperbolic equation is a frequently used mathematical model not only in the seismic ground response analysis but also in the many analysis of ground (Kondner 1963). This model is the same with the skeleton curve explained as the Hardin–Drnevich model in Sect. 6.6.1. The stress–strain curve is already shown as Eq. (6.7), which is rewritten as

$$\tau = \frac{G_0\gamma}{1+\gamma/\gamma_r} = \frac{G_0\gamma}{1+G_0\gamma/\tau_f} \tag{6.7}$$

where G_0 denotes an elastic shear modulus, τ denotes shear strength, and γ_r denotes reference strain. Shear stress reaches shear strength as the shear strain increases to infinity. Shear stress at $\gamma = \gamma_r$ is a half of the shear strength and the shear moduli ratio becomes one half at this strain.

Since this model uses only two parameters, G_0 and τ_f (or γ_r), the mathematical treatment is easy.

Characteristics of the hyperbolic model can be seen from Fig. 8.7. At first, shear stress does not reach shear strength at the finite strain; it reaches shear strength at infinite strain in the laboratory test. On the other hand, shear stress reaches shear strength at a finite strain. In addition, increase of the shear stress is very slow at strains larger than the reference strain. Therefore, applicability of this model may not be good at large stresses. One of the solutions of this limitation is proposed by Duncan–Chang's model (1970), in which τ_f is set a little larger than the actual shear strength.

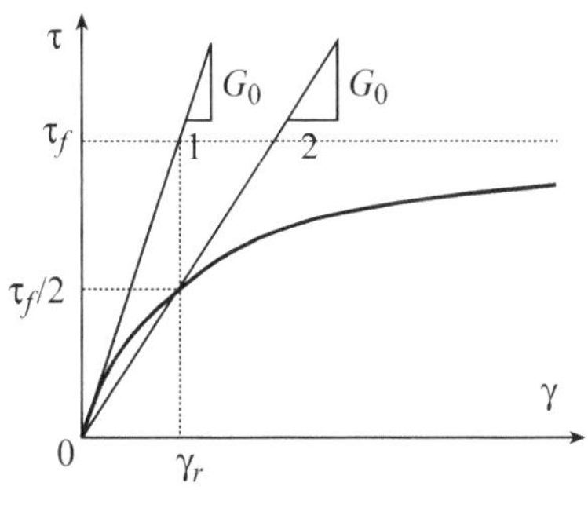

Fig. 8.7 Skeleton curve of the hyperbolic model

This limitation indicates that agreement at small strains becomes bad when parameters are evaluated so that agreement at the large strain becomes well. Therefore, in many practical calculations, shear strength is evaluated by the product of the elastic shear modulus and the reference strain. Then a good fitting can be expected at strains lower than the reference strain although τ_f is not an actual strength.

The hysteresis curve of the hyperbolic model is obtained by applying Masing's rule. Then the damping characteristics become

$$h = \frac{4}{\pi}\left(1 + \frac{\gamma_r}{\gamma}\right)\left[1 - \frac{\gamma_r}{\gamma}\ln\left(1 + \frac{\gamma}{\gamma_r}\right)\right] - \frac{2}{\pi} \tag{8.7}$$

Two examples of the fitting are shown in Figs. 8.8 and 8.9. The Toyoura sand is tested under the undrained test in Fig. 8.8 (Yamashita 1992), whereas it is tested

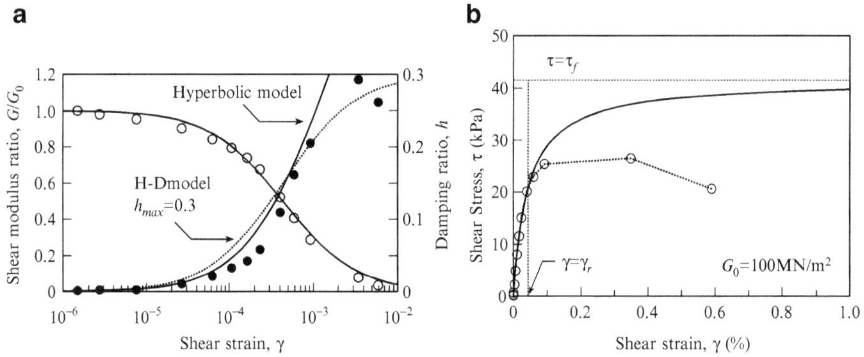

Fig. 8.8 Applicability of hyperbolic model (undrained condition test). (**a**) Cyclic shear deformation characteristics. (**b**) Stress–strain curve

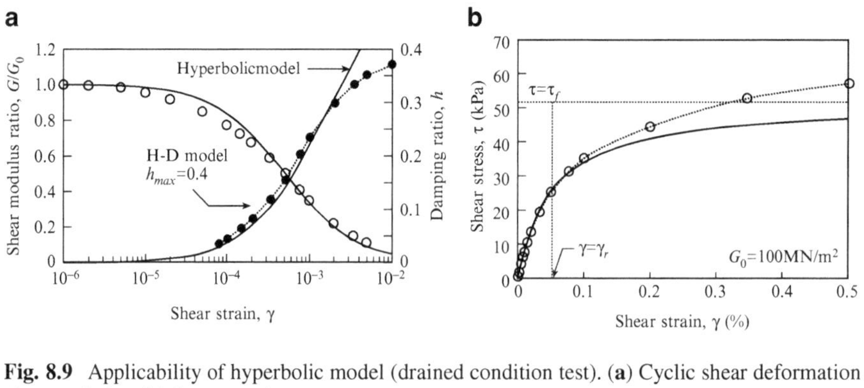

Fig. 8.9 Applicability of hyperbolic model (drained condition test). (**a**) Cyclic shear deformation characteristics. (**b**) Stress–strain curve

under the drained condition in Fig. 8.9 (PWRI 1982). An initial shear modulus $G_0 = 100$ MN/m^2 is assumed in drawing the stress–strain curve. Agreement of the G–γ relationships seems good in Fig. 8.8a. On the other hand, agreement of the damping characteristics also seems good, but it is overestimated at large strains as discussed in the previous subsection.

Agreement of the shear modulus ratio seems good in Fig. 8.8a, but different feature is seen when looking at the stress–strain curve in Fig. 8.8b. Error of the shear stress at large strains is fairly large. As absolute value of G/G_0 is small at large strains, error seems small in the shear modulus reduction curve. This limitation can be solved by using stress–strain curves instead of modulus reduction curves. Behavior at large strain is important as the error of the shear stress at large strains relates error of upper bound acceleration.

As the shear strength decreases rapidly at large strains under the undrained test due to the excess porewater pressure generation as described in Sect. 6.7.2, the shear stress cannot become large. Therefore, the hyperbolic model overestimates shear stress. On the other hand, the hyperbolic model underestimates the shear stress at large strains from 10 % in the drained test.

In order to examine the applicability on the hyperbolic model to a variety of soils, the cyclic shear deformation characteristics test results shown in Sect. 7.2. 12 are normalized by the strain γ_r at $G/G_0 = 0.5$ (reference strain) and are shown in Fig. 8.10. When looking at the stiffness reduction curve shown in Fig. 8.10a, agreement seems better, but when looking at the stress–strain curve in Fig. 8.10b, significant underestimation of the shear stresses at large strains is clearly seen, where shear stress is normalized by the shear strength $\tau_f = G_0\gamma_r$ and shear strain is normalized by the reference strain γ_r. These observations indicate that there is no method by which parameters can be evaluated automatically, and the engineer needs to evaluate the parameter case by case. Since there is only one parameter that can adjust the nonlinear characteristics, applicability on the damping characteristics is not discussed here; it will be discussed in Sect. 8.2.4.

The hyperbolic model was employed at first in the computer program DESRA (Finn et al. 1976) as far as the author knows. In Japan, however, this model is frequently called a modified Hardin–Drnevich model. This naming came from

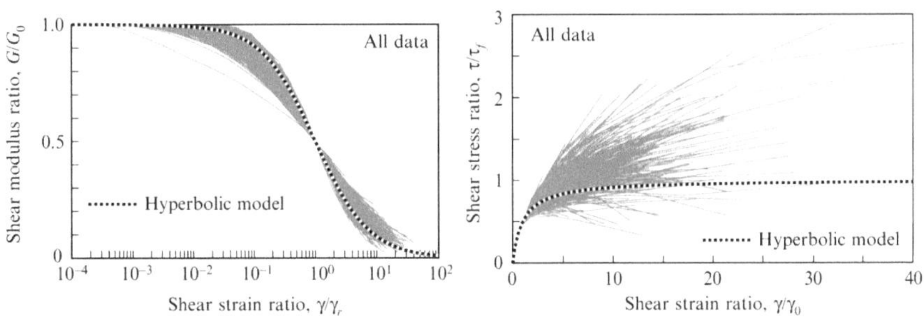

Fig. 8.10 Applicability of hyperbolic model. (**a**) Stiffness reduction curve. (**b**) Stress–strain curve

technical paper by Kokusho et al. titled "On the Modified Hardin–Drnevich Model" (Kokusho and Sakurai 1979). Here the meaning of "modified" is that they used the hyperbolic model for the skeleton curve and Masing's rule to build the hysteresis curve instead of the original equation by Hardin and Drnevich (1972) (see Sect. 6.6). This rule is exactly the same with the hyperbolic model described above. Therefore, the name modified Hardin–Drnevich (H–D) model is not used but the hyperbolic model is used in this book.

Constitutive models are frequently improved even by the same researchers. In order to distinguish from original model, the term "modified" is frequently added in front of the name of the original constitutive models. Therefore, the meaning of the "modified" may not have the same meaning for all constitutive models. It is recommended to confirm which modification is made from the original model when using these constitutive models.

Next, applicability of the Hardin–Drnevich model is examined. The skeleton curve is already examined because it has the same skeleton curve with the hyperbolic model. The damping characteristics are shown in Figs. 8.8a and 8.9a by dotted line. Here, $h_{max} = 0.3$ is used in the undrained test, whereas $h_{max} = 0.4$ in the drained condition. The damping characteristics in general agree well with the test result in Fig. 8.8 (undrained condition) although damping ratio is overestimated up to 10^{-4} and decreasing characteristics cannot be explained at strains larger than 10^{-3}. Agreement becomes better in Fig. 8.9 (drained condition) as hysteresis loop does not become inverse-S shape.

8.2.4 Ramberg–Osgood Model

The Ramberg–Osgood model (R–O model) was developed for metal that does not have clear yield stress such as aluminum (Ramberg and Osgood 1943). This model was used for the cyclic behavior by Jennings (1964), with Masing's rule for the behavior after unload.

The stress–strain relationships in the original paper are replaced by the shear stress τ–shear strain γ relationships, yielding.

$$\gamma = \frac{\tau}{G_0} + \alpha \gamma_r \left(\frac{\tau}{\tau_f} \right)^{\beta} \tag{8.8}$$

where α and β are parameters. The stress is expressed as a function of the strain in many mathematical models, but vice versa in the Ramberg–Osgood model. This equation is rewritten as

$$\frac{\gamma}{\gamma_r} = \frac{\tau}{\tau_f} + \alpha \left(\frac{\tau}{\tau_f} \right)^{\beta} \quad \text{or} \quad \gamma = \frac{\tau}{G_0} \left\{ 1 + \alpha \left(\frac{\tau}{\tau_f} \right)^{\beta - 1} \right\} \tag{8.9}$$

The first term of this equation is an elastic strain. The second term has a nature to increase more as stress increases, and it corresponds to a plastic strain.

The damping characteristics associated with Masing's rule become

$$h = \frac{2}{\pi}\frac{\beta - 1}{\beta + 1}\left(1 - \frac{G}{G_0}\right) \tag{8.10}$$

Here, if $h_{\max}$ is defined as

$$h_{\max} = \frac{2}{\pi}\frac{\beta - 1}{\beta + 1}, \tag{8.11}$$

then Eq. (8.10) becomes the same with the proposal by Hardin and Drnevich (1972) described in Sect. 6.6. A combination of Eq. (8.9) and Masing's rule is called the Ramberg–Osgood model in this book.

The shear modulus reduction curve is obtained by substituting $\tau = G_\gamma$ and $\tau_f = G_0\gamma_r$ into Eq. (8.9) as

$$1 = \frac{G}{G_0}\left\{1 + \alpha\left(\frac{G}{G_0}\frac{\gamma}{\gamma_r}\right)^{\beta-1}\right\} \tag{8.9}$$

There are four parameters G_0, τ_f, α, β in this model, but if a new parameter defined as $\overline{\alpha} = \alpha/\tau_f^{\beta-1}$ is employed, Eq. (8.9) yields

$$\gamma = \frac{\tau}{G_0}\left\{1 + \overline{\alpha}\tau^{\beta-1}\right\} \tag{8.12}$$

This equation indicates that there are only three independent parameters G_0, $\overline{\alpha}$, and β. As the number of independent parameters increases to three from the hyperbolic model that has only two parameters, the Ramberg–Osgood model has more flexibility than the hyperbolic model. Damping characteristic, for example, may be controlled, whereas it cannot in the hyperbolic model. Good agreement is expected since the damping characteristics are identical to the Hardin–Drnevich model.

The meaning of each parameter becomes clear when equation is expressed with four parameters although there are only three independent parameters. The skeleton curve is shown schematically in Fig. 8.11 to recognize mechanical meaning of the parameters. At first, τ_f is set to indicate the shear strength. Then the reference strain defined in the preceding can be used in the R–O model, too. It is noted that this τ_f need not be an actual shear strength but a value evaluated by $G_0\gamma_r$.

Next, strain at $\tau = \tau_f$ is $(1 + \alpha)\gamma_r$, which is another feature of this model. This indicates that the skeleton curve always passes the state point $((1 + \alpha)\gamma_r, \tau_f)$. Finally, β controls curvature of the skeleton curve. The stress–strain curve becomes a linear line at $\beta = 1$, and it becomes elastic–perfectly plastic model when $\beta = \infty$. The damping ratio increases with increasing β as can be seen from Eq. (8.10).

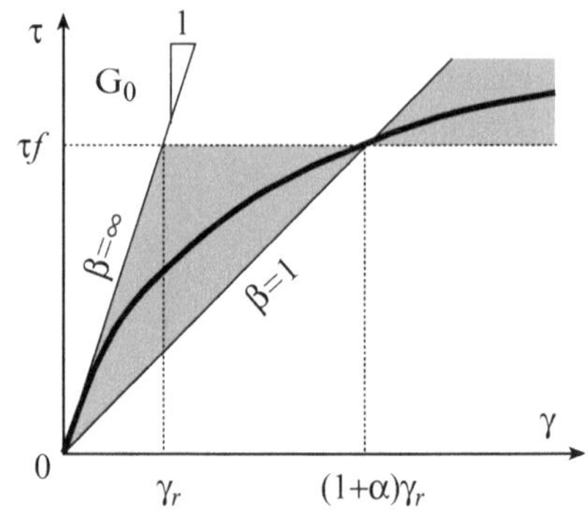

Fig. 8.11 Skeleton curve of Ramberg–Osgood model

Finally, the shear stress tends to infinity when the shear strain becomes infinite. One may feel that this is a significant limitation of this model as this nature does not agree with test results. In practice, it is not a limitation because the strain never reaches infinity, which will be discussed at the end of this section.

Values of G_0 and τ_f can be evaluated by the same procedure with the hyperbolic model. Then only one parameter left is to be determined. Several methods have been proposed to evaluate it and introduce hereafter.

If, by some method, β is determined, the value of α is calculated by substituting $\gamma = \gamma_r$ and $\tau = \tau_f/2$ into Eq. (8.9) as

$$\alpha = 2^{\beta-1} \tag{8.13}$$

8.2.4.1 Method 1: Curve Fitting

Only shear modulus reduction curves are interested in by this method. A linear equation is convenient to apply the least square method in order to get better set of parameters. The shear modulus reduction curve is obtained from Eq. (8.8) as

$$1 = \frac{G}{G_0}\left[1 + \alpha\left(\frac{G}{G_0}\cdot\frac{\gamma}{\gamma_r}\right)^{\beta-1}\right] = \frac{G}{G_0} + \alpha''\left(\frac{G}{G_0}\right)^{\beta}\gamma^{\beta-1} \tag{8.14}$$

where

$$\alpha'' = \frac{\alpha}{\gamma_r^{\beta-1}} \tag{8.15}$$

This equation is modified to

$$\ln\ (1 - G/G_0) + \ln\ \gamma = \ln\ \alpha' + \beta\,(\ln\ G/G_0 + \ln\ \gamma) \tag{8.16}$$

Then the relationship between $\ln(1 - G/G_0) + \ln\gamma$ and $\ln G/G_0 + \ln\gamma$ becomes linear.

On the other hand, the reference strain $\gamma_r = \gamma_{0.5}$ is a convenient parameter as it can be used in the hyperbolic model. If one uses this as one parameter, only one parameter remains undefined. The original equation is modified to

$$\log\left(1 - \frac{G}{G_0}\right) - \log\frac{G}{G_0} = (\beta - 1)\log\left(2\frac{G}{G_0}\frac{\gamma}{\gamma_r}\right) \tag{8.17}$$

Then the relationship between $\log(1 - G/G_0) - \log\ G/G_0$ and $\log(2G/G_0 \cdot \gamma/\gamma_r)$ is linear.

The better fit curve may be found from the figure as an alternate method. If one uses $\gamma = \gamma_r$ as one parameter, he/she can draw parametric figures by changing the value of β and chooses the better fitting curve (Shamoto and Mori 1980).

8.2.4.2 Method 2: Maximum Damping Ratio

When h_{max} is evaluated by some ways such as the Hardin–Drnevich model (Sect. 6. 6.1), then β can be calculated from Eq. (8.11) as

$$\beta = \frac{2 + \pi h_{max}}{2 - \pi h_{max}} \tag{8.18}$$

8.2.4.3 Method 3: Failure Strain

The failure strain γ_f is defined as the strain at failure. Substitution of $\tau = \tau_f$ into Eq. (8.9) yields

$$\gamma_f = (1 + \alpha)\,\gamma_r \ \text{ or } \ \alpha = \gamma_f/\gamma_r - 1 \tag{8.19}$$

It is more practical to determine the point (γ_f, τ_f) to be that the skeleton curve is desired to pass. Description in Sect. 6.7.6 may give an idea on the failure strain.

As described in the preceding, the shear stress tends to infinity with increasing shear strain. In the actual analysis, however, strain never reaches infinity. Therefore, this is not a limitation of the Ramberg–Osgood model in the practical use. If the engineer does not like this feature, shear stress at strains larger than failure strain is set constant as

$$\tau = \tau_f \ldots \gamma \geq \gamma_f = (1 + \alpha)\,\gamma_r \tag{8.20}$$

This method is called Hara's modification (Ishihara 1982) and is an extension of the method 3.

The R–O model has been modified to various ways (Ishihara 1982). For instance, the expression as Eq. (8.12) emphasizes that there are three independent parameters. Next, the exponent $\beta - 1$ in this equation is replaced to β (Adachi and Tatsuoka

1986). These models are called modified Ramberg–Osgood model. The term "modified" just indicates that the expression is different from the original paper (Adachi and Tatsuoka 1986), although all modified models are essentially identical. These modifications may give some confusion to engineers, because expressions employed in various computer programs are not identical. Therefore, one needs to recognize which modification is used in the particular program, and the value of the parameters should be determined based on that equation. Typical methods are shown in the preceding, and conversion is not difficult even if they are not shown.

Result of the simulation for the same material used in the hyperbolic model is shown in Figs. 8.12 and 8.13. Values of the parameters are determined based on the reference strain γ_r and maximum damping ratio $h_{\max}$.

In Fig. 8.12 (undrained test), $h_{\max}$ is set to 0.3. Agreement seems better in general although G/G_0 is overestimated at large strains. However, when looking at the stress–strain curve in Fig. 8.12b, shear stress is overestimated significantly. As described frequently, behavior at large strains is important in the seismic ground

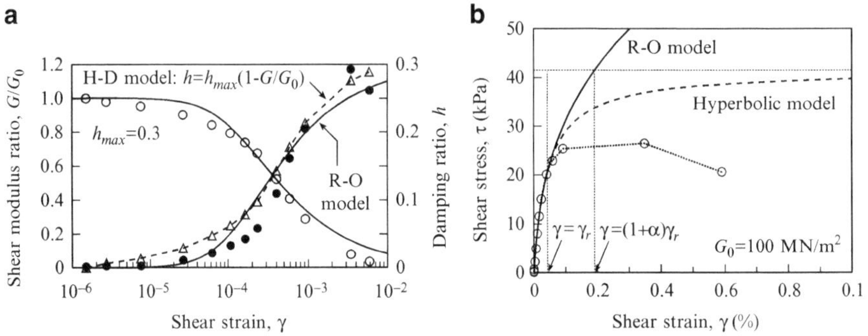

Fig. 8.12 Applicability of the R–O model (undrained test). (**a**) Cyclic shear deformation characteristic. (**b**) Stress–strain curve

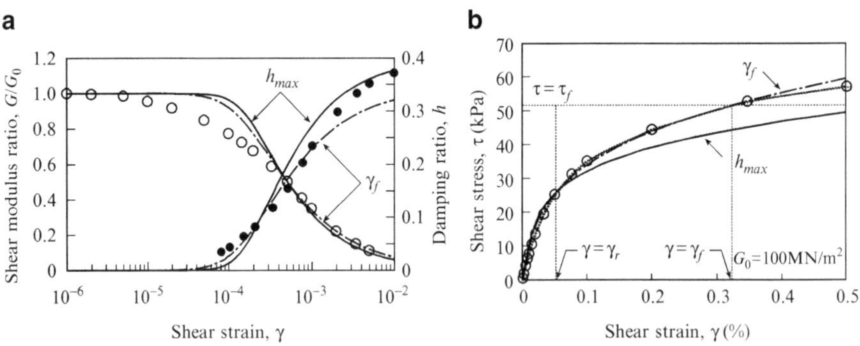

Fig. 8.13 Applicability of the R–O model (drained test). (**a**) Cyclic shear deformation characteristics. (**b**) Stress–strain curve

response analysis under large earthquakes, in which sense this simulation is not satisfactory. Furthermore, same as the hyperbolic model, importance to look at the stress–strain curve is emphasized or it is more important than to see the shear modulus reduction curve.

Getting good agreement in the damping characteristics is difficult in general because damping ratio decreases at large strains. Result of the H–D model evaluated from h_{max} and shear modulus ratio of test is shown by symbol Δ in Fig. 8.12a. Agreement of damping ratio is nearly the same with the H–D model in Fig. 8.8a.

In the drained test in Fig. 8.13, G/G_0 is overestimated at strains less than the reference strain. The same feature is also observed in Fig. 8.12a. As the R–O model was originally proposed for metal, it has large quasi-linear regions, which is the reason why G/G_0 is overestimated at small strains. On the other hand, shear stress is underestimated in Fig. 8.13b. In order to improve this limitation, failure strain is defined to be a strain at $\tau = \tau_f$ and determine values of parameters based on method 3 above. Then agreement of the stress–strain curve becomes much better than before. However, damping ratio becomes smaller as shown in Fig. 8.13a. This error may be allowable compared with the error by other factors. As can be seen in this example, some effort is necessary to get a better fitting in this model. It is noted that this kind of effort is not necessary in the hyperbolic model; these jobs are required when the number of parameters is greater than the number of conditions.

Data in Sect. 7.2.12 is again used to investigate the applicability of the Ramberg–Osgood model. If the value of one parameter is determined from the reference strain, the resultant stress–strain curves scatter as already shown in Fig. 8.10. Therefore, two parameters are used in order to get a better agreement of the stress–strain curve. Stiffness reduction curve is normalized so that it agrees with the Ramberg–Osgood model at $G/G_0 = 0.5$ and 0.8. Comparison between the test data and the Ramberg–Osgood model is shown in Fig. 8.14a, and the stress–strain relationships are compared in Fig. 8.14b. The Ramberg–Osgood model shows larger stress, whereas the hyperbolic model shows smaller stress at large strains in general.

One of the advantages of the Ramberg–Osgood model is that damping ratio can be expressed well. Figure 8.14c compares the relationship between the normalized damping ratio h/h_{max} and shear modulus ratio G/G_0. Here h_{max} is not actual maximum damping ratio but the damping ratio at $\gamma = 0.01$ (1 %). The relation becomes linear line connecting two intersects according to Eq. (8.10), but data scatters very much. Therefore, again engineering judgment is required to evaluate relevant values for the parameters even if this model is used.

8.2.5 *Yoshida's Model*

Models introduced above cannot simulate test results perfectly, and, at the same time, one can understand that perfect simulation is impossible by using ordinary mathematical equation. In contrary, this model aims to get perfect agreement with the test result. The model is composed of two fundamental concepts.

8.2.6 *Modified GHE Model*

The GHE model is already explained in Sect. 6.6. It considers from small strain to large strain up to the failure, whereas the Hardin–Drnevich model considers relatively small strains with 10^{-3} or a little large. Same with other models, this model cannot be applied to the complicated behavior such as undrained test described in Sect. 6.7.2. It uses eight parameters, which may be too many in the practical sense.

This model is proposed only for the skeleton curve. The hysteresis loop can be obtained by applying Masing's rule. The model that uses hysteresis rule in Sect. 8.2.2.2. is proposed by Murono (1990) and is called the modified GHE model. The meaning of modified here is to add the hysteresis loop for the cyclic loading.

8.3 Constitutive Models for Multidimensional Analysis

Constitutive models for a one-dimensional analysis explained in the previous sections are expressed by a relatively simple mathematical equation, and its mechanical meaning is easily understood. These constitutive models, however, cannot be used simply in the multidimensional analysis. The world of the general constitutive models is very wide and cannot be explained in the limited pages of this book. Sometimes it exceeds the knowledge level of the reader expected in this book. Therefore, only limited amount of the constitutive models is introduced. Constitutive models that are extended from the one-dimensional analysis are explained in detail and only the fundamental concepts are explained in the multidimensional analysis.

8.3.1 *Extended Model from One Dimension*

Although stress and strain are tensors which have magnitude, direction, and working plane, the latter is frequently not recognized well in the one-dimensional analysis because direction is expressed only by either plus or minus sign. However, in the multidimensional space, there are more directions and it is impossible to express it by an index. When focused on the shear deformation, the equivalent stress and the equivalent strain explained in Eq. (4.4) are indices to express shear deformation. As the deviatoric stress and strain correspond to the second invariants of the tensor, they are always of plus value, which makes modifications of the one-dimensional model into multidimensional space models difficult.

Figure 8.16 shows this problem schematically. An element is subjected to a simple shear deformation, in which case the independent stress is only τ_{xy}. Therefore, equivalent stress σ_e and shear stress τ_{xy} have the same meaning except the difference of the coefficient in Eq. (4.4). Two loading cases are considered. The one is a path A → B → C → D under which τ_{xy} takes both positive and negative

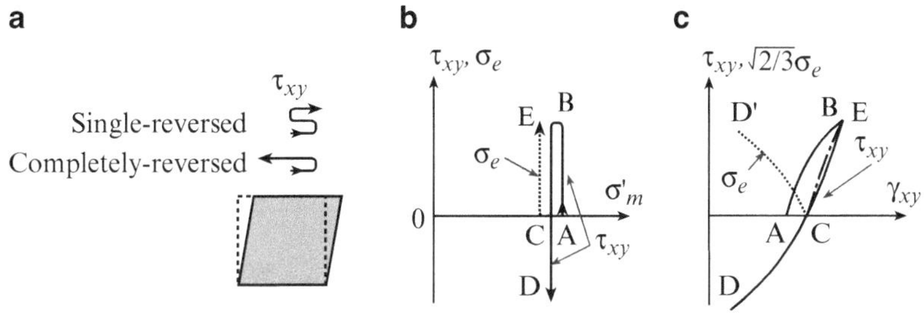

Fig. 8.16 Shear stress τ_{xy} and equivalent stress σ_e under simple shear. (**a**) Element in consideration. (**b**) Stress path. (**c**) Stress–strain curve

signs. The other is a path A → B → C → E under which τ_{xy} takes only positive value. The paths C → D and C → E are clearly distinguished in Fig. 8.16b, distinction between the two processes is impossible by looking at σ_e. The τ_{xy}–γ_{xy} relationships of the former case are drawn by a solid line. Here, τ_{xy} is negative in the path C → D, but it is expressed by a dotted line path C → D′ if σ_e is used as ordinate. On the other hand, path C → E is expressed by a chained line, which is quite different from C → D′. This is the difficulty to use the constitutive models for the one-dimensional space into the multidimensional space. Several methods have been proposed to solve this problem.

8.3.1.1 Behavior Under Simple Shear

Behavior under the S-wave propagation is important in the engineering practice. When earthquake wave propagates in the vertical direction, simple shear deformation such as Fig. 8.16 occurs. Therefore, the tangent shear modulus G to be used in Eq. (4.8) can be evaluated from the τ_{xy}–γ_{xy} relationships (Fukutake and Ohtsuki 1989). Although Eq. (4.8) is elastic relationships, it is used as incremental tangent relationships in the nonlinear analysis.

Limitation of this method is that the shear stress caused by the deviator stress cannot be considered. Therefore, effect of initial K_0 stress state and deviator stress caused by the rocking of a structure cannot be considered.

Two methods are considered to extend this method.

The one can be used for a three-dimensional analysis, in which there are three shear components. Therefore, one-dimensional models are used for τ_{xy}–γ_{xy}, τ_{yz}–γ_{yz}, and τ_{zx}–γ_{zx} relationships individually. The tangent modulus G_{xy}, G_{yz}, and G_{zx} and its average value $G = (G_{xy}+G_{yz}+G_{zx})/3$ are used in building the material stiffness matrix (Eq. 4.8); G is used for the normal stress term, and other G's are used for shear stress term (Fukutake 1997).

Predominant shear deformation occurs in the tilted direction in the irregular ground such as an earth dam or an embankment (Kuwano and Ishihara 1988). As an extension of this method, the shear modulus can be evaluated by this plane instead of the horizontal plane.

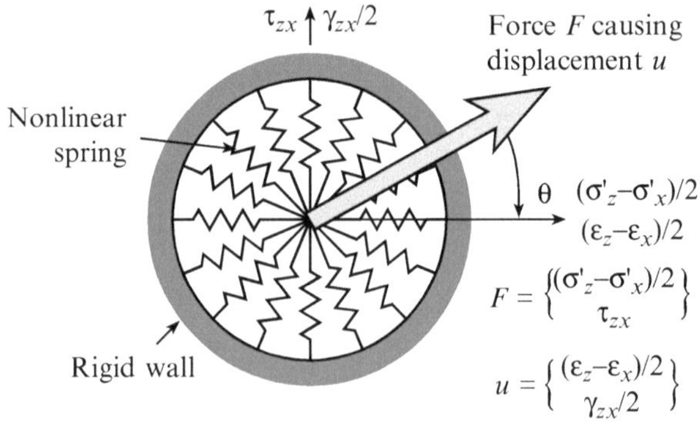

Fig. 8.17 Schematic figure of multispring model

8.3.1.2 Multispring Model

The method described above is not applicable for the deviator stress. The multispring model is proposed to improve this limitation. This model is composed of number of springs set in a rigid circular wall as schematically shown in Fig. 8.17 (Towhata and Ishihara 1985) for a simple two-dimensional space on z–x plane. The simple shear (relationships between τ_{zx} and $\varepsilon_{xy}=\gamma_{zx}/2$) and deviator (relationships between $(\sigma_z^{'}-\sigma_x^{'})/2$ and $(\varepsilon_z-\varepsilon_x)/2$) behaviors are extreme behaviors and are expressed by the springs in z- and x-axis, respectively. Other multidirectional shear behavior is expressed by the corresponding spring. Stress–strain behavior is described by the behavior at the center of the model.

All the springs are connected in one point in this model; this point locates the center for zero stress or initial state. It is difficult to make initial state corresponding to the given stress in this model because nonlinear equation is to be solved. Therefore, initial stresses are usually obtained by a switch-on-gravity analysis or a layer construction analysis starting zero stress. It indicates that past stress history described in Sect. 6.7.8 is not considered, which may result in overestimation of the shear deformation.

8.3.1.3 Duncan–Chang's Model and Yoshida's Model

Duncan and Chang proposed a simple model applicable to the multidimensional analysis based on the triaxial test (Duncan and Chang 1970). The model uses maximum and minimum principal stresses, σ_1 and σ_3, and the shear stress is expressed by a deviator stress σ_1–σ_3. As this stress state is somewhat different from the state during earthquake, τ–γ relationships are used here. It is noted that, as seen in Fig. 8.18a, deviator stress σ_1–σ_3 corresponds to the diameter of Mohr's circle, whereas τ corresponds to the radius of the circle.

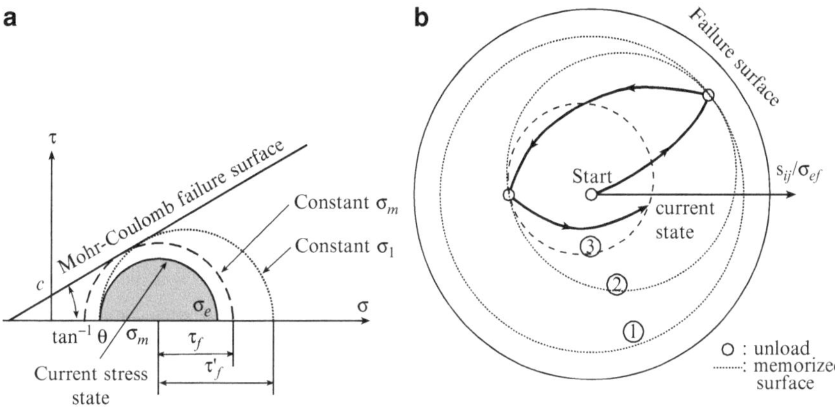

Fig. 8.18 Schematic figure showing concept of Yoshida's model. (**a**) Shear strength and Mohr's circle. (**b**) Hysteresis rule in multidimension

When the hyperbolic model in Eq. (6.7) is used, stress–strain behavior was expressed as

$$\tau = \frac{G_0\gamma}{1 + G_0\gamma/\tau_f}, \quad \text{or} \quad \gamma = \frac{\tau}{G_0\left(1 - \tau/\tau_f\right)} \tag{6.7}$$

Here, the shear stress is expressed as a function of strain in the first equation, whereas strain is expressed as a function of stress in the second equation. The tangent modulus G_t is calculated by differentiating the first equation by γ. Substitution of G_t into the second equation to eliminate γ results in

$$G_t = \frac{G_0}{\left(1 + \frac{\tau}{\tau_f\left(1 - \tau/\tau_f\right)}\right)^2} \tag{8.23}$$

This equation has an advantage that the tangent shear modulus is evaluated from the current shear stress. Nonlinear analysis becomes possible by using G_t into the incremental equation of Eq. 4.8.

Shear strength τ_f, i.e., radius of Mohr's circle, is necessary to use Eq. (8.23). Since Duncan and Chang considered the triaxial test, they enlarged Mohr's circle keeping the minimum principal stress σ_3 constant (dotted circle in Fig. 8.18a). Then the shear strength is evaluated as

$$\tau_f = \frac{c\cos\phi + \sigma_3\sin\phi}{1 + \sin\phi} \tag{8.24}$$

In addition, they propose to use the shear strength a little larger than the actual shear strength, which will improve the limitation of the hyperbolic model discussed in Sect. 8.2.3. This model has been used in many field in the soil mechanics.

Although Duncan–Chang's model is an easy handling constitutive model, it deals with monotonic loading only. Yoshida et al. improved this model into general use and into the cyclic loading (Yoshida and Tsujino 1993). It is composed of three concepts. As the radius of Mohr's circle is equivalent to the equivalent stress, all one-dimensional models can be used to the multidimensional analysis by replacing the shear stress into equivalent stress. In addition, minimum principle stress is assumed to keep constant in evaluating shear strength in Duncan–Chang's model, whereas confining stress σ_m is kept constant in Yoshida's proposal so as to be applied in the general cases. The shear strength is evaluated by

$$\tau_f = c \cos\phi + \sigma_m \sin\phi \tag{8.25}$$

It is noted that τ_f evaluated from Eq. (8.24) is constant in whole analysis, whereas τ_f evaluated from Eq. (8.25) changes with σ_m.

The third one is the hysteresis rule under the cyclic loading. At first, τ–γ curve is evaluated by the method in Sect. 8.2.2.2. The deviatoric stress is normalized by the confining stress as schematically shown in Fig. 8.18b in order to consider the effective stress dependency of the mechanical properties. The subsequent yield surface after unloading can be drawn by two conditions that the circle is tangent to previous yield surface and it passes current stress point. The stress ratio can be either radius of the yield surface or a half of the distance between the unloaded point and the current point, among which the latter is suggested (Yoshida 1996).

8.3.2 *Plasticity Theory*

Many constitutive models based on the plasticity theory have been proposed. They are summarized in, for example, JSSMFE (1983), Pande and Zienkiewicz (1982), and Murayama (1985), Editorial committee on elastic-plastic constitutive model of soil (2009), Research committee on effect of constitutive model on numerical analysis (2002), etc. However, there is no constitutive model that is installed in many computer programs. Therefore, engineers need to study based on, for example, the user's manual or technical papers on the particular constitutive model although fundamental concept is introduced in this book following the JGS (2007). A discussion is made by Tobita and Yoshida (2004) on the practical applicability.

The plasticity theory is composed of five equations:

1. Elastic stress–strain relationship such as Eq. 4.8.
2. Yield function f: Mohr–Coulomb's failure criteria or its scaled down surface is usually used. For example,

$$f = \eta - h \tag{8.26}$$

where $\eta = \sigma_e/\sigma'_m$ denotes shear stress ratio and h denotes a hardening parameter. Yielding continues when $f = 0$ and direction of the stress increment is outside to the surface; otherwise, unload occurs.

3. Plastic potential function g: Direction of the strain increment is normal to g. Potential function g is set equals to yield function f in many constitutive models, but $g \neq f$ for soil in order to express dilatancy behavior well. A Cam-clay-type stress–dilatancy relationship is frequently used for soil:

$$\frac{\mathrm{d}\varepsilon_v^p}{\mathrm{d}\gamma^p} = \mu_m - \eta \tag{8.27}$$

where $\mathrm{d}\varepsilon_v^p$ denotes plastic volumetric strain, γ^p denotes plastic shear strain, and μ_m denotes stress ratio at phase transform. This equation is originally proposed for clay and a little modification may be done for sand. Then plastic potential function is obtained from Eq. (8.27) and the condition that direction of strain increment is normal to g, resulting in

$$g = \eta + \mu_m \ \ln\left(\frac{\sigma'_m}{\sigma'_m}\right) \tag{8.28}$$

4. Flow rule: Relationships between stress increment and plastic strain increment are defined. The rule is called as associated flow rule when $g = f$, whereas it is called as non-associate flow rule when $g \neq f$.
5. Hardening rule: Size of h or yield surface is defined. A hyperbolic equation is frequently used:

$$h = \frac{G_{p0}\gamma^p}{1 + G_{p0}\gamma^p/\mu_f} \tag{8.29}$$

where G_{p0} denotes initial modulus of the hardening function, γ^p denotes plastic shear strain, and μ_f denotes stress ratio at failure.

In order to recognize how these equations are used, let us consider a very simple case that there are only two stress components σ'_m and τ. Then non-associate flow rule is written as

$$\begin{Bmatrix} \mathrm{d}\gamma^p \\ \mathrm{d}\varepsilon_v^p \end{Bmatrix} = \mathrm{d}\lambda \begin{Bmatrix} \partial g/\partial\tau \\ \partial g/\partial\sigma'_m \end{Bmatrix} = \mathrm{d}\lambda \begin{Bmatrix} 1/\sigma'_m \\ (\mu_m - \eta)/\sigma'_m \end{Bmatrix} \tag{8.30}$$

where $\mathrm{d}\lambda$ is a scalar that defines magnitude of the plastic strain increment and the vector denotes direction normal to the plastic potential function. Here, $\eta = \tau/\sigma'_m$ in this simple model. Then one can confirm that Eq. (8.27) holds. Yielding continues when $\mathrm{d}\lambda > 0$. Other conditions that $f = 0$ and $\mathrm{d}f = 0$ are also to be satisfied when yielding continues:

$$\mathrm{d}f = \{\partial f/\partial\tau \ \ \partial f/\partial\sigma'_m\} \begin{Bmatrix} \mathrm{d}\tau \\ \mathrm{d}\sigma'_m \end{Bmatrix} + \frac{\partial f}{\partial h}\frac{\partial h}{\partial\gamma^p}\mathrm{d}\gamma^p = 0 \tag{8.31}$$

Stress increments $d\gamma$ and $d\varepsilon_v$ are the sum of the elastic and plastic components as

$$\begin{Bmatrix} \mathrm{d}\gamma \\ \mathrm{d}\varepsilon \end{Bmatrix} = \begin{Bmatrix} \mathrm{d}\gamma^e + \mathrm{d}\gamma^p \\ \mathrm{d}\varepsilon_v^e + \mathrm{d}\varepsilon_v^p \end{Bmatrix} = \begin{Bmatrix} \dfrac{\mathrm{d}\tau}{G_{e0}\sigma'_m} + \dfrac{\mathrm{d}\lambda}{\sigma'_m} \\ \dfrac{\mathrm{d}\sigma'_m}{K_{e0}\sigma'_m} + \dfrac{\mathrm{d}\lambda\left(\mu_m - \eta\right)}{\sigma'_m} \end{Bmatrix} \tag{8.32}$$

where superscript e denotes elastic component and Ge_0 and K_{e0} are elastic shear and bulk moduli normalized by σ'_m so as to consider the confining stress dependency of the elastic moduli. On the other hand, the constitutive model for the elastic component is written as

$$\begin{Bmatrix} d\tau \\ d\sigma'_m \end{Bmatrix} = \begin{bmatrix} G_{e0}\sigma'_m & 0 \\ 0 & K_{e0}\sigma'_m \end{bmatrix} \begin{Bmatrix} d\gamma - d\gamma^p \\ d\varepsilon_v - d\varepsilon_v^p \end{Bmatrix} \tag{8.33}$$

It is noted that the confining stress dependency is considered in Eq. (8.33) so that elastic modulus is proportional to the effective confining stress. Eliminating the plastic strain components in the above equations, one obtains

$$\begin{aligned} \mathrm{d}\lambda &= \frac{\begin{Bmatrix} \dfrac{\partial f}{\partial \tau} & \dfrac{\partial f}{\partial \sigma'_m} \end{Bmatrix} \begin{bmatrix} G_{e0}\sigma'_m & 0 \\ 0 & K_{e0}\sigma'_m \end{bmatrix} \begin{Bmatrix} \mathrm{d}\gamma - \mathrm{d}\gamma^p \\ \mathrm{d}\varepsilon_v - \mathrm{d}\varepsilon_v^p \end{Bmatrix}}{\begin{Bmatrix} \dfrac{\partial f}{\partial \tau} & \dfrac{\partial f}{\partial \sigma'_m} \end{Bmatrix} \begin{bmatrix} G_{e0}\sigma'_m & 0 \\ 0 & K_{e0}\sigma'_m \end{bmatrix} \begin{Bmatrix} \partial g/\partial \tau \\ \partial g/\partial \sigma'_m \end{Bmatrix} - \dfrac{\partial f}{\partial h}\dfrac{\partial h}{\partial \gamma^p}\left|\dfrac{\partial g}{\partial \tau}\right|} \\ &= \frac{G_{e0}\mathrm{d}\gamma - K_{e0}\eta \mathrm{d}\varepsilon_v}{\left\{h' + G_{e0} - K_{e0}\eta\left(\mu_m - \eta\right)\right\}/\sigma'_m} \\ &= \frac{G_{e0}\mathrm{d}\gamma - K_{e0}\eta \mathrm{d}\varepsilon_v}{D} \end{aligned} \tag{8.34}$$

where h' denotes partial differential of h with respect to γ^p:

$$h' = \frac{G_{p0}}{\left(1 + G_{p0}\gamma^p/\mu_f\right)^2} = G_{p0}\left(1 - \frac{\eta}{Gm_f}\right) \tag{8.35}$$

Finally, the stress increment is obtained from the strain increment as

$$\begin{aligned} \begin{Bmatrix} \mathrm{d}\tau \\ \mathrm{d}\sigma'_m \end{Bmatrix} &= \left(\begin{bmatrix} G_{e0}\sigma'_m & 0 \\ 0 & K_{e0}\sigma'_m \end{bmatrix} - \frac{1}{D} \begin{bmatrix} G_{e0}\sigma'_m & 0 \\ 0 & K_{e0}\sigma'_m \end{bmatrix} \begin{Bmatrix} \partial g/\partial \tau \\ \partial g/\partial \sigma\sigma'_m \end{Bmatrix} \right. \\ &\quad \times \left. \begin{Bmatrix} \dfrac{\partial f}{\partial \tau} & \dfrac{\partial f}{\partial \sigma\sigma'_m} \end{Bmatrix} \begin{bmatrix} G_{e0}\sigma\sigma'_m & 0 \\ 0 & K_{e0}\sigma\sigma'_m \end{bmatrix} \right) \begin{Bmatrix} \mathrm{d}\gamma \\ \mathrm{d}\varepsilon \end{Bmatrix} \\ &= \begin{bmatrix} G_{e0}\sigma'_m - \frac{G_{e0}^2}{D} & \dfrac{G_{e0}K_{e0}\eta}{D} \\ -\dfrac{G_{e0}K_{e0}\left(\mu_m - \eta\right)}{D} K_{e0}\sigma'_m & \dfrac{K_{e0}^2\left(\mu_m - \eta\right)\eta}{D} \end{bmatrix} \begin{Bmatrix} \mathrm{d}\gamma \\ \mathrm{d}\varepsilon \end{Bmatrix} \end{aligned} \tag{8.36}$$

The stiffness matrix is unsymmetrical unless $g = f$, which makes use of this type of constitutive model difficult. At the beginning of loading, $\varepsilon_v = 0$ and plastic shear strain is nearly zero. Then one obtains

$$\mathrm{d}\tau = \frac{G_{e0} G_{p0} \sigma'_m}{G_{e0} + G_{p0}} \mathrm{d}\gamma \tag{8.37}$$

This equation indicates that initial stiffness is not the elastic modulus $G_{e0}\sigma'_m$ unless $G_{p0} = \infty$. Therefore, elastic region is to be considered in order to get elastic modulus at the beginning of loading.

Since non-associated flow rule is employed in the soil, theoretical confliction may occur in the extreme stress or loading condition, but it does not affect the practical problem (Tobita and Yoshida 2004). A simple constitutive model that the elastic modulus depends on the confining stress also creates theoretical problem such that energy is created by a circle of lading, but it does not affect the practical problem, too (Tobita and Yoshida 2004). As such, there are some distance between the theory and practical behavior in the case of soil.

Constitutive models based on the plasticity theory are the most rigorous in theory, but it does not indicate that result by means of these models is accurate. Actually, there is an evaluation that the most successful constitutive model is Duncan–Chang's model (Towhata 1989). It indicates that a constitutive model that is theoretically rigorous does not achieve good results in the practical problem (Tobita and Yoshida 2004).

There are a few fundamental differences between the extended models and the constitutive models based on the plasticity theory.

One of the big differences relates direction of the strain increment. Let us consider, for simplicity, an element with K_0 initial stress state is subjected to the simple shear load. It is clear from Eq. (4.8) that only the shear deformation is produced under the simple shear load. On the other hand, direction of the strain increment is constrained by the potential function in the plasticity theory. Generally, as seen in Fig. 8.19, not only the shear deformation but also the volume change occurs. It indicates that only the horizontal displacement occurs in the extended

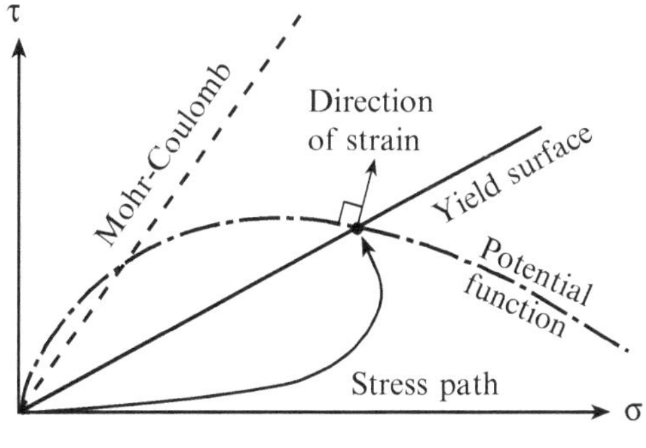

Fig. 8.19 Schematic figure showing image of plasticity theory

model, whereas settlement as well as shear displacement occurs in the plasticity model. The only exception is the multispring model; it shows behavior similar to the plasticity theory (Towhata 1992).

Another difference is the definition of strain. The shear stress is defined as a function of strain. Therefore, strain is to be calculated from the zero stress state in the extended model. If the confining stress changes during the loading, for example, the strain is to be converted to new stress state. On the other hand, this procedure is not necessary in the plasticity theory because strain increment is defined to be a function of stress increment and current stress. Note that the same procedure becomes possible in the extension model if the tangent modulus is written by the stress state such as Eq. (8.23).

The final difference is the stiffness matrix. Stiffness matrix is symmetrical in the extended model and in the plasticity model with associate flow rule. However, it becomes unsymmetrical in the plasticity model with non-associate flow rule. It causes some difficulty in the numerical analysis. At first, it requires more memory than the symmetric stiffness matrix. Then time required for solving simultaneous equation increases. The symmetric stiffness matrix has a positive definite nature, which makes numerical analysis easy, but this nature is lost in the unsymmetrical stiffness matrix.

8.4 Choice of Models and Evaluation of Parameters

It is very important to choose a constitutive model in the practical analysis. Engineers cannot use all models, but can use only constitutive models installed in the computer program they have. Hyperbolic model and/or Ramberg–Osgood model is frequently installed in the one-dimensional computer program, whereas Yoshida's model and the GHE model are installed in only limited computer programs. Choice of the constitutive models and the method how to evaluate the parameters are discussed in this section by considering these situations.

There are two types of constitutive models. The one is a constitutive model where stress–strain curve is expressed by mathematical equations. Hyperbolic model and Ramberg–Osgood model are examples of this type. On the other hand, some constitutive models require values of shear moduli and damping ratios at several strains such as Yoshida's model. They are called a mathematical model and a table type model, respectively, in the following:

Two situations are considered when seismic ground response analysis is made. The one is analyzing a particular site in order to, for example, construct a structure, and the other is analyzing many sites to make the risk analysis. Detailed soil test including the cyclic shear deformation characteristics test will be made in the former case, whereas the cyclic shear deformation characteristics will be obtained from the empirical equations because laboratory test is practically impossible. Cyclic shear test may not be carried out even in the previous case. The strategy to choose the constitutive model and to evaluate values of the parameters is different in these two situations.

If the engineer has a table type test data and the computer program allows table type models, it is the best way to use the table type model. On the other hand, a mathematical model has an advantage that mechanical meaning of some parameters is clear. Affecting factors such as the confining stress dependency may be taken into account easily. In addition, frequently used mathematical models have lots of past experiences, which makes understanding of the behavior easy. Therefore, it is convenient in the risk analysis because many sites are analyzed under the same concepts.

If the computer program does not have a table type model but has another model such as hyperbolic model and/or Ramberg–Osgood model, the engineer needs to choose one of them. Generally speaking, a model that has more parameters is better because it has more flexibility. This is, however, not always true because the basic functional shape also affects the accuracy. Determining many variables also requires many efforts.

When a mathematical model is used, the value of the parameters is to be evaluated so that the stress–strain curve agrees with test data as much as possible by using, for example, a least square method. In this case, it is recommended to focus on the shear stress, but not on shear modulus ratio, because, as frequently explained in this book, absolute value of the shear modulus ratio is small at large strain. It indicates that contribution of the behavior at large strain is small, but the behavior at large strain is important in the seismic ground response analysis. It is also noted that the behavior at strains larger than the maximum shear strain during the seismic ground response analysis need not be focused because they are never used.

Parameters that have clear mechanical meaning, such as elastic modulus and shear strength, are to be determined at first. Then other parameters that do not have mechanical meaning are evaluated. Parameters α and β in the Ramberg–Osgood model are examples of this type. Finally, the parameters that have mechanical meaning are modified regardless of the mechanical meaning. If, for example, shear stress does not reach the shear strength, disagreement of the shear strength does not affect the result of the analysis. Therefore, it can be treated as an adjusting parameter. Evaluation of τ_f as a product $G_0\gamma_r$ is an example of this type.

When the engineer does not have a test data, he/she should use empirical equations. In this situation, it is important to consider the factors that affect the cyclic shear deformation characteristics as much as possible. The confining stress and the plasticity index are examples of this kind. Soil classification is also an important factor. Here, an empirical equation that is based on similar soils is preferable to use because the cyclic shear deformation characteristics cannot always be distinguished by the physical test result; there are factors such as sediment conditions that do not appear in the physical test result. After determining the cyclic shear deformation characteristics, he/she should move to choose the constitutive models as shown in the preceding.

8.5 Complex Modulus

Complex modulus is a quite different concept from aforementioned constitutive models. This model is developed from the needs of equivalent linear method that will be described in Sect. 11.4, which are linear stress–strain relationship and hysteretic damping. It cannot be realized in the world of real numbers, but becomes possible by introducing a complex number.

Let us start a model composed of a spring and a dashpot as shown in Fig. 8.20. The spring and dashpot are connected in serial in the Maxwell model, whereas they are connected in parallel in the Voigt model (Kelvin model or Kelvin–Voigt model). Behavior of the Voigt model is investigated here, but similar behavior is obtained from Maxwell model.

Let us denote spring constant as G and viscous coefficient C of the model in Fig. 8.20b. Then the stress–strain relationships are expressed as

$$\tau = G\gamma + C\dot{\gamma} \tag{8.38}$$

When this model is subjected to a sinusoidal loading, both τ and γ vibrate sinusoidal at the steady state. This behavior can be expressed by introducing the complex amplitude τ_0 and γ_0 as

$$\gamma = \gamma_0 e^{i\omega t}, \quad \tau = \tau_0 e^{i\omega t} \tag{8.39}$$

Here, τ and γ are also complex numbers and $e^{i\omega t}$ is given from Euler's formula

$$e^{i\omega t} = \cos\ \omega t + i\ \sin\ \omega t \tag{8.40}$$

The real part of τ and γ is actual vibration. Substitution of τ and γ into Eq. (8.38) results in

$$\tau_0 = G\gamma_0 + i\omega C\gamma_0 = (G + i\omega C)\gamma_0 \tag{8.41}$$

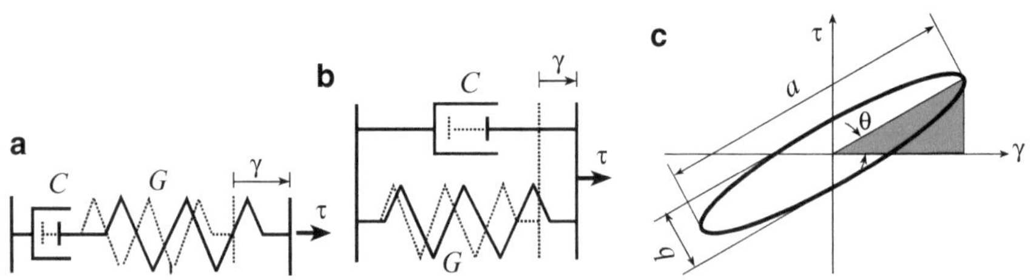

Fig. 8.20 Mechanical models for complex modulus. (**a**) Maxwell model. (**b**) Voigt model. (**c**) Hysteresis loop

This equation indicates that shear stress amplitude depends on frequency. On the other hand, stress–strain relationships of soil are independent from frequency as described in Sect. 6.7.3. In order to obtain the frequency independent hysteresis loop, viscous coefficient of the dashpot is assumed to be a function of circular frequency ω as

$$C = \frac{2hG}{\omega} \tag{8.42}$$

where h denotes damping constant. Then Eq. (8.41) yields

$$\tau_0 = G\,(1 + 2ih)\,\gamma_0 = \overline{G}\gamma_0 \tag{8.43}$$

Here

$$\overline{G} = G\,(1 + 2ih) \tag{8.44}$$

is called a complex modulus. By introducing this complex modulus, the τ_0–γ_0 relationships become linear. Stress–strain curve obtained as a real part of the complex stress and strain becomes an elliptic shape as shown in Fig. 8.20c, which enable to consider the hysteretic damping. If one uses a Maxwell model, the same procedure above results in

$$\overline{G} = \frac{G}{1 - 2ih} \approx (1 + 2ih)\,G \tag{8.45}$$

Displacement amplitude u of a single-degree-of-freedom system subjected to sinusoidal input motion with amplitude $\ddot{u}_g$ is evaluated as

$$u = \frac{1}{\sqrt{\left(1 - (\omega/\omega_0)^2\right)^2 + (2h)^2}} \frac{\ddot{u}_g}{\omega_0^2} \tag{8.46}$$

On the other hand, the displacement amplitude of the same system except that viscous damping is considered instead of the hysteretic damping is expressed as

$$u = \frac{1}{\sqrt{\left(1 - (\omega//\omega_0)^2\right)^2 + (2\omega h/\omega_0)^2}} \frac{\ddot{u}_g}{\omega_0^2} \tag{8.47}$$

where ω and ω_0 denote circular frequency of the input motion and the natural circular frequency of the system, respectively. There are some differences in amplification. This difference can be improved by setting the complex modulus as (Lysmer 1973; Desai and Christian 1977)

$$\overline{G} = G\left(1 - 2h^2 + 2ih\sqrt{1 - h^2}\right) \tag{8.48}$$

Two complex moduli (Eqs. 8.44 and 8.48) are nearly equal for $h < 0.3$. Therefore, Eq. (8.44) is sufficient in the ordinary case, but Eq. (8.48) seems better when $h > 0.3$. Therefore, recent computer program uses complex modulus defined in Eq. (8.48). Although the amplitudes agree to each other by employing Eq. (8.48), phases do not agree.

All calculations are made in the complex plane, and its real part is output as results of the analysis when using the complex modulus. Here, stress–strain curve becomes elliptic shape as Fig. 8.20c when both τ and γ are calculated from the real part of the complex number. However, shear stress can also be calculated as a product of the real part of the strain and the shear modulus, in which case stress–strain curve becomes linear and the hysteretic damping cannot be seen although it is considered in the analysis. Original SHAKE (Schnabel et al. 1972) used the latter method. Although the same calculations are made, difference appears in the response of the shear stress between two methods. For example, times at which stress and strain become maximum value are different in the former method as the hysteresis loop is elliptic in shape, whereas they are identical in the latter method. Maximum shear stresses in two methods are not the same because of the same reason.

References

Adachi N, Tatsuoka F (1986) Soil mechanics III- consolidation, shear and dynamic analysis-. Gihodo, Tokyo, p 339

Desai CS, Christian JT (1977) Numerical methods in geotechnical engineering, McGraw Hill, New York, pp. 696–699

Duncan JM, Chang CY (1970) Nonlinear analysis of stress and strain in soils. Proc ASCE J SM 96(No. SM5):1629, 1653

Editorial Committee on Dynamic Analysis of Ground -from Foundation to Application (2007) Dynamic analysis of ground -from fundamental to application, JGS, Tokyo, 152 pp

Editorial committee on elastic-plastic constitutive model of soil (2009) Elastic-plastic constitutive model of soil, JGS, Tokyo (in Japanese)

Finn WDL, Byrne PL, Martin GR (1976) Seismic response and liquefaction of sands. J GED ASCE 102(GT8):841–856

Fukutake K (1997) Study of three dimensional liquefaction analysis of soil-structure system considering multi-directional nature of shear deformation. Ohsaki research report, No. 97-03, Shimizu Corp, Tokyo. (in Japanese)

Fukutake Y, Ohtsuki A (1989) Analysis by ALISS. In: Proceedings of symposium on behavior of ground and soil-structures during earthquake, JGS, Tokyo, pp 125–134 (in Japanese)

Hardin BO, Drnevich VP (1972) Shear modulus and damping in soils: design equations and curves. Proc Am Soc Civ Eng 98(SM7):667–692

Ishihara K (1982) Evaluation of soil properties for use in earthquake response analysis. In: International symposium on numerical models in geomechanics, Zurich, pp 237–259

Ishihara K, Yoshida N, Tsujino S (1984) Modelling of hysteresis rule of stress-strain model. In: Proceedings of the 39th annual conference of the Japan society of civil engineering, vol I, pp 735–736 (in Japanese)

Ishihara K, Yoshida N, Tsujino S (1985) Modelling of stress-strain relations of soils in cyclic loading. In: Proceedings of the 5th international conference for numerical method in geomechanics, Nagoya, vol 1, pp 373–380

Jennings PC (1964) Periodic response of a general yielding structure. Proc ASCE EM2:131–163

JSSMFE (1983–1984) Introduction of constitutive model of soil. Tsuchi-to-Kiso, JSSMFE, Tokyo, vol 31–32 (in Japanese)

Kokusho T (1980) Cyclic triaxial test of dynamic soil properties for wide strain range. Soils Found 20(2):45–60

Kokusho T, Sakurai A (1979) On modified Hardin-F\Drnevich model. In: Proceedings of the 33rd Japan national conference on geotechnical engineering, JSSMFE, Tokyo, vol III, pp 1181–1184

Kondner RL (1963) Hyperbolic stress-strain response; cohesive soils. Proc ASCE SM1:115–143

Kuwano J, Ishihara K (1988) Analysis of permanent deformation of earth dams due to earthquakes. Soils Found 28(1):41–50

Lade PV, Nelson RB (1987) Modelling the elastic behavior of granular materials. Int J Numer Anal Methods Geomech 11:521–542

Lysmer J (1973) Modal damping and complex stiffness. University of California Note, University of California, Berkeley

Lysmer J, Udaka T, Tsai C-F, Seed HB (1975) FLUSH a computer program for approximate 3-D analysis of soil-structure interaction problems, Report No. EERC75-30. University of California, Berkeley

Murayama S (1985) Constitutive laws of soils. Report of ISSMFE subcommittee on constitutive laws of soils and proceedings of discussion session IA XI ICSMFE, San Francisco

Murono Y (1990) A study on the seismic design methodology of pile foundation considering the non-linear dynamic interaction caused by strong ground motions. RTRI Report, Special No. 32, Railway Technical Research Institute (in Japanese)

Nakagawa K, Soga K, Mitchell JK (1997) Observation of Biot compressional wave of the second kind in granular soils. Geotechnique 47:133–147

Pande GN, Zienkiewicz OC (1982) Soil mechanics-transient and cyclic loads constitutive relations and numerical treatment. Wiley, Chichester

PWRI (1982) Numerical analysis of seismic response of ground -SHAKE, DESRA-, Report of PWRI, No. 1778, pp 1–47 (in Japanese)

Ramberg W, Osgood WT (1943) Description of stress-strain curves by three parameters, Tech Note 902, NACA, Washington, DC

Research committee on effect of constitutive model on numerical analysis (2002) Report of the committee on effect of characteristics of constitutive model on numerical analysis, Japanese Geotechnical Society, Tokyo, 81 pp (in Japanese)

Sato M, Yasuda S, Yoshida N, Masuda T (1998) Simplified method for estimating maximum shear stress in the ground during earthquakes. J Geotechn Eng. Proc JSCE, 610/III-45:83–96 (in Japanese)

Schmertmann JH (1991) The mechanical aging of soils. J GT ASCE 117(9):1288–1330

Schnabel PB, Lysmer J, Seed HB (1972) SHAKE a computer program for earthquake response analysis of horizontally layered sites, Report no. EERC72-12, University of California, Berkeley

Shamoto, Y., Mori, N (1980) Applicability of Ramberg-Osgood model on hard clayey soil and silt. In: Proceedings of the 15th Japan national conference of soil mechanics and foundation engineering, JGS, Tokyo, pp 597–600 (in Japanese)

Tobita Y, Yoshida N (2004) State of arts and problems of constitutive model of geomaterial. Tsuchi-to-Kiso 50(8):1–6 (in Japanese)

Towhata I (1989) Models for cyclic loading, mechanics of granular materials, report of ISSMFE technical committee on mechanics of granular materials, ISSMFE, Rio de Janeiro, pp 80–90

Towhata I (1992) Dynamics deformation characteristics model of geo-material. Tsuchi-to-Kiso 40(5):75–82 (in Japanese)

Towhata I, Ishihara K (1985) Modelling soil behavior under principal stress axes rotation. In: Proceedings of the 5th international conference for numerical method in geomechanics, Nagoya A.A. Balkema, Torrerdam, vol 1, pp 523–530

Yamashita S (1992) A study on effect of variety of factors that affect cyclic shear deformation characteristics of sand and its application. A thesis in the Department of Civil engineering presented in partial fulfillment of the requirements for the degree of Doctor of Engineering, Hokkaido University, 258 pp (in Japanese)

Yoshida N (1996) Initial stress effect on response of level ground. In: Proceedings of the 11th world conference on earthquake engineering, Acapulco, Mexico, Paper No. 1023

Yoshida N, Tsujino S (1993) A simplified practical stress-strain model for the multi-dimensional analysis under repeated loading. In: Proceedings of the 28th Japan national conference of soil mechanics and foundation engineering, pp 1221–1224 (in Japanese)

Yoshida N, Sawada S, Takeshima Y, Mikami T, Sawada S (2003) Effect of hysteretic damping characteristics on earthquake response of ground, JSCE J Earthq Eng Symp, vol 28, Paper no. 158 (in Japanese)

Chapter 9
Equation of Motion

9.1 Equation of Motion and Wave Propagation

Governing differential equation to control the vibration of the ground is called a wave equation, an equation of vibration, or an equation of motion. They are essentially equivalent but appearance may be different depending on the way to derive it.

Let us consider the simplest model, i.e., one-degree-of-freedom system shown in Fig. 9.1a. Equilibrium equation yields

$$m\ddot{u} + c\dot{u} + ku = -m\ddot{u}_g \tag{9.1}$$

where m denotes mass, c denotes damping coefficient, k denotes spring constant (stiffness), u denotes displacement, and u_g denotes displacement of the base. Dot indicates differentiation with respect to time. As this equation expresses the vibration of the structure above the foundation, it is called an equation of vibration.

Equation (9.1) is modified by dividing both terms by m, resulting in

$$\ddot{u} + 2h\omega_0\dot{u} + \omega_0^2 u = -\ddot{u}_g \tag{9.2}$$

where

$$\omega_0 = \sqrt{\frac{k}{m}}, \quad h = \frac{c}{2\sqrt{km}} \tag{9.3}$$

Here ω_0 denotes natural circular frequency of this system and h denotes damping constant. Natural period T and natural frequency f of this system are evaluatedfrom ω_0 as

N. Yoshida, *Seismic Ground Response Analysis*, Geotechnical, Geological and Earthquake Engineering 36, DOI 10.1007/978-94-017-9460-2_9

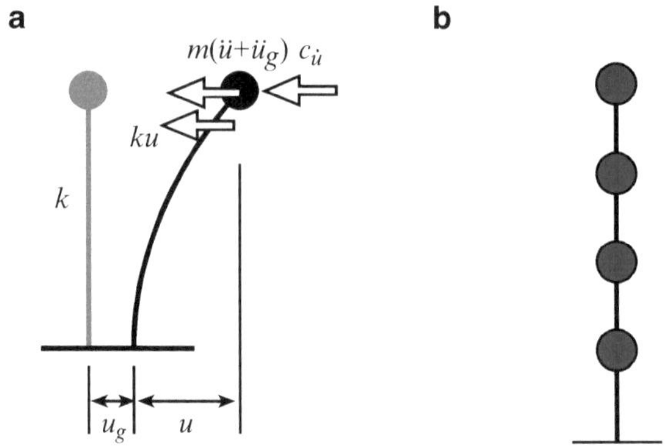

Fig. 9.1 Lumped mass model. (**a**) One-degree-of-freedom system. (**b**) Multi-degree-of-freedom system

$$T = \frac{2\pi}{\omega_0}, \quad f = \frac{1}{T} = \frac{\omega_0}{2\pi} \tag{9.4}$$

Next, let us consider the wave propagating in the ground. Forces working in a small element under one-dimensional wave propagation are already shown in Fig. 1.15 and equilibrium condition is shown in Eq. (1.8), which was

$$\frac{d\tau}{dz} = \rho\ddot{u} \tag{1.8}$$

The τ–γ relationships under the elastic behavior were shown in Eq. (4.1) as

$$\tau = G\gamma \tag{4.1}$$

Substitution of this equation and strain–displacement relationships

$$\gamma = \frac{du}{dz} \tag{9.5}$$

into Eq. (1.8) yields

$$G\frac{\partial^2 u}{\partial z^2} = \rho\frac{\partial^2 u}{\partial t^2} \tag{9.6}$$

where G denotes elastic shear modulus and ρ denotes density. Since displacement u is a function of both time t and space z, Eq. (9.6) is a partial differential equation. This equation has a characteristic that only displacement appears. This expression is called wave equation.

Both Eqs. (9.6) and (9.1) are essentially equivalent, but there are two differences. Equation (9.6) does not include viscous damping term, but it can be added if necessary by adding a relevant force shown in Fig. 1.15. The second difference is that Eq. (9.6) uses an absolute displacement, whereas displacement in Eq. (9.1) is a relative displacement from the base.

The term "wave equation" is used for the wave traveling in the media, whereas "equation of vibration" is used for the vibrating system, but it is equivalent. This can be understood by considering that the one-dimensional wave traveling in the ground is modeled as Fig. 9.1b as an extension from Fig. 9.1a and that a multi-degree of structure is also modeled as Fig. 9.1b.

The behavior of the multi-degree-of-freedom system such as Fig. 9.1b can be expressed as a sum of the one-degree-of-freedom system by means of eigenvalue problem, which is not shown in this book as it appears in almost all textbooks on vibration and as it is hardly used in the nonlinear analysis of ground.

9.2 Multidimensional Analysis

Difference between the wave equation and the equation of vibration is hardly seen in the multidimensional analysis. For simplicity, let us consider an element with width dx and height dy in the two-dimensional condition. Forces working in this element in the x-direction are shown in Fig. 9.2. It is noted that tension is positive in this figure. Then equilibrium condition leads to

$$\frac{d\sigma_x}{dx} + \frac{d\tau_{yx}}{dy} - \rho\ddot{u}_x = 0 \tag{9.7}$$

An equation same with Eq. (9.6) is obtained by applying the linear stress–strain relationships and strain–displacement relationship to Eq. (9.7).

In the same manner, equilibrium conditions in three-dimensional are written as

$$\begin{aligned}
&\frac{d\sigma_x}{dx} + \frac{d\tau_{xy}}{dy} + \frac{d\tau_{zx}}{dz} - \rho\ddot{u}_x = 0 \\
&\frac{d\tau_{xy}}{dx} + \frac{d\sigma_y}{dy} + \frac{d\tau_{yz}}{dz} - \rho\ddot{u}_y = 0 \\
&\frac{d\tau_{zx}}{dx} + \frac{d\tau_{yz}}{dy} + \frac{d\sigma_z}{dz} - \rho\ddot{u}_z = 0
\end{aligned} \tag{9.8}$$

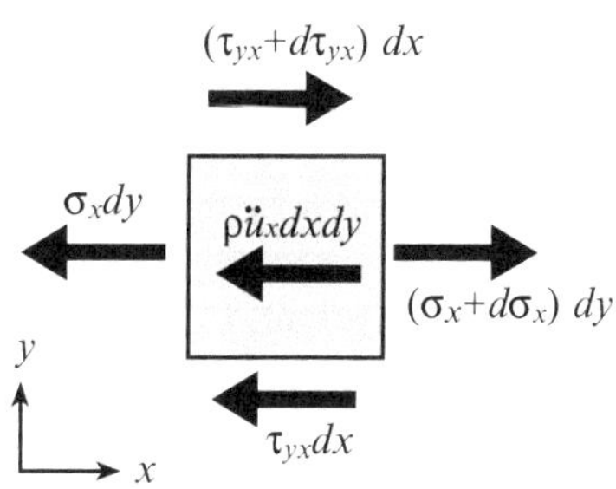

Fig. 9.2 Force acting on infinitesimal small element in x-direction

This equation is the governing differential equation. It is noted that Eq. (9.8) is simplified by using the relations $\tau_{yx} = \tau_{xy}$, $\tau_{yz} = \tau_{zy}$, and $\tau_{zx} = \tau_{xz}$. If, for example, finite element method (FEM) is applied to Eq. (9.8), one obtains

$$[m]\{\ddot{u}\} + [c]\{\dot{u}\} + [k]\{u\} = \{0\} \tag{9.9}$$

where $[m]$, $[c]$, and $[k]$ are mass, damping, and stiffness matrices, respectively. This equation is a wave equation and also an equation of vibration. They are called an equation of motion in this book.

A term that is not included in Eq. (9.8) is added when Eq. (9.9) is derived from Eq. (9.8). It is the damping term $[c]\{\dot{u}\}$. This is similar to the relation between Figs. 1.15 and 9.1a or between Eqs. (9.6) and (9.1). This may come from the easiness to image damping. If, for example, a structure vibrates in the water, the velocity proportional damping is easily recognized such as in Fig. 9.1a. On the other hand, damping force as a body force is difficult to image in Figs. 1.15 and 9.2. The former damping is called an external damping which is caused by an interaction to the outside media. However, in the actual situation, the internal damping that is caused by the relative velocity difference is prone. Discussions on damping are made in Chap. 12 in detail.

Displacement in Eq. (9.8) is an absolute displacement. As the absolute displacement is not important in the engineering practice, a relative displacement is important. It is separated into two components, i.e., displacement at the base u_b and relative displacement u_r from the base

$$\{u\} = \{u_r\} + \{I\}\,u_b \tag{9.10}$$

where $\{I\}$ is a vector whose component is unity when the corresponding freedom agrees with the direction of loading and is zero for other freedom. Then, Eq. (9.9) is rewritten as

$$[m]\{\ddot{u}_r\} + [c]\{\dot{u}_r\} + [k]\{u_r\} = -[m]\{I\}\,\ddot{u}_b \tag{9.11}$$

Since u_b is a rigid displacement in the region of interest, then

$$[k]\{u_b\} = \{0\} \tag{9.12}$$

If damping is caused by the internal damping, the damping term becomes $\{0\}$ against the rigid body motion; these terms are omitted in Eq. (9.11). Then damping term remains under the external damping. It may not be eliminated when the structure vibrates in the large viscous liquid such as water. However, this term is very small under the vibration in the air and is usually neglected in the actual situation. Therefore, Eq. (9.11) is valid in almost all situations. The subscript r is omitted for the simplicity under the definition that u indicates relative displacement in the following. Next, displacement at the base or at the referent point from which relative displacement is measured is denoted as u_g. Then Eq. (9.11) yields

$$[m]\{\ddot{u}\} + [c]\{\dot{u}\} + [k]\{u\} = -[m]\{I\}\ddot{u}_g \tag{9.13}$$

Differences between Eqs. (9.1) and (9.13) are that the latter is expressed by means of matrix and vector. Matrix expression is not used except for the case that it is necessary in this book for simplicity. Then Eq. (9.13) is written the same with Eq. (9.11) as

$$m\ddot{u} + c\dot{u} + ku = -m\ddot{u}_g \tag{9.14}$$

This is the equation of motion used in this book.

It is noted that an external damping may become an important issue in the practical situation, which again will be explained in Chap. 12.

When using a computer program on the seismic ground response analysis, one should be careful on the output of acceleration, velocity, and displacement. They are frequently output simply as acceleration, velocity, and displacement. Here, the acceleration is always absolute acceleration. On the other hand, displacement is usually relative displacement; the referent point is either base, outcropped base, or node specified by the user. The problem is velocity; many computer programs output relative velocity because it is naturally obtained as a solution of the equation of motion (Eq. 9.13). On the other hand, in the field of earthquake engineering, velocity is obtained by integrating the observed acceleration. Since acceleration measured by a seismograph is an absolute acceleration, this velocity is an absolute velocity, and there is no way to separate relative velocity from the obtained record. It is easy to distinguish absolute and relative velocity by looking at the velocity at the base or referent point; it is zero when the relative velocity is output, whereas it is not zero under the absolute velocity output.

As inertia force is a product of the mass and the absolute acceleration, the output by the absolute acceleration has mechanical meaning. In addition, acceleration measured by a seismograph is the absolute acceleration. On the other hand, the absolute displacement does not have sense and the displacement relative to the referent point is important. The velocity is important as an index of the earthquake motion for considering the long period structure such as bridges and high-rise structures. The absolute velocity is better in this sense. The relative velocity appears in the velocity response spectrum, but this may be an exception for using velocity values. Absolute velocity is to be calculated by the user when necessary because many computer programs output relative velocity. Here results are affected by the method to integrate it.

This book used relative velocity. If absolute velocity is necessary, it is written as the absolute velocity. The velocity in Fig. 1.5 is an absolute velocity.

9.3 Back Analysis

Earthquake motion at the base $\ddot{u}_g$ is usually given in the seismic ground response analysis, in which case the equation motion is given by Eq. (9.14).

There are cases that the earthquake motion at the ground surface or at a particular depth is known. Let us consider, then, the case that $\ddot{u}_a$ at a particular point is known and $\ddot{u}_g$ is unknown. In the one-dimensional analysis, equation of motion is written emphasizing it as

$$\begin{bmatrix} m_a & 0 \\ 0 & \mathbf{m}_b \end{bmatrix} \begin{Bmatrix} \ddot{u}_a \\ \ddot{\mathbf{u}}_b \end{Bmatrix} + \begin{bmatrix} c_{aa} & \mathbf{c}_{ab} \\ \mathbf{c}_{ba} & \mathbf{c}_{bb} \end{bmatrix} \begin{Bmatrix} \dot{u}_a \\ \dot{\mathbf{u}}_b \end{Bmatrix} + \begin{bmatrix} k_{aa} & \mathbf{k}_{ab} \\ \mathbf{k}_{ba} & \mathbf{k}_{bb} \end{bmatrix} \begin{Bmatrix} u_a \\ \mathbf{u}_b \end{Bmatrix} = -\begin{bmatrix} m_a & 0 \\ 0 & \mathbf{m}_b \end{bmatrix} \begin{Bmatrix} 1 \\ \mathbf{I} \end{Bmatrix} \ddot{u}_g \tag{9.15}$$

Here, bold character indicates matrix or vector and subscript a is the node where the earthquake motion is specified. All the components of $\mathbf{I}$ is unity in the one-dimensional analysis.

The following equation holds at the node where earthquake motion is specified:

$$\ddot{v}_a = \ddot{u}_a + \ddot{u}_g \tag{9.16}$$

where $\ddot{v}_a$ denotes an absolute acceleration and is a known quantity because it is specified. Then, the acceleration at the base is given as

$$\ddot{u}_g = \ddot{v}_a - \ddot{u}_a \tag{9.17}$$

Substituting this acceleration into Eq. (9.15), one obtains

$$\begin{bmatrix} m_a & 0 \\ 0 & \mathbf{m}_b \end{bmatrix} \begin{Bmatrix} \ddot{u}_a \\ \ddot{\mathbf{u}}_b \end{Bmatrix} + \begin{bmatrix} c_{aa} & \mathbf{c}_{ab} \\ \mathbf{c}_{ba} & \mathbf{c}_{bb} \end{bmatrix} \begin{Bmatrix} \dot{u}_a \\ \dot{\mathbf{u}}_b \end{Bmatrix} + \begin{bmatrix} k_{aa} & \mathbf{k}_{ab} \\ \mathbf{k}_{ba} & \mathbf{k}_{bb} \end{bmatrix} \begin{Bmatrix} u_a \\ \mathbf{u}_b \end{Bmatrix} = -\begin{bmatrix} m_a & 0 \\ 0 & \mathbf{m}_b \end{bmatrix} \begin{Bmatrix} 1 \\ \mathbf{I} \end{Bmatrix} (\ddot{v}_a - \ddot{u}_a) \tag{9.18}$$

Since $\ddot{u}_a$ in the right-hand side is unknown, it is moved to the left-hand side:

$$\begin{bmatrix} m_a & 0 \\ 0 & \mathbf{m}_b \end{bmatrix} \begin{Bmatrix} \ddot{u}_a \\ \ddot{\mathbf{u}}_b \end{Bmatrix} - \begin{bmatrix} m_a & 0 \\ 0 & \mathbf{m}_b \end{bmatrix} \begin{Bmatrix} 1 \\ \mathbf{I} \end{Bmatrix} \ddot{u}_a + \begin{bmatrix} c_{aa} & \mathbf{c}_{ab} \\ \mathbf{c}_{ba} & \mathbf{c}_{bb} \end{bmatrix} \begin{Bmatrix} \dot{u}_a \\ \dot{\mathbf{u}}_b \end{Bmatrix} + \begin{bmatrix} k_{aa} & \mathbf{k}_{ab} \\ \mathbf{k}_{ba} & \mathbf{k}_{bb} \end{bmatrix} \times \begin{Bmatrix} u_a \\ \mathbf{u}_b \end{Bmatrix} = -\begin{bmatrix} m_a & 0 \\ 0 & \mathbf{m}_b \end{bmatrix} \begin{Bmatrix} 1 \\ \mathbf{I} \end{Bmatrix} \ddot{v}_a$$

or

$$\begin{bmatrix} 0 & 0 \\ \mathbf{m}_b\mathbf{I} & \mathbf{m}_b \end{bmatrix} \begin{Bmatrix} \ddot{u}_a \\ \ddot{\mathbf{u}}_b \end{Bmatrix} + \begin{bmatrix} c_{aa} & \mathbf{c}_{ab} \\ \mathbf{c}_{ba} & \mathbf{c}_{bb} \end{bmatrix} \begin{Bmatrix} \dot{u}_a \\ \dot{\mathbf{u}}_b \end{Bmatrix} + \begin{bmatrix} k_{aa} & \mathbf{k}_{ab} \\ \mathbf{k}_{ba} & \mathbf{k}_{bb} \end{bmatrix} \begin{Bmatrix} u_a \\ \mathbf{u}_b \end{Bmatrix}$$
$$= -\begin{bmatrix} m_a & 0 \\ 0 & \mathbf{m}_b \end{bmatrix} \begin{Bmatrix} 1 \\ \mathbf{I} \end{Bmatrix} \ddot{v}_a \tag{9.19}$$

Equation (9.19) is the final equation under given condition.

This governing equation is inconvenient because mass matrix is unsymmetric. Furthermore, this equation is difficult to solve from stability point. In the past researches, $\beta = 3$ or $\theta = 10$ by Newmark's beta method and Wilson's θ method (Sect. 11.1.1), respectively, were used (Sakai et al. 1997). These values are never used in the ordinary seismic ground response analysis.

9.4 Multiple-Support Excitation

Different earthquake motions work at different nodes in the multiple-support excitation problem. This case hardly occurs in the seismic ground response analysis, but it may occur in analyzing a soil-structure system. There are two cases in the multiple-support excitation problem.

The first case is to analyze a long bridge, for example, in which earthquake motions of the foundations are different from that of other foundations. The evaluation of the inertia force becomes problem in this case. A method is shown, for example, by Clough and Penzien (1975), but this method is hardly used in the analysis of the ground.

The second case appears in analyzing the pile in the ground, for example. There will not become a problem when solving a whole soil-structure system in one time. However, geomaterial uses special constitutive models that are hardly used in the analysis of structures. On the other hand, constitutive models used in the structural element may not be installed in a computer code for ground. Analysis of a whole soil-structure system is not possible in these cases. The multiple-excitation formulation can be used to solve the structure and the ground separately (Tanaka et al. 1983). Figure 9.3 is a schematic figure showing the multiple-support excitation problem under consideration. In this case, a free-field ground is solved separately and its effect is considered in the pile-structure system as multiple-support excitation problem. If they are solved simultaneously, the problem yields the Penzien model (Penzien et al. 1964) or similar (AIJ 2006).

Equation of motion is already shown in Eq. (9.9) for the absolute displacement. In order to emphasize that the displacement is an absolute displacement, it is represented by u^t:

$$[m]\{\ddot{u}^t\} + [c]\{\dot{u}^t\} + [k]\{u^t\} = \{0\} \tag{9.20}$$

Fig. 9.3 Example of multiple-support excitation

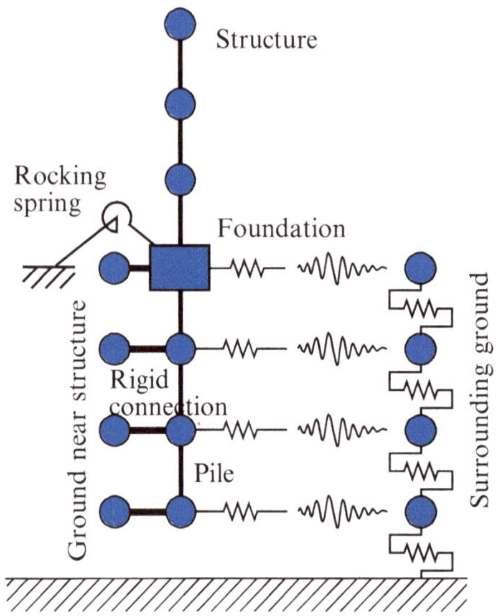

Then, u^t is separated to be the displacement at the referent point u^R and the relative displacement u from the referent point. The displacement u^R is a rigid displacement. The referent point is usually base layer, but it need not be a base layer but can be arbitrary point.

$$\{u^t\} = \begin{Bmatrix} u_a^t \\ u_b^t \end{Bmatrix} = \underset{\text{Rigid}}{\begin{Bmatrix} u_a^R \\ u_b^R \end{Bmatrix}} + \underset{\text{Relative}}{\begin{Bmatrix} u_a \\ u_b \end{Bmatrix}} \begin{matrix} \text{free} \\ \text{support} \end{matrix} \tag{9.21}$$

Here, subscripts a and b denote the free node and the excitation support, respectively. In the ordinal one-directional shaking problem, the relative displacement at the excitation support is zero, but it is not zero in this case.

Substitution of this equation into Eq. (9.20) yields

$$\begin{bmatrix} m_a & 0 \\ 0 & m_b \end{bmatrix} \begin{Bmatrix} \ddot{u}_a + \ddot{u}_a^R \\ \ddot{u}_b + \ddot{u}_b^R \end{Bmatrix} + \begin{bmatrix} c_{aa} & c_{ab} \\ c_{ba} & c_{bb} \end{bmatrix} \begin{Bmatrix} \dot{u}_a + \dot{u}_a^R \\ \dot{u}_b + \dot{u}_b^R \end{Bmatrix} + \begin{bmatrix} k_{aa} & k_{ab} \\ k_{ba} & k_{bb} \end{bmatrix} \times \begin{Bmatrix} u_a + u_a^R \\ u_b + u_b^R \end{Bmatrix} = \begin{Bmatrix} 0 \\ 0 \end{Bmatrix} \tag{9.22}$$

Retrieving free component from this equation results in

$$[m_a]\{\ddot{u}_a\} + [c_{aa}]\{\dot{u}_a\} + [k_{aa}]\{u_a\} = -[m_a]\{I_a\}\ddot{u}_a^R - [c_{ab}]\{\dot{u}_b\} - [k_{ab}]\{u_b\} \tag{9.23}$$

where $\{I_a\}$ denotes a vector whose component is unity in the direction of earthquake motion and zero in other component.

Same as the procedure in Sect. 9.2, trivial terms under the rigid body displacement and the velocity proportional terms that are small compared with inertia and stiffness terms are omitted. This equation is the same with the ordinary equation of motion except the last two terms appear in the right side.

This method is proved to give equivalent solution with the whole analysis (Yoshida and Shirato 2004). In addition, significant error is shown to be caused when velocity term input is neglected (Yoshida et al. 2005).

References

AIJ (2006) Response analysis considering dynamic interaction between building and ground, Gihodo, Tokyo, pp 379 (in Japanese)

Clough RW, Penzen J (1975) Dynamics of structures. McGraw-Hill Kogakusha, Tokyo

Penzien J, Scheffey CF, Parmelee A (1964) Seismic analysis of bridges on long piles. J Eng Mech Div (EM), Proc ASCE, vol 90, No. EM3, pp 223–254

Sakai H, Sawada S, Toki K (1997) A basic study to identify the incident seismic wave on the base layer in time domain. J Struct Mech Earthq Eng Proc JSCE 557/I-41:53–64

Tanaka T, Yoshida N, Kameoka Y, Hasegawa Y (1983) Multiple excitation analysis of underground structures. In: Proceedings of the 38th annual conference of the Japan society of civil engineering, vol I, pp 49–50 (in Japanese)

Yoshida N, Shirato M (2004) Multiple excitation analysis of underground structure. In: Proceedings of the 59th annual conference of the Japan society of civil engineering, vol I, pp 601–602 (in Japanese)

Yoshida N, Funahara H, Kobayashi Y, Kobayashi K (2005) Finite element earthquake response analysis, its advantage and disadvantage from case study. Tsuchi-to-Kiso 53(8):4–6 (in Japanese)

Chapter 10
Equation of Motion: Spatial Modeling

The equation of motion is a partial differential equation with respect to space and time. Therefore, it is to be solved against both variables. Among them, a method to solve with respect to space is shown in this chapter and that with respect to time is explained in the next chapter.

10.1 Modeling of Analyzed Region

The most accurate solution of the differential equation is an analytical solution. It is, however, very difficult or impossible to get them with increasing degrees of freedom of a system. An analytical solution is obtained in a limited case such as a multiple reflection theory (see Sect. 11.3) or a few degree-of-freedom systems.

All other methods are approximate methods. A finite element method (FEM) is the most commonly used method. Since there are so many textbooks on the FEM, details of the method are not treated in this book.

The final form of the equation of motion by the finite element formulation is shown in Eq. (9.11). An element stiffness matrix $[K]$ is obtained by a volume integral of the material stiffness. This volume integral is not calculated analytically, but is calculated numerically by

$$[K] = \int [B]^{\mathrm{T}} [D] [B] \,\mathrm{d}v = \int\int\int [B]^{\mathrm{T}} [D] [B] \,\mathrm{d}x\mathrm{d}y\mathrm{d}z \approx \sum_{i=1}^{N}\sum_{j=1}^{N}\sum_{k=1}^{N} w_i w_j w_k \left[B_{ijk}\right]^{\mathrm{T}} \left[D_{ijk}\right] \left[B_{ijk}\right] \tag{10.1}$$

where $[D]$ denotes a material stiffness matrix which is already explained in Sect. 4.2.5, $[B]$ is obtained by differentiating a shape function (the interpolation function to obtain the displacements from the nodal displacements) in space and is

N. Yoshida, *Seismic Ground Response Analysis*, Geotechnical, Geological and Earthquake Engineering 36, DOI 10.1007/978-94-017-9460-2_10

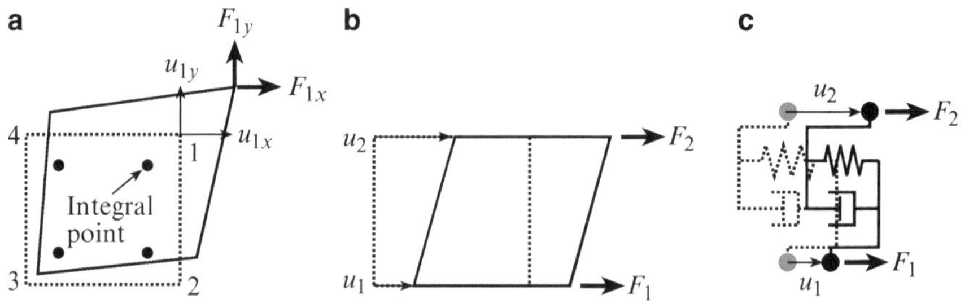

Fig. 10.1 Modeling of ground. (**a**) FEM model. (**b**) 1D model. (**c**) Spring-mass model

simply called a B-matrix, w denotes a weighted function, and subscripts correspond to the coordinate component. There are many quadratures to obtain the element stiffness matrix, but the Gauss quadrature is the most frequently used method. Values of [B] and [D] are calculated only at a few points, called integral points, and they are summed up in an element after multiplying the weight at this point. Figure 10.1a shows an element with the nodal force, the nodal displacement, and the integral point.

When N integral points are considered in an axis (usually called N-point Gauss quadrature), the Gauss quadrature gives the exact integral when the order of integrant is less than or equals to ($2N$-1)th polynomials. Therefore, the value of N is automatically determined from the shape function. However, in actual situation, values smaller than this requirement are frequently used, in which case it is called a reduced integration.

In the one-dimensional problem, a simpler element is used because only the shear deformation shown in Fig. 10.1b is interested in. As a liner shape function is employed in this element, the shear stress and shear strain are constant in whole element. This element is modeled as schematically shown in Fig. 10.1c, which is called a lumped mass–spring model. This model is equivalent to the model shown in Fig. 10.1b although appearance is significantly different.

Various conditions such as regions to be analyzed, boundary conditions, size of each element, number of integral points, etc. are to be determined in the finite element analysis of ground.

10.2 Irregularity of Ground

10.2.1 Dimension

Methods to model grounds are schematically shown in Fig. 10.2. Figure 10.2a shows horizontally layered ground, in which the same soil profile is assumed to expand in the horizontal direction. Therefore, only one seismic ground response analysis is enough to obtain the behavior of this ground. On the other hand, the ground is

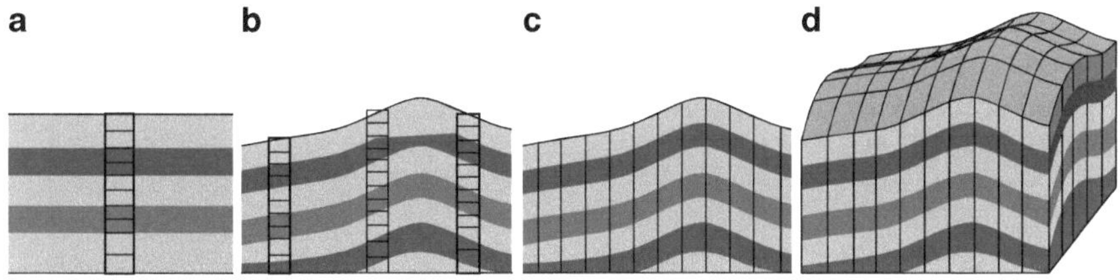

Fig. 10.2 Various modeling of ground. (**a**) Horizontal. (**b**) One dimension. (**c**) Two dimension. (**d**) Three dimension

not considered to be horizontally layered in the one-dimensional model shown in Fig. 10.2b, but behavior of the ground can be grasped by several one-dimensional analyses. The horizontally layered model and the one-dimensional model are not usually distinguished in the engineering practice. Therefore, they are simply called one-dimensional model hereafter. The two- and three-dimensional models are used in order to grasp the behavior of ground considering the irregular topography. Behavior obtained by one-dimensional analyses is sometimes different from the one by the multidimensional analysis because soil column is constrained to move free by neighboring soil columns in the multidimensional analysis, whereas it can move free under the one-dimensional analysis. It means that the one-dimensional analysis gives larger displacement than the multidimensional analysis in many cases when the ground is close to the horizontal ground. If, on the other hand, the ground is tilted, multidimensional analysis gives larger displacement because soils become weak when initial shear stress works. Two-dimensional analysis is sometimes carried out, and three-dimensional analysis is hardly made in the engineering practice.

In the practical engineering, the degree of irregularity that requires a multidimensional analysis may be one of the primary interests. Difference between one- and two-dimensional analyses may appear near the horizontally layered ground if soil liquefaction is considered (Yoshida 1988). In the case of total stress analyses, the following reports may give a good guide.

Building Research Institute, Japan, reported that behavior of ground is affected near the cliff and correction factor is to be considered from the result of one-dimensional analysis (Building Research Institute and Japan Society of Construction Engineering 1997). According to the result of the analysis, correction factor need not be considered in the following cases:

1. Slope angle of a cliff is less than or equal to 15°.
2. Height of a cliff is less than or equal to 4 m.
3. Site is apart from the cliff by six times of the height of the cliff.

Effect of irregularity on the ground surface appears when the slope angle of a base layer is larger than or equal to 5° (Design specification of high pressure gas pipeline (Committee of gas facility standard 2000)). Effect of soft layers does not appear even if angle of tilts of ground is less than 5° (Sekiba et al. 2009).

The criteria are different depending on researches, probably because of the difference of the analyzed ground. It is also noted that these criteria depend on the required accuracy of the analysis. If required accuracy becomes high, the criteria must become stricter.

10.2.2 Lens Shape Irregularity

Lens shape layers are frequently observed in the cross section of the ground. A schematic example is shown in Fig. 10.3. As layer ① appears in all borehole data, this layer is expressed as continuum although it tilts. On the other hand, there is no layer that corresponds to layer ② in the right and left side of the borehole data. Therefore, it is drawn as a lens shape layer. The same discussion can be made on layers ③ and ④ although layer ④ is not a lens shape because thickness of the layer is too thick to draw a lens shape. The exact shape of these layers is not known because there is no data in the neighboring borehole.

The practical problem occurs when these lens shape layer is a soft clayey soil, a liquefiable soil, or a humus soil in the old riverbed, for example. As explained in Sect. 1.1.2, acceleration response at the ground surface is controlled by the existence of the weak layer in the one-dimensional analysis. Therefore, if one-dimensional analysis is made using a soil profile that includes lens shape weak layers, PGA may be underestimated because lens shape layer exists only locally and the soil profile is not representative at the site.

There is few report on this topic; among them Yoshida et al. (1999) pointed out that use of the one-dimensional analysis is justified as the effect of local weak layer such as a lens shape layer appears at the ground surface. However, it is noted that the soil profile is not a representative soil profile at the site. In this sense, it is recommended to perform the analysis by neglecting the lens shape weak layer by considering the layer to be the same with upper or lower layer.

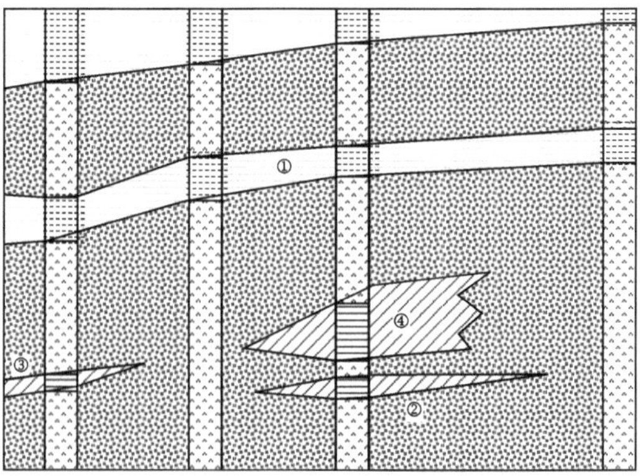

Fig. 10.3 Example of lens shape irregularity

As mentioned, effect of a soft weak layer affects results of the one-dimensional analysis significantly and response attenuates. Therefore, it may be preferable to neglect the weak layer so as to make the design conservative. The similar problem may occur when there is tilted weak layer (Yoshida et al. 1999).

10.3 Size of Layer Thickness and Mesh

Subdivision of layers in the one-dimensional analysis and determination of mesh size in the finite element analyses are to be checked in two viewpoints: wave propagation and representation of material properties that changes continuously in a layer. The latter appears, for example, as a result of the confining stress dependency of elastic modulus and shear strength. This topic is already described in Sect. 5.5.

In a finite element method or similar, displacements between nodes are obtained from a shape function, i.e., interpolation of the nodal displacements. In the one-dimensional analysis, for example, interpolation function is a linear function as shown in Fig. 10.1b. This indicates that displacement is constrained from free deformation and that high-frequency wave cannot propagate the element. The size of an element is to be less than one fifth to one sixth of the wavelength at the desired highest frequency.

The wavelength λ is calculated from the S-wave velocity V_s as

$$\lambda = \frac{V_s}{f} \tag{10.2}$$

where f denotes the highest frequency to be considered. Result of the seismic ground response analysis is used not only to grasp the behavior of the ground but also the design earthquake motion of the superstructure. Therefore, the highest frequency is determined by how the result will be used. In many practical problems, frequency between 10 and 15 Hz is sufficient as conservative design.

Let us consider a soft ground with $V_s = 100$ m/s and the highest frequency of 15 Hz, and then the wavelength yields

$$\lambda = 100/15 \approx 6.7 \text{ m} \tag{10.3}$$

This indicates that thickness of the layer is about 1.3 m at maximum. When nonlinear behavior takes place, the wave velocity becomes smaller. When shear modulus becomes a half of the elastic modulus, for example, S-wave velocity becomes $1/\sqrt{2} = 0.707$ times smaller than the initial velocity as the S-wave velocity is proportional to square root of the shear modulus (Eq. 5.8). Then wavelength, hence maximum thickness of the layer, becomes 0.707 times smaller, resulting in about 1 m. If shear modulus becomes much smaller than this, maximum allowable thickness becomes much smaller, too. However, high-frequency wave

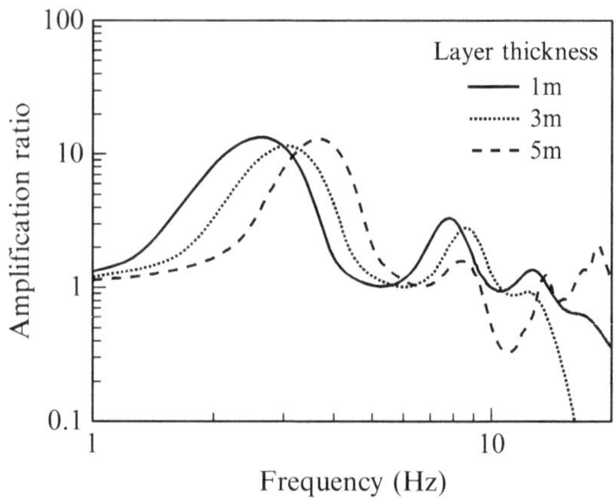

Fig. 10.4 Effect of layer division

attenuates under significant nonlinear behavior as seen in the example in Fig. 1.16. Therefore, finer layer thickness than this may not be necessary in the practice. In conclusion, maximum allowable thickness is about 1/100 of the S-wave velocity.

Finer layer thickness is required depending on the problem. For example, natural frequency of the equipment can become larger than the frequency considered above. Earthquake resistant design of the equipment is based on the floor response. Therefore, if result of the seismic ground response analysis of the ground is used as an input motion of the structure, and if floor response is calculated from the response analysis of the structure, layer thickness is to be smaller than the above example. Since small change of mechanical property also affects high-frequency response, their modeling is to be more detailed to reflect the effective stress dependency, for example, although there is no research on this topic.

An example of Fig. 10.4, where a ground with 15 m thick, $V_s = 150$ m/s, and unit weight $\gamma_t = 18$ kN/m^3, is analyzed under the El Centro NS wave input. The thickness of the layer is set either 1, 3, or 5 m. Theoretical natural frequency of this system is 2.5 Hz (natural period $T = 4H/V_s = 0.4$ s). Predominant frequency of the analysis with 1 m layer thickness agrees with the theoretical natural frequency. However, it becomes larger with increasing layer thickness. This increase of predominant frequency occurs as apparent stiffness increases because of the constraint of displacement as mentioned above.

This example shows importance of setting the relevant layer thickness by two points, i.e., to pass high-frequency wave and to get reasonable natural frequency.

Another issue on the layer division comes from the modeling of the mechanical property. They change even in the same layer because stiffness and strength change with depth, for example. Therefore, it is necessary to change material property layer by layer in the analysis in a precise sense. In the engineering practice, however, several layers are frequently treaded as a same group and the same mechanical property is used in one group. This is quite a different issue from the aforementioned mesh division problem.

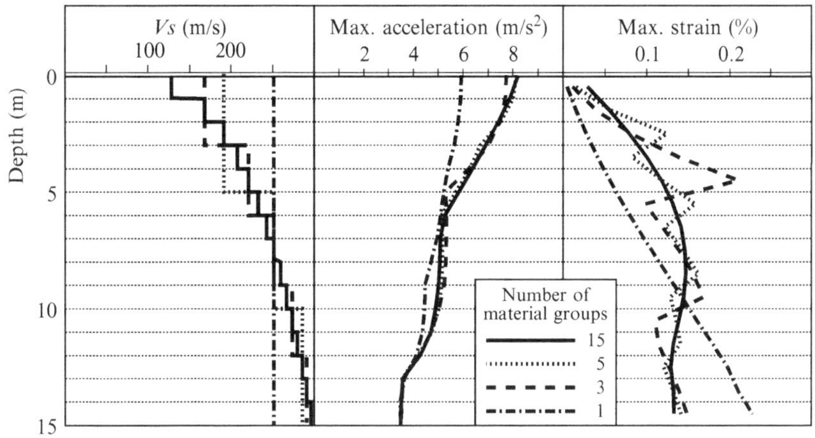

Fig. 10.5 Effect of grouping of mechanical property

The ground same with Fig. 10.4 is analyzed again. Confining stress dependency is considered for the elastic modulus by setting $G_0 = 10,000\sigma'^{0.5}_v$ [kPa] ($V_s = 130{\sim}300$ m/s) and nonlinear behavior is considered by using the H–D model (the reference strain $\gamma_r = 10^{-3}$ and the maximum damping ratio $h_{\max} = 0.25$). Thickness of the layer is set to 1 m; the total number of the layer is 15. Four parametric studies are carried out by changing the representative elastic modulus in each layer. Sublayers are grouped every 1, 3, 5, and 15 layers by dividing the ground by 15, 5, 3, and 1. Elastic modulus is set to be the one at the middle of each group. The result is shown in Fig. 10.5. Distributions of maximum acceleration are similar except the case with the same elastic modulus in all sublayers. However, distributions of maximum strain scatter very much. Maximum strain is the largest at the bottom of each group. Since the elastic modulus and the reference strain are identical in each group, the shear strength is also the same. On the other hand, inertia force becomes large with depth. This is the reason why the shear strain is larger in the lower layer in each group. As seen in this example, constant S-wave velocity or elastic shear modulus results in constant shear strength. Therefore, effective stress dependency is important especially under large earthquake motion because shear strength relates upper bound acceleration at the ground surface.

Layer subdivision from the different viewpoint is necessary in the analysis based on multiple reflection theory (Sect. 11.3). Since equation of motion is solved analytically in space in the multiple reflection theory, the result of the analysis is independent from layer subdivision in the elastic problem. In the nonlinear problem, however, mechanical properties change during earthquake because the degree of nonlinear depends on the depth. Since mechanical property at the mid-depth of the layer is used as representative values of that layer, thickness of the sublayer is to be determined from this viewpoint. As shown in the example in Sect. 11.3 and based on the author's experience, thickness can be much larger than the thickness discussed above.

In the multidimensional analysis, element size in the horizontal direction is also to be determined under the same consideration. However, it increases the number of element very much because the width of the analyzed region is usually larger than the thickness of the ground. One may be able to assume that waves traveling in the horizontal direction are mainly surface wave. Then he/she can use wide and thin elements because wavelength of the surface wave is larger than that of the body wave.

Discussion in this subsection is focused on S-wave. The same discussion can be made on the P-wave, but is not considered in the practice, partly because predominant frequency of P-wave is usually larger than that of S-wave although V_p is larger than V_s and partly because behavior under the P-wave is not much interested in.

10.4 Boundary Conditions

In the ordinary mechanics and physics, a boundary is defined not to be affected by the behavior in the media. In the earthquake geotechnical engineering, however, the boundary is to be defined in the finite depth although the ground is semi-infinite. The same discussion is made on the lateral boundary. Therefore, engineering judgment is required to define boundaries, and special treatment may be required.

Generally, an analyzed region is set to have a rectangular shape as shown in Fig. 10.6 in the seismic ground response analysis. Earthquake motion enters to the analyzed region through the bottom boundary (base). On the other hand, lateral boundary absorbs waves going out of the region without reflection.

Boundary conditions for the discrete system such as a finite element method and a mass–spring system are described in this section. Boundary conditions are defined differently in the multiple reflection theory described in Sect. 11.3.

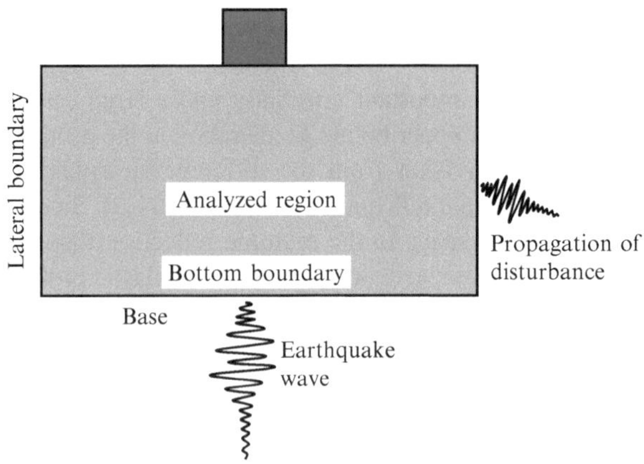

Fig. 10.6 Schematic figure showing boundaries for the seismic ground response analysis

10.4.1 Lateral Boundary

Interaction between the analyzed region and the region outside occurs in the ground. It is known as dynamic interaction problem and there are a great number of publications on it. The theory is difficult and exceeds the level of this book. In addition, these theories are hardly used in the practical computer program because they are difficult to be used in the nonlinear analysis. Only boundary conditions frequently used in the practical analysis are introduced in this book.

In the practice of the seismic ground response analysis, a horizontally layered ground is assumed to continue semi-infinitely in the lateral direction outside the analyzed region. Functions required for the lateral boundary are mainly to absorb the wave going out of the region. Horizontally layered ground that is not affected by the behavior of analyzed region or is located far from the analyzed region is called a free field. A free field can be modeled to one-dimensional model. Therefore, disturbance from the free-field behavior in the analyzed region is expected to be absorbed on the lateral boundary.

Lateral boundaries frequently used in the practice are schematically shown in Fig. 10.7. It is noted that the effect of the boundary can be erased if sufficient width is modeled as analyzed region, because waves reflected at the lateral boundary are absorbed by the internal damping before reaching the actually interested region. Width several times larger than the interested region is believed to be sufficient in this purpose, but it depends on the type of the boundary:

(a) Free Boundary

Lateral boundary is set to behave free in the free boundary. As nothing constrains the displacement outside the region, analyzed region may expand laterally or bending deformation of the whole analyzed region appears. In this sense, it is not a good boundary.

(b) Roller Boundary

The bending deformation of the whole analyzed region causes vertical movement at the lateral boundary. This vertical movement can be constrained by constraining

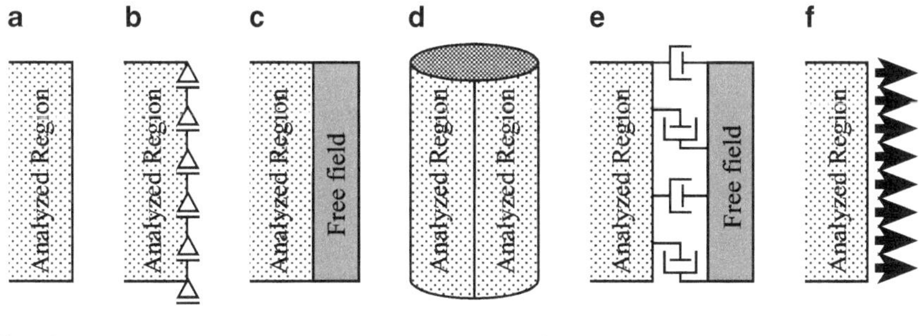

Fig. 10.7 Various lateral boundaries (**a**) Free boundary, (**b**) Roller boundary, (**c**) Lateral boundary, (**d**) Cyclic boundary, (**e**) Viscous boundary, (**f**) Other boundary

the vertical displacement at the boundary. This boundary is called a roller boundary because it allows free horizontal displacement. In addition, since vertical displacement is constrained, it is not relevant when the actual vertical displacement occurs such as the liquefaction problem or the vertical input problem. Expansion toward the outside region may occur by this boundary, which may not be good.

(c) Lateral Free Field

The two aforementioned boundaries have limitations in the practical situation. In order to improve them, free-field elements can be compulsory added at the end of the analyzed region. This can be done be setting the displacement at two adjacent nodes composing free field to be same. Since gradual transition from the analyzed region to the free field is neglected, it is not a perfect lateral boundary, but this boundary seems much better than the previous boundaries. The width of the free field must be sufficiently large such as the large mass explained in (e) so that behavior of the analyzed region does not affect the free-field behavior.

(d) Cyclic Boundary

This boundary can be used only when there are same horizontal layers at both sides of the analyzed region, and the displacements at both ends at the same depth are set identical. It may be difficult to image this boundary in Fig. 10.7d. Figure 10.8 may give more ideas to recognize this boundary; same analyzed regions are connected sequentially. Figure 10.7d shows this situation by connecting both boundaries by rolling the analyzed region.

This boundary is perfect when analyzing one-dimensional ground in the two-dimensional analysis. It shows good-looking displacement field as the extension into the outside direction observed in the free and roller boundaries does not occur. It is, however, noted that this is not a perfect boundary because effect of the analyzed region affects neighboring region and vice versa.

Node number is to be numbered in the horizontal direction so that maximum band width becomes small in the cyclic boundary because size of the maximum band width affects computing time significantly.

(e) Viscous Boundary

When the wave goes out of the analyzed region, vibration of the analyzed region attenuates as kinematic energy goes out of the region with the wave. This attenuation is called a radiation damping. Viscous boundary proposed by Lysmer

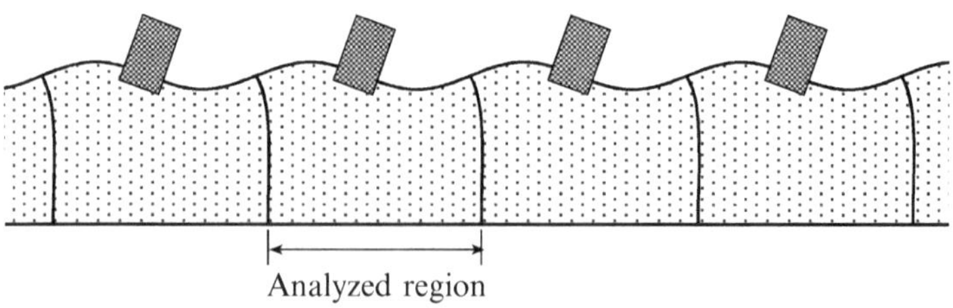

Fig. 10.8 Schematic figure showing cyclic boundary

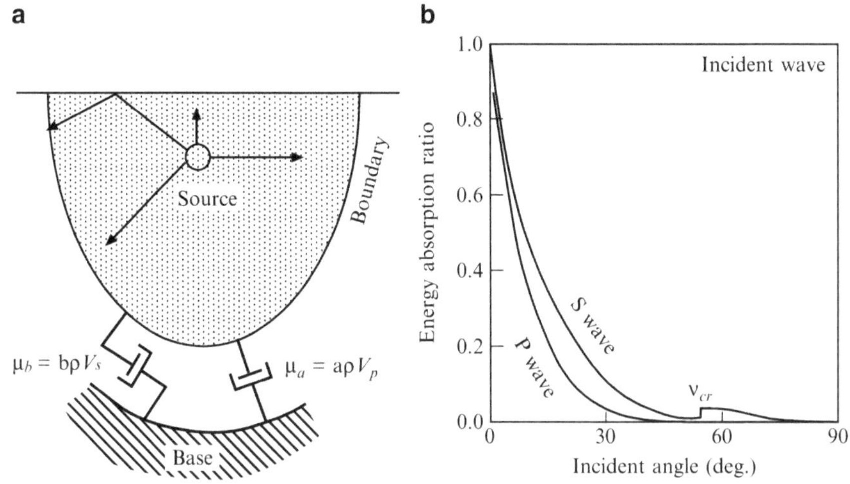

Fig. 10.9 Viscous boundary. (**a**) Dashpot and viscous coefficient. (**b**) Energy absorption ratio

and Kuhlemeyer (1969) can consider this radiation damping. Figure 10.9a shows viscous boundary schematically; they are composed of dashpots that are parallel and perpendicular to the boundary. Viscous coefficients are product of density and wave velocity in the direction of consideration. In addition, coefficients a and b are multiplied in each direction, respectively. Energy absorption ratio of the dashpot is shown in Fig. 10.9b, under the condition $a = b = 1$. When body wave goes perpendicular to the boundary, viscous boundary works a perfect boundary as all the energy is absorbed. The boundary is also known to work perfectly against the surface wave, in which case coefficients a and b are not constant but functions of frequency. They become unity at zero frequency (static case) and are called a standard boundary. If the coefficients are frequency dependent, this boundary cannot be used in the nonlinear method by the step-by-step time integration (see Sect. 11.1). In this situation, standard boundary is usually used.

Viscous boundary absorbs the energy of the vibration produced at a source in the region and going out of the region. On the other hand, the region out of the analyzed region also vibrates in the seismic ground response analysis. Therefore, as shown in Fig. 10.7e, dashpots are placed between the analyzed region and the free field, and it has a function to absorb the energy of the wave different from the free field. As the boundary lies along the vertical direction, viscous coefficients per unit length are ρV_p against the horizontal direction and ρV_s for the vertical direction.

By the way, behavior of the analyzed region may affect the behavior of the free field when a dashpot is installed, but it should be avoided. Two methods are known to avoid it:

1. Large Mass Method

   Effect to the free field can be negligibly small if mass of the free field is sufficiently large compared with the mass in the analyzed region. It can be made by setting the width large; a width of 1,000~10,000 times larger than that of the analyzed region is known to be relevant.

The shape of the free-field elements becomes wide and thin. This element shape is suggested to be avoided in the finite element analysis, but it does not produce problem because wave propagation in the horizontal direction is not considered in the free field. However, too large width may cause accuracy of the analysis because of the effective digit number of the computer. The recommended value above is determined based on these considerations.

2. Single Work Dashpot

A specially designed element damping matrix is used in this method; force which appears in the dashpot as a result of relative velocity between analyzed region and free field works only to the analyzed region but not to the free field. It becomes possible to set

$$\begin{Bmatrix} F_a \\ F_b \end{Bmatrix} = \begin{bmatrix} C & -C \\ 0 & 0 \end{bmatrix} \begin{Bmatrix} \dot{u}_a \\ \dot{u}_b \end{Bmatrix} \tag{10.4}$$

where C denotes the viscous coefficient, F denotes the equivalent nodal force, and $\dot{u}$ denotes the velocity. Subscripts a and b indicate the analyzed region and the free field. As coefficient matrix becomes unsymmetrical, numerical analysis requires more time than the case with symmetric coefficient matrix. If, however, constitutive models based on the plasticity theory with a non-associated flow rule are used, unsymmetric nature of Eq. (10.4) is no more a limitation because coefficient matrix is already unsymmetrical.

According to the theory of the dynamic interaction between the structure and the ground, the excavation force is to be considered in addition to the previous dashpot. Just before the earthquake, forces are in equilibrium state at the boundary. However, equilibrium state is not hold during analyses because stress working along the boundary changes or in the free field, unbalanced force should be applied along the boundary to keep equilibrium, which force is called an excavation force. However, this excavation force is frequently neglected in the practical analysis, which may come from the following reasons:

1. Initial stresses are in equilibrium state between the analyzed region and the outside region, and they do not affect the result of analysis during the seismic ground response analysis.
2. Difference of stresses along the boundary is essentially small because boundary of the analyzed region behaves similar to free field.
3. Forces that act on inner and outer side of the boundary are to be the excavation force. Change of stress in free field, however, does not become the excavation force as dashpots connect the analyzed region and the ground that is infinitely far from the analyzed region.

It is also noted that the shear force in the free field is not considered at the beginning of the earthquake. A report pointed out that excavation forces as well as velocity proportional term absorb good amount of energy (Miura and Okinaka 1989).

(f) Other Boundaries

There are boundaries that absorb the energy by some treatment at the boundary and are called energy-transmitting boundary, silent boundary, no reflection boundary, boundary element method, etc. Boundary shown in Fig. 10.7e can absorb surface wave. It is formulated in the frequency region and is called an energy-transmitting boundary (Lysmer and Drake 1971). Reflections at the fixed boundary and that of the free boundary have phase difference of 180°. Therefore, sum of them becomes zero, which indicated that reflection wave is not observed (Smith 1974). Other methods are briefly shown by, for example, Achenbach (1983). These boundaries are hardly used in the practical calculation partly because they are not installed in the practical computer programs and partly because they can be applicable only in an elastic problem.

(g) Summary

Either the cyclic boundary or the viscous boundary is frequently used in the practical calculation.

Back to the beginning, the region outside the analyzed region is a horizontally layered deposit. However, this assumption is not the actual situation. If irregular ground configuration out of the interested region exists, soil investigation in these regions is also required, which requires huge cost in addition to the computer power. Considering these situations, there is no perfect lateral boundary that can be used in the practical analysis.

10.4.2 Base Boundary

Earthquake motion is usually defined at an engineering seismic base layer as described in Sect. 3.1. Correspondingly, a boundary is set there, and the region above the engineering seismic base layer is analyzed. This boundary is frequently called simply as "base" or "base layer." It is also noted that a term "base" does not always indicate an engineering seismic base layer. It is also used in order to indicate at the bottom of the analyzed region from where earthquake motion works.

There are two types of base layer, i.e., a rigid base and an elastic base. When the base layer is assumed to be rigid in the rigid base, all the waves that propagate toward the rigid base are reflected at the boundary. Vibration of the base layer is specified as the input motion in the rigid base. On the other hand, base layer is assumed to be a semi-infinite elastic medium in the elastic base. Therefore, the wave that propagates to the boundary transmits into the semi-infinite region as well as reflects at the boundary. When the wave enters the base layer, energy in the analyzed region decreases, which results in attenuation of the vibration in the analyzed region. This attenuation is called a radiation damping and is explained in Sect. 12.4. Since the radiation damping cannot be considered in the rigid base, internal damping in the analyzed region may be added as an alternate damping.

Rigid base is used when the observed earthquake motion in the underground is used as input motion. Observed motion is to be specified as a composite wave (=sum of incident and reflected wave) in this case.

Theoretical background of the elastic base was developed by Joyner (1975). Let us consider the case an earthquake incident wave into the analyzed region perpendicular to the base in the one-dimensional problem. Displacement u is a function of time t and depth x as

$$u_I = u_I\left(x + V_s t\right) \text{ and } u_R = u_R\left(x - V_s t\right) \tag{10.5}$$

where subscripts I and R indicate an incident and a reflected wave, respectively, and V_s denotes the S-wave velocity of the base (media just below the analyzed region). Then the shear stress at the base τ_B becomes

$$\tau_B = G_0\left(\frac{\partial u_I}{\partial x} + \frac{\partial u_R}{\partial x}\right) \tag{10.6}$$

Relationships between the shear strain and the velocity are obtained by differentiating Eq. (10.5) with respect to x and t, resulting in

$$\frac{\partial u_I}{\partial x} = V_s\frac{\partial u_I}{\partial t} = V_s v_I, \qquad \frac{\partial u_R}{\partial x} = -V_s\frac{\partial u_R}{\partial t} = -V_s v_R \tag{10.7}$$

where v_I and v_R are velocities of incident and reflected waves, respectively. The total velocity v_B of the base is the sum of these two velocities as

$$v_B = v_I + v_R \tag{10.8}$$

Substitution of Eqs. (10.7) and (10.8) into Eq. (10.6) yields

$$\tau_B = \frac{G_0}{V_s}\left(2v_I - v_B\right) = \rho V_s\left(2v_I - v_B\right) \tag{10.9}$$

In order to consider the meaning of this equation, let us consider two grounds shown in Fig. 10.10a. The subsurface layers exist on the base layer in the left

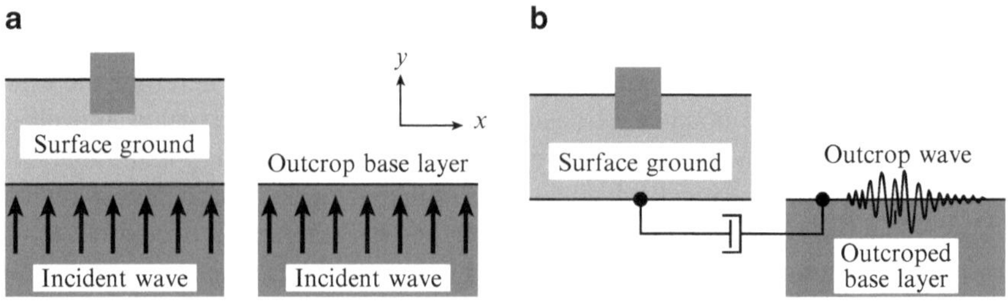

Fig. 10.10 Treatment of incident wave in elastic base. (**a**) Image of ground. (**b**) Modeling

figure, whereas the base layer outcrops in the right figure. Earthquake motion is generally defined at the surface of this ground, whereas the engineer should analyze the ground in the left figure. The incident wave below the base layer is the same, but reflection waves are different; it is affected by the surface layer in the left figure, whereas it is identical to the incident wave in the right figure. The ground in the left figure is modeled as shown in Fig. 10.10b.

According to Eq. (10.9), the ground below the base layer is omitted and dashpots are installed. The other ends of the dashpot are placed on the outcropped base layer. The end of the dashpot is shaken by the outcrop motion, i.e., double of the incident wave.

Considering this dashpot, the equation of motion (9.9) is rewritten as

$$[m]\{\ddot{u}\} + [c]\{\dot{u}\} + [\mu]\left(\{\dot{u}\} - \{I_b\}\dot{u}_f\right) + [k]\{u\} = \{0\} \tag{10.10}$$

where $[\mu]$ is a matrix composed of the dashpot. In other words, $[\mu]$ is a diagonal matrix whose component is $\rho V_s l$, where l is the length that the dashpot covers, at the freedom where dashpot is set. The vector $\dot{u}_f$ is the velocity at the outcropped base layer. This equation is inconvenient because the input earthquake is given by the velocity of the outcropped base layer. Then a new relative displacement is introduced as

$$\{u\} = \{u_r\} + \{I_b\}u_f \tag{10.11}$$

where $\{u_r\}$ denotes displacements relative to the outcropped base. Substituting Eq. (10.11) into Eq. (10.10) and following the same procedure from Eqs. (9.9) to (9.11), one obtains

$$[m]\{\ddot{u}\} + [c']\{\dot{u}\} + [k]\{u\} = -[m]\{I\}\ddot{u}_f \tag{10.12}$$

where

$$[c'] = [c] + [\mu] \tag{10.13}$$

Again, subscript r is not written because it is clear.

Equation (10.12) is the same form with the equation of motion except that the reference displacement is not the base but the outcropped base. Therefore, displacement at the base is not zero.

Discussion above is made against the S-wave. If dashpots are installed against the P-wave, the corresponding term is also to be considered.

Equation (10.12) is not an approximate method but the exact method in the one-dimensional analysis because the incident and the reflected waves are always perpendicular to the base. On the other hand, reflected wave may not be perpendicular to the boundary in the multidimensional analysis, in which case this dashpot is an approximation.

A numerical problem may occur when dashpots are installed at the base boundary as well as the lateral boundary. If all boundaries are supported only by the dashpots, the analyzed region is not supported statically. Therefore, it may move under static or quasi-static load, resulting in the settlement or differential settlement of the analyzed region. The analyzed region is in equilibrium state at the beginning of the earthquake, but this equilibrium state cannot maintain during the earthquake, which causes these settlements. If only the horizontal waves are considered, the roller boundary works well, but there is no relevant method to stop these deformations against the input in both the horizontal and vertical directional input. If displacement is constrained at some places, earthquake motion propagates through these points.

Since many computer programs are based on the small deformation theory, the load acting in the element is not affected by these deformations because they are determined referring to the location before the earthquake. Therefore, if the engineer does not like large subsidence or tilt of the analyzed region, he/she can correct it by rotating or moving so that, for example, base boundary becomes horizontal or average rotation angle becomes zero.

10.5 Multidimensional Analysis

Descriptions in the preceding are mainly focused on the one-dimensional analysis because many seismic response analyses are the one-dimensional analyses. Several additional problems appear in the multidimensional analysis although the basic concepts are the same. They are explained in this section.

10.5.1 Consideration of Vertical Motion

Only the horizontal input motion is usually considered even in the multidirectional analysis, but both horizontal and vertical input motions are sometimes considered as input motion. Additional difficulties may appear in the latter cases as will be discussed in the following. Same as in the preceding, horizontal input is considered as the S-wave and vertical input as the P-wave.

Predominant period of P-wave is usually shorter than that of S-wave. Therefore, response at the higher frequency than S-wave is to be interested in the analysis of the P-wave. This difference affects mesh size and time interval, etc., which are already discussed in Sect. 10.3 or will be discussed later in Sect. 11.1.

Another problem is a bottom boundary (base boundary). Dashpot is usually used in order to consider the elastic base or when the input motion is defined as incident or outcrop motions. As discussed in Sect. 10.4.2, settlement or differential settlement of the analyzed region may occur as the dashpot does not resist static or quasi-static load.

A computer program FLUSH (Lysmer et al. 1975) which is frequently used in the practice same as SHAKE (Schnabel et al. 1972) has another problem, which comes from the way to consider the nonlinear behavior. As explained in Sect. 10.4.2, both SHAKE and FLUSH evaluate the shear modulus from the equivalent shear strain.

There is no other modulus in SHAKE, but one more modulus, bulk modulus, is required in the multidirectional analysis. It is calculated by assuming that Poisson's ratio keeps constant in FLUSH. It indicates that the bulk modulus depends on shear strain, which is against the observation.

This problem occurs against the horizontal input, too. However, as described in the preceding, the bulk modulus is related to the P-wave velocity and behavior under the P-wave propagation is not very important under the horizontal input. However, input of the vertical motion indicates that the behavior under the P-wave propagation is to be interested in. Therefore, relevant evaluation of the bulk modulus is important, and the method used in FLUSH is inappropriate.

Additional procedures are necessary in order to improve this situation. The procedures are as follows:

1. Analyze under only horizontal input.
2. Evaluate Poisson's ratio ν from the converged shear modulus G and original bulk modulus K_0.
3. Analyze considering both the horizontal and the vertical input by using G and ν obtained before. No iteration is necessary in this process.
4. The analysis ends in item (3). If error seems large, iterative process can be done from item (2) by using the new moduli.

10.5.2 Mass Distribution

One of the frequently discussed problems in the finite element modeling is a mass matrix. There are two methods to develop the mass matrix.

In the finite element formulation, displacement in an element is interpolated from the nodal displacements. The same interpolation function is used in the velocity terms. This procedure is common in almost all the finite element formulations.

Difference appears in developing a mass matrix. The same interpolation function is used in developing the mass matrix in one method, which matrix is called a consistent mass matrix. On the other hand, a more simplified interpolation function is used, in which mass is concentrated at the nodes, and it is called a lumped mass matrix. Since the lumped mass matrix is a diagonal matrix, treatment is easier.

Let us consider a simple square element with a total mass of 4. Lumped mass and consistent mass matrices of this element become as follows:

$$\text{Lumped mass} \begin{bmatrix} 1 & 0 & 0 & 0 \\ 0 & 1 & 0 & 0 \\ 0 & 0 & 1 & 0 \\ 0 & 0 & 0 & 1 \end{bmatrix}, \text{ consistent mass} \begin{bmatrix} 1/2 & 1/8 & 1/8 & 1/4 \\ 1/8 & 1/2 & 1/4 & 1/8 \\ 1/8 & 1/4 & 1/2 & 1/8 \\ 1/4 & 1/8 & 1/8 & 1/2 \end{bmatrix}$$

The consistent matrix was considered to be the only method in the old day (Zienkiewicz 1971), but it has changed because of several reasons. Lumped mass can be derived theoretically by using the interpolation function different from the

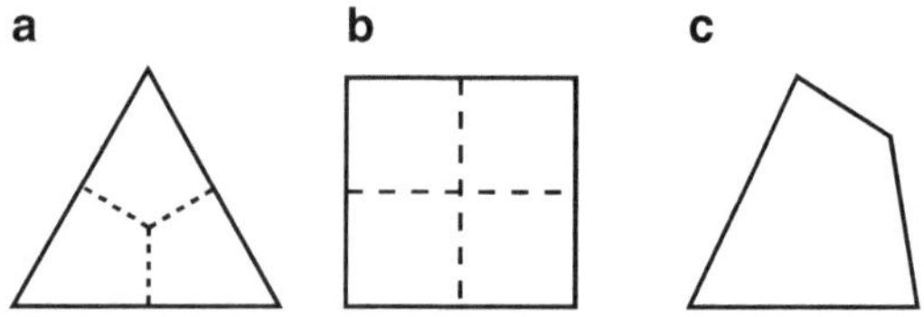

Fig. 10.11 Element shape for considering lumped mass (**a**) Triangular element, (**b**) Rectangular element, (**c**) Quadrilateral element

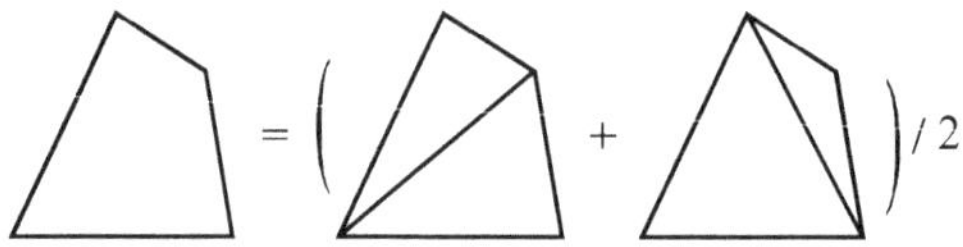

Fig. 10.12 Example of distribution of lumped mass

one used for displacement. The consistent mass is not always accurate compared with the lumped mass (Togawa 1975). In addition, mesh size becomes finer because of the improvement of computers. Therefore, at present, choice of mass matrix is not a big issue in the practice. If the engineer can choose, computing time can be reduced a little by using a lumped mass.

There remains, however, an issue of how to distribute lumped mass into each node. There is no argument in Fig. 10.11a, b; one third and a quarter are relevant value, respectively. However, several methods can be listed in Fig. 10.11c. For example, an element is divided into four triangular elements as shown in Fig. 10.12 (Association for Disaster Prevention Research 1990) by a method because lumped mass is uniquely determined for triangular element. The mass in each node is weighted by the diagonal term of the lumped mass by a method (Cook et al. 1989). Other methods are also possible. However, the engineer cannot choose the method because the computer program usually prepares one method.

10.5.3 Shape and Configuration of Element

Figure 10.13 shows a rectangular region divided by two triangular elements. Deformation of the triangular element is also shown under the horizontal load. Area of the element A does not change, whereas area changes in the element B. This change of area causes change of the confining stress in element B. Since mechanical properties such as the stiffness and the strength depend on the (effective) confining stress, they change only in the element B, in which behavior is not natural.

An example is shown in Fig. 10.14, in which tensile failure of an embankment is examined by using triangular elements (Sakai et al. 2000). Elements with tensile failure and those with shear failure are located side by side in the horizontal direction at the lower portion of the embankment, which probably comes from the above reason.

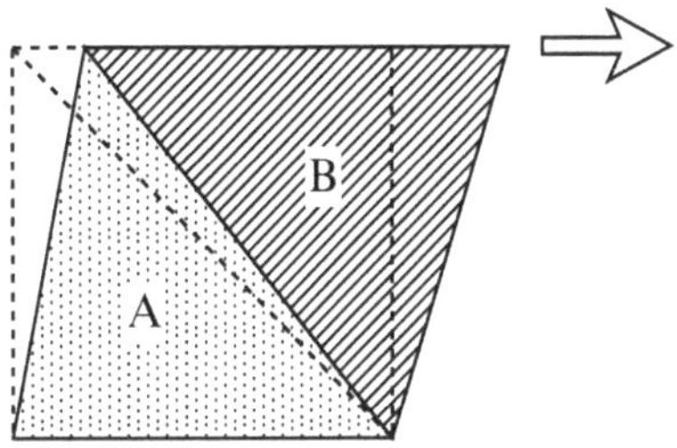

Fig. 10.13 Area change of triangular elements

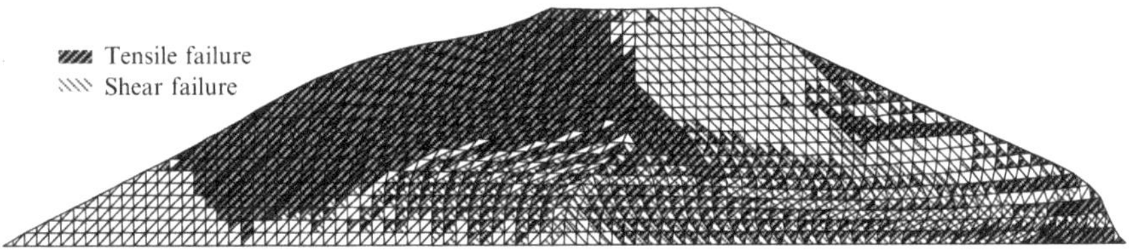

Fig. 10.14 Example of analysis by using triangular element

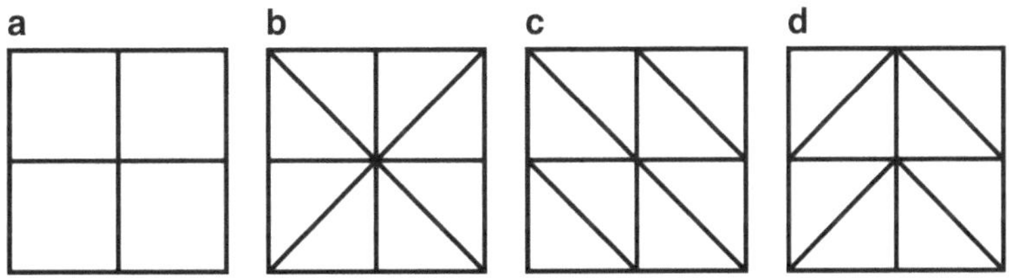

Fig. 10.15 Example of mesh arrangement (**a**–**d**) Typical mesh arrangement

Concentration of the mass to the particular elements must be avoided in making the meshes. Several mesh patterns are shown in Fig. 10.15 in which a rectangular area is divided into meshes. Rectangular elements are used in Fig. 10.15a, in which no problem occurs. In Fig. 10.15b, on the other hand, masses from eight elements concentrate at the center node, whereas mass from two elements concentrates in the side note. Even if this rectangular element is repeatedly arranged, mass from four elements concentrates. Difference of the mass causes irregular inertia force, which may create local irregular behavior. Therefore, this kind of mesh configuration is to be avoided. Mass distribution is balanced in Fig. 10.15c, but anisotropic deformation may occur because of the anisotropic configuration of triangular element. If triangular elements are to be used, Fig. 10.15d is preferable because of the balanced configuration.

If rectangular element is used in the problem such as Fig. 10.15a, no problem occurs. In this sense, triangular element is difficult to use. Considering the discussion in Fig. 6.15, in addition, it may be better to avoid a triangular element.

10.5.4 Integral Points, Volume Locking, and Hourglass Instability

Gauss quadrature is used to obtain an element stiffness matrix in the finite element method. Stiffnesses at integral points (Eq. 10.1) are used to evaluate an element stiffness matrix. A reduced integration scheme in which the number of integral points is smaller than that required from the shape function as described in Sect. 10.1 is usually used partly because it is easy to handle and partly because accuracy may be good by the reduced integration. The number of integral point is 3 for general quadrilateral element for exact evaluation, but 2 is frequently used as in Fig. 10.16.

Geomaterial is composed of a soil particle and the porewater under the water table. Behavior of the soil particle and the water is separately considered in the effective stress analysis, but they are treated as one body in the total stress analysis. Poisson's ratio of this geomaterial becomes close to 0.5, or no volume change occurs partly because the water does not resist against the shear and partly because the bulk modulus of water is much larger than that of soil skeleton. Same situation also occurs in the effective stress analysis because movement of the porewater into the neighboring elements is negligibly small as duration of the earthquake is too short. This condition is called undrained condition, in which apparent bulk modulus is the same with the one in the total stress analysis.

Then let us consider a quadrilateral element shown in Fig. 10.17a subjected to a pair of a coupled force. The element deforms as a trapezoid shape. It means that the volume contracts in the upper integral points, whereas the volume expands in the lower integral points. If Poisson's ratio is 0.5, however, volume change is not allowed. As a result, apparent stiffness against trapezoid shape deformation becomes very large, resulting in underestimation of the displacement. This behavior is called a volume locking. The volume locking does not occur against the shear deformation in Fig. 10.17b as there is no volume change. Therefore, it does not occur in the one-dimensional analysis.

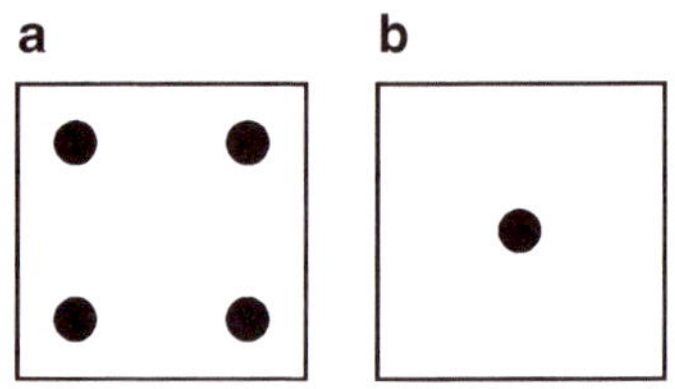

Fig. 10.16 Integral points for Gauss quadrature. (**a**) Two-point integral. (**b**) One-point integral

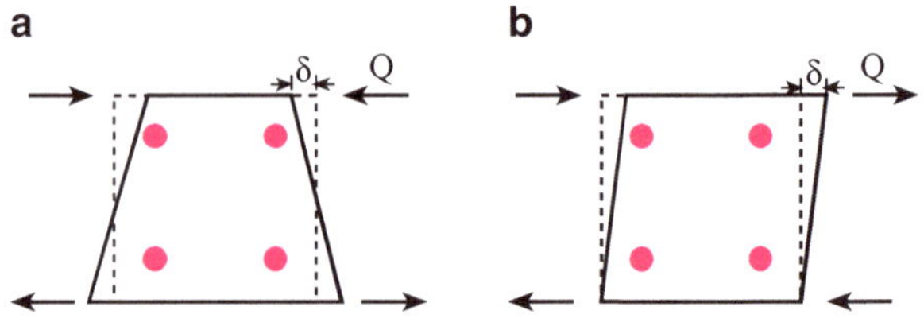

Fig. 10.17 Hourglass and simple shear deformations (**a**) Hourglass deformation, (**b**) Shear deformation

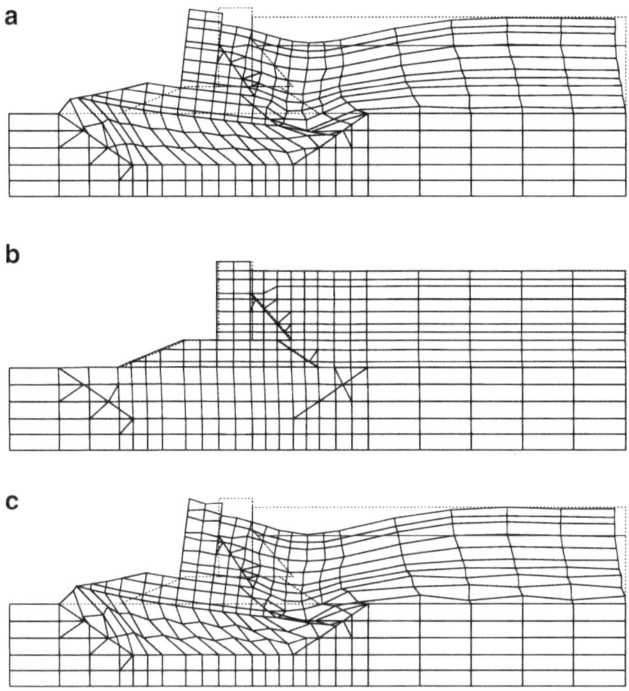

Fig. 10.18 Example of volume locking and hourglass instability (**a**) One-point integral + anti-hourglass stiffness. (**b**) Two-point integral. (**c**) One-point integral

An example is shown in Fig. 10.18 in which a quay wall is analyzed statically when liquefaction occurs (Yasuda et al. 1999). Here Fig. 10.18a is a relevant deformation or the solution of this problem. On the other hand, displacement in Fig. 10.18b is significantly small because of the volume locking (two-point Gauss quadrature).

Excess reduced integration is known to be effective to avoid the volume locking. The one-point integral shown in Fig. 10.17b is an example of excess reduced integral. Volume locking does not occur in this case because there is no volume change at the center of the element or the integral point under the couple of forces in Fig. 10.17a.

Excess reduced integral, however, may cause another problem called hourglass instability. Neither volume change nor shear deformation occurs at the center of the element against the couple of force shown in Fig. 10.17a. It indicates that a trapezoid shape deformation can occur without any force. Figure 10.18c, for example, is a result by using the one-point integral, and vertical displacement is not smooth in the horizontal direction in the surface layer as well as beneath the caisson; vertical heave up and down displacement is repeated. This deformation pattern is called an hourglass deformation or an hourglass instability. Similar behavior is confirmed to occur in the seismic ground response analysis (Ohya 2009) too in Fig. 10.19.

Fig. 10.19 Schematic deformation pattern caused by hourglass instability

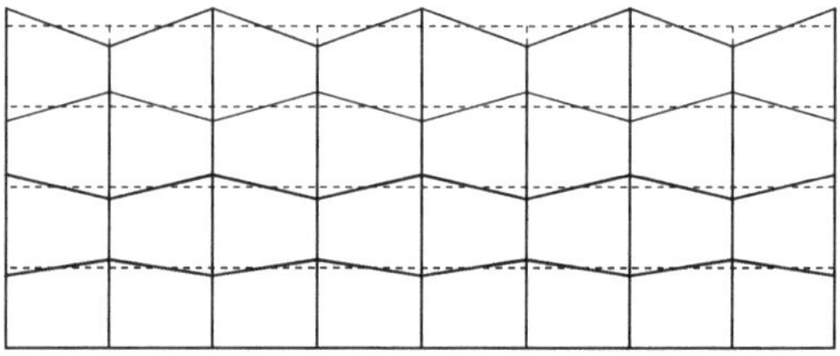

Several methods have been proposed to avoid the volume locking. Anti-hourglass stiffness, special stiffness to resist deformations such as in Fig. 10.17a, is installed in addition to the ordinary stiffness matrix obtained by the one-point integration (Flanagan and Belytschko 1981; Yoshida et al. 1989). The two-point integration is used for the shear deformation and the one-point integration is used for volume change in the SRI (selective reduced integration) method. The apparent Poisson's ratio becomes close to 0.5 because soil skeleton and water behave as one body. Poisson's ratio of soil skeleton is about 0.3 or less in the ordinary cases. Therefore, two-point integration is used for soil skeleton and one-point integration is used for water (Ohya and Yoshida 2008). However, a computer program may not always prepare several choices to avoid the volume locking. Therefore, from the point of view as a user, it is suggested to use a method that can avoid volume locking if the computer program has the function. Otherwise one-point integration with hourglass deformation is to be accepted, because the order of the displacements is the same under the one-point integration and the improved methods as can be seen in Fig. 10.18c as an example.

10.6 Initial Conditions

Equation of motion can be solved when the initial conditions and the boundary conditions are specified. Here, boundary conditions are already described as the lateral and the base boundary as well as the earthquake motion. The initial conditions are described in this subsection, although they are not a modeling of space, because they are related to it.

The initial conditions are defined as the displacement, the velocity, the acceleration, and the initial stress just before the earthquake. Among them, the velocity and the acceleration are set to zero. The displacement is also set to zero in many cases, but the displacement during the construction may be considered. The treatment is not difficult just to add the displacement by two analyses. Therefore, displacement can also be zero at the beginning of the earthquake.

On the other hand, relevant evaluation of the initial stress is not an easy task. The behavior of soil should be followed in order to obtain the accurate initial stress since it depends on the past loading histories, but it is practically impossible. Effect during the construction may sometimes be traced, but it is an exception.

In addition, even if the past loading history is traced, only the stresses can be passed to the seismic ground response analysis, because constitutive models for the seismic ground response analysis are usually different from that of the static and consolidation analyses. Considering these situations, it may be practical to consider them simply.

The analysis to obtain an initial stress state is called a switch-on-gravity analysis, a layer construction analysis, an excavation analysis, etc. The weight of the soil and the structure is applied in one time in the switch-on-gravity analysis, whereas the elements and the loads are added or removed following the construction process in the layer construction and the excavation analyses, respectively.

Since the elastic moduli are obtained by the ground survey, a switch-on-gravity analysis based on these elastic moduli has been used in the engineering practice. Since the equivalent linear analysis has been prone, the initial stress hardly affects the result of the seismic ground response analysis because of the following reasons:

1. Initial stress cannot be considered in the equivalent linear analysis.
2. As elastic moduli are obtained from the ground survey, initial stress is not required to evaluate.
3. Stres-strain relationships are also obtained by the laboratory test or empirical equations.

An example of the switch-on-gravity analysis is shown in Fig. 10.20, in which a retaining wall is analyzed. The settlement of the ground under the retaining wall is larger than that at other places because the retailing wall is heavy. The ground near the retaining wall is sheared significantly following the settlement of the retaining wall. However, as easily understood, this displacement has no meaning by considering the practical construction. The ground near the retaining wall is once excavated, for example, and filled after constructing the retaining wall. Therefore, it is not sheared due to the settlement of the ground. As the ground is leveled after the construction, settlement of the ground surface such as in Fig. 10.20 is not realistic. Figures such as Fig. 10.20 appear as a result of the switch-on-gravity analysis. It, however, just shows that the switch-on-gravity analysis is carried out, but the deformation does not affect the result of the seismic ground response analysis.

Situation changes in the nonlinear method because of two reasons. Firstly, the stiffness and the strength are evaluated from the existing stress (the principle of the

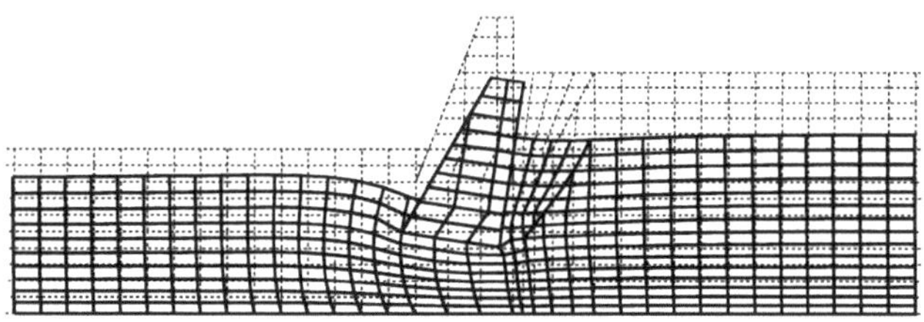

Fig. 10.20 Example of displacement of the gravity wall by switch-on-gravity analysis

effective stress). Secondly, if the initial shear exists, soil shows nonlinear behavior at the early stage of loading. For example, since soil elements near the retaining wall in Fig. 10.20 are subjected to significant shear, they will fail quickly, but it is not realistic. Therefore, rational evaluation of the initial stress is important although exact evaluation is difficult or impossible.

The switch-on-gravity analysis is the most popular analysis. Since the soil shows nonlinear behavior, the gravity force is applied incrementally. Here, if the stiffnesses are not given a priori, or they are calculated based on the effective stress, difficulty appears in the numerical calculation. The stiffness and the strength increase very rapidly as the mean stress increases. In addition the stiffness and the strength are zero at the beginning of the analysis as there is no stress. If the initial moduli are specified based on the ground survey, the switch-on-gravity analysis becomes much easier.

Displacement is described to have no meaning as explained in the preceding. A simple example is shown in order to recognize it. Figure 10.21 shows the result of a switch-on-gravity analysis and the layer construction analysis of the horizontally layered or the one-dimensional ground with three layers. Displacement in Fig. 10.21a is obtained when weight is applied in one time (switch-on-gravity analysis). On the other hand, Fig. 10.21b–d is the result of the layer construction analysis in which elements are added layer by layer. When the first layer is developed, no element exists above the first layer. In the same manner, there is no surface layer when the second layer is placed. Therefore, displacement of the ground surface starts when the third layer is placed. The resultant displacement by the layer construction analysis becomes as in Fig. 10.21d. Displacements in Fig. 10.21a, d are quite different to each other, but resultant stresses are the same.

As shown in this example, the displacement depends on the method to evaluate the initial stress. Therefore, it is natural to neglect the displacement and to set zero in the seismic ground response analysis. Of course, when displacement by the switch-on-gravity analysis or the layer construction analysis has mechanical meaning, it can be considered.

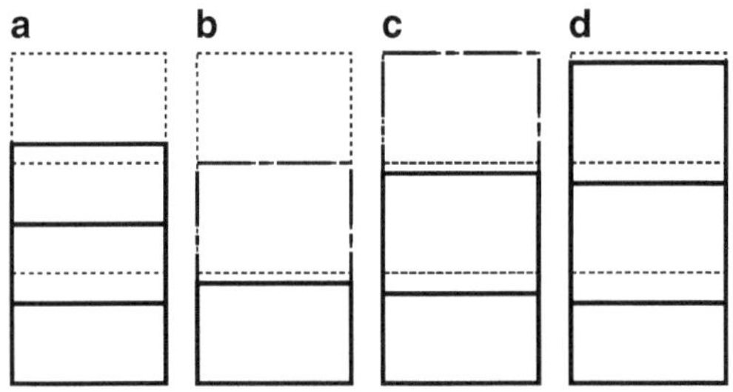

Fig. 10.21 Displacement by switch-on-gravity analysis and layer construction analysis. (**a**) Switch-on-gravity analysis. (**b**) Construction of first layer. (**c**) Construction of second layer. (**d**) After layer construction

Another problem may occur in the analysis to obtain the initial stress; resultant initial stress may violate the failure condition or the yield condition. This violation hardly occurs in the horizontally layered ground but easily occurs in the embankment because the stress along the surface is zero. It is necessary to avoid this phenomenon in the nonlinear analysis.

The switch-on-gravity analysis and the layer construction analysis are difficult if process of the sedimentation is followed exactly as shown in the preceding. Even if this problem is solved, it is still not perfect. Past history of the earthquakes is also traced in order to follow the loading history. It is easily imagined that the exact initial stress is almost impossible to obtain considering these situations. Therefore, various methods are used in the practice, which are much simpler than the above.

There are some methods by which the initial stress state is obtained simpler. First is an elastic analysis or a simple nonlinear analysis in which both the elastic and the nonlinear constants do not vary.

(a) Load shedding analysis in which the stresses exceeding the stresses associated from the current strain are released.
(b) K_0 stress state, which is obtained by putting $\sigma_h = K_0\sigma_v$, $\tau_{vh} = 0$

In the second case, the assumption that $K_0 = 1.0$ is frequently assumed. The equivalent linear method uses this assumption.

It is noted that effective stress analysis is necessary in the initial stress analysis even in the total stress analysis if the effective stress dependency of the elastic modulus and the shear strength is considered. This analysis can be made easily by using the buoyancy unit weight instead of the total unit weight.

References

Achenbach JD (ed) (1983) Computational methods for transient analysis, vol 1, Mechanics and mathematical methods. Elsevier Science Publications, Amsterdam

Association for Disaster Prevention Research (1990) Report on investigation and research on development of two-dimensional liquefaction analysis computer program (in Japanese)

Building Research Institute and Japan Society of Construction Engineering (1997) Sectional committee report on risk evaluation of earthquake motion amplification, Building Research Institute Technical Project "Development of earthquake disaster prevention in urban area", 381 pp (in Japanese)

Committee of gas facility standard (2000) Design specification of high pressure gas pipeline. Japan Gas Association (in Japanese)

Cook RD, Malkus DS, Plesha ME (1989) Concepts and applications of finite element analysis, 3rd edn. Wiley, New York

Flanagan DP, Belytschko T (1981) A uniform strain hexahedron and quadrilateral with orthogonal hourglass control. Int J Numer Methods Eng 17:679–706

Joyner WB (1975) A method for calculating nonlinear response in two dimensions. Bull Seismol Soc Am 65(5):1337–1357

Lysmer J, Drake LA (1971) Chapter 6: A finite element method for seismology. In: Alder B, Fernbach F, Boit BA (eds) Methods in computational physics, vol 11, Seismology. University of California, Berkeley

Lysmer J, Kuhlemeyer RL (1969) Finite dynamic model for infinite media. J Eng Mech Div Proc ASCE 95(EM4):859–877
Lysmer J, Udaka T, Tsai C-F, Seed HB (1975) FLUSH a computer program for approximate 3-D analysis of soil-structure interaction problems, Report no. EERC75-30. University of California, Berkeley
Miura F, Okinaka H (1989) Dynamic analysis method for 3-D soil-structure interaction systems with the viscous boundary based on the principle of virtual work. J Geotech Eng Proc JSCE 404(I–11):395–404 (in Japanese)
Ohya Y (2009) A study on numerical analysis of liquefaction and liquefaction-induced flow for practical use, theses submitted to Tohoku Gakuin University, 180 pp (in Japanese)
Ohya Y, Yoshida N (2008) FE formulation of effective stress analysis free from volume locking and hourglass instability. J Struct Eng 54B:45–50 (in Japanese)
Sakai H, Sawada S, Toki K (2000) Non-linear analyses of dynamic behavior of embankment structures considering tensile failure. In: Proceedings of the 12th WCEE, New Zealand, No. 678/5/A
Schnabel PB, Lysmer J, Seed HB (1972) SHAKE a computer program for earthquake response analysis of horizontally layered sites, Report no. EERC72-12. University of California, Berkeley
Sekiba Y, Chiba Y, Yoshida N (2009) Effect of soft layer on earthquake response of irregular ground. In: Proceedings, Meeting of Tohoku Branch of JSCE, Tagajo, pp 319–320 (in Japanese)
Smith WD (1974) A non reflecting boundary for wave propagation problems. J Comput Phys 15:492–503
Togawa H (1975) Vibration analysis by finite element method. Science-sha (in Japanese)
Yasuda S, Yoshida N, Adachi K, Kiku H, Gose S, Masuda T (1999) A simplified practical method for evaluating liquefaction-induced flow. J Geotech Eng Proc JSCE 638/III-49:71–89 (in Japanese)
Yoshida N (1988) Effectiveness of the conventional effective stress analysis to predict liquefaction-induced permanent ground displacement. In: Proceedings, First Japan-US workshop on liquefaction, large ground deformation and their effects on lifeline facilities, Tokyo, pp 57–65
Yoshida N (1989) Large deformation analysis of liquefaction-induced ground displacement by reduced integral method. In: Proceedings of the 3rd symposium on computational mechanics, Japan Union of scientists, Tokyo, pp 391–396 (in Japanese)
Yoshida N, Abo H, Yasuda S, Higuchi K (1999) Case study on applicability of one-dimensional analysis. In: Proceedings of the 54th annual conference of the Japan Society of Civil Engineering, vol 1(A), Hiroshima, pp 216–217 (in Japanese)
Zienkiewicz OC (1971) The finite element method in engineering science. McGraw-Hill, London, 521 pp

Chapter 11
Solution in Time

Solution in space is described in the previous chapter, whereas solution in time is described in this section. There are two methods. One is made in the time domain by solving the step-by-step time integration scheme, and the other is made by converting the time domain into the frequency domain.

The basic equation of motion is already shown in Chap. 9, where it is written as

$$m\ddot{u} + c\dot{u} + ku = -m\ddot{u}_g \tag{9.1}$$

or

$$\ddot{u} + 2h\omega_0\dot{u} + \omega_0^2 u = -\ddot{u}_g \tag{9.2}$$

This equation is developed for a single-degree-of-freedom system, but, as described in the preceding, it can be extended to multi-degree-of-freedom system by employing vector or matrix. Various solution methods are explained in this section based on these equations.

Different governing equation is used in the multiple reflection theory in Sect. 11.3, which was shown as wave equation in Chap. 9.

11.1 Time Domain Analysis

The equation of motion is solved in the time domain by integrating in a small time increment, which is called a step-by-step time integration method. A small time interval is used because the change of the mechanical property and/or the external load is small, in which case they can be approximated by the linear or low degree of polynomial equation.

N. Yoshida, *Seismic Ground Response Analysis*, Geotechnical, Geological and Earthquake Engineering 36, DOI 10.1007/978-94-017-9460-2_11

When the nonlinear behavior is considered, for example, Eq. (9.1) is to be written as

$$m\ddot{u} + c\dot{u} + f(u) = -m\ddot{u}_g \tag{11.1}$$

in which the restoring force term is changed to express nonlinear behavior. However, during a small time interval, Eq. (11.1) can be written as follows by using an incremental expression:

$$m\mathrm{d}\ddot{u} + c\mathrm{d}\dot{u} + k\mathrm{d}u = -m\mathrm{d}\ddot{u}_g \tag{11.2}$$

where k is a representative stiffness in this duration. It is usually tangential stiffness at the beginning of the increment. This equation is formally the same with Eq. (9.1) except that "*d*" is added indicating the incremental value.

11.1.1 Numerical Integration Scheme

Both initial and boundary conditions are to be specified to solve the equation of motion, among which the initial condition and part of the boundary conditions were already described in Chap. 10. Remaining boundary condition is an earthquake motion.

Generally, earthquake motion is specified by digital values at the small time interval. Thus earthquake motion in the time interval is not specified. In other words, the problem is not perfectly defined. Therefore, there are infinite numbers of solutions in the equation of motion by assuming different interpolations in the time interval.

When, however, continuum function such as sinusoidal function is specified, the problem is perfectly defined and there is a unique solution. Therefore, one of the best methods is to define the earthquake motion as continuum function by using relevant interpolation functions. A linear interpolation may be acceptable in this problem. Nigam and Jennings (1964) showed a method to obtain the exact solution of this problem by which single-degree-of-freedom system is solved very easily.

The equation of motion between times t_n and t_{n+1} is written as

$$\ddot{u} + 2h\omega_0\dot{u} + \omega_0^2 u = -\ddot{u}_{g,n} - \frac{\Delta\ddot{u}_g}{\Delta t}(t - t_n) \tag{11.3}$$

where $\Delta\ddot{u}_g = \ddot{u}_{g,n+1} - \ddot{u}_{g,n}$ is acceleration increment in this duration. Solution of this equation under the initial condition that

$$u = u_n \quad \text{and} \quad \dot{u} = \dot{u}_n \quad \text{at} \quad t = t_n \tag{11.4}$$

is obtained analytically as

$$y(t) = \mathrm{e}^{-h\omega(t-t_n)}\left\{a_1 \cos\omega\,(t-t_n) + a_2 \sin\omega\,(t-t_n)\right\}$$
$$-\frac{\ddot{u}_{g,n}}{\omega_0^2} + \frac{2h}{\omega_0^3}\frac{\Delta\ddot{u}_g}{\Delta t} - \frac{1}{\omega_0^2}\frac{\Delta\ddot{u}_g}{\Delta t}(t-t_n) \tag{11.5}$$

where

$$\omega = \omega_0\sqrt{1-h^2}$$
$$a_1 = u_n - \frac{2h}{\omega_0^3}\frac{\Delta\ddot{u}_g}{\Delta t} + \frac{\ddot{u}_{g,n}}{\omega_0^2}$$
$$a_2 = \frac{1}{\omega}\left(h\omega_0 u_n + \dot{u}_n - \frac{2\eta^2-1}{\omega_0^2}\frac{\Delta\ddot{u}_g}{\Delta t} + \frac{h}{\omega_0}\ddot{u}_{g,n}\right) \tag{11.6}$$

Then the recurrent relation is obtained between the start and end of the increments as

$$\begin{Bmatrix} u_{n+1} \\ \dot{u}_{n+1} \end{Bmatrix} = \begin{bmatrix} b_{11} & b_{21} \\ b_{21} & b_{22} \end{bmatrix}\begin{Bmatrix} u_n \\ \dot{u}_n \end{Bmatrix} + \begin{bmatrix} c_{11} & c_{21} \\ c_{21} & c_{22} \end{bmatrix}\begin{Bmatrix} \ddot{u}_{g,n} \\ \ddot{u}_{g,n+1} \end{Bmatrix} \tag{11.7}$$

where

$$b_{11} = \mathrm{e}^{-h\omega_0\Delta t}\left(\frac{h}{\sqrt{1-h^2}}\sin\ \omega\Delta t + \cos\ \omega\Delta t\right), \quad b_{12} = \frac{\mathrm{e}^{-h\omega_0\Delta t}}{\omega}\sin\ \omega\Delta t$$
$$b_{21} = -\frac{h\mathrm{e}^{-h\omega_0\Delta t}}{\sqrt{1-h^2}}\sin\ \omega\Delta t, \quad b_{22} = \mathrm{e}^{-h\omega_0\Delta t}\left(\cos\ \omega\Delta t - \frac{h}{\sqrt{1-h^2}}\sin\ \omega\Delta t\right)$$
$$c_{11} = \mathrm{e}^{-h\omega_0\Delta t}\left\{\left(\frac{2h^2-1}{\omega_0^2\Delta t} + \frac{h}{\omega_0}\right)\frac{\sin\ \omega\Delta t}{\omega} + \left(\frac{2h}{\omega_0^3\Delta t} + \frac{1}{\omega_0^2}\right)\cos\ \omega\Delta t\right\} - \frac{2h}{\omega_0^3\Delta t}$$
$$c_{12} = -\mathrm{e}^{-h\omega_0\Delta t}\left(\frac{2h^2-1}{\omega_0^2\Delta t}\frac{\sin\ \omega\Delta t}{\omega} + \frac{2h}{\omega_0^3\Delta t}\cos\ \omega\Delta t\right) - \frac{1}{\omega_0^2} + \frac{2h}{\omega_0^3\Delta t}$$
$$c_{21} = \mathrm{e}^{-h\omega_0\Delta t}\left\{\left(\frac{2h^2-1}{\omega_0^2\Delta t} + \frac{h}{\omega_0}\right)\left(\cos\ \omega\Delta t - \frac{h}{\sqrt{1-h^2}}\sin\ \omega\Delta t\right)\right\}$$
$$-\left(\frac{2h}{\omega_0^3\Delta t} + \frac{1}{\omega_0^2}\right)(\omega\ \sin\omega\Delta t + h\omega\cos\omega\Delta t) + \frac{1}{\omega_0^2\Delta t}$$
$$c_{22} = -\mathrm{e}^{-h\omega_0\Delta t}\left\{\frac{2h^2-1}{\omega_0^2\Delta t}(\omega\sin\ \omega\Delta t + h\omega\ \cos\ \omega\Delta t)\right.$$
$$\left.-\frac{2h}{\omega_0^3\Delta t}(\omega\ \sin\ \omega\Delta t + h\omega\ \cos\ \omega\Delta t)\right\} - \frac{1}{\omega_0^2\Delta t}$$

Since b_{ij} and c_{ij} are constants, response can be obtained successively by using Eq. (11.7) by specifying an initial condition such as $u_0 = \dot{u}_0 = \ddot{u}_0 = 0$.

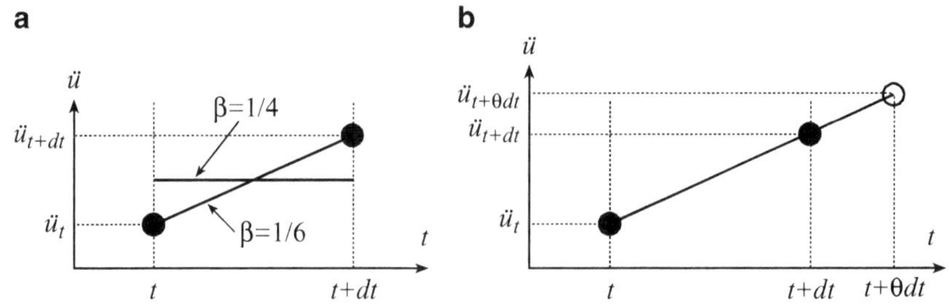

Fig. 11.1 Interpolation of acceleration in time interval. (**a**) Newmark's β method. (**b**) Wilson's θ method

The acceleration at the end of the interval $\ddot{u}_{n+1}$ is calculated from Eq. (11.3). Integration by this scheme does not have instability problem described in Sect. 11.1.2 because the solution is obtained analytically. Because of the stability and fast evaluation, this method is frequently used in calculating the response spectrum (Shibata 2010; Ohsaki 1994).

It becomes, however, rapidly difficult to get the analytical solution with the increase of degrees-of-freedom, and it is practically impossible to solve differential equation analytically.

Response of the system is interpolated instead of interpolating the input earthquake motion in the next method. Figure 11.1 shows the typical interpolation functions for the acceleration response.

11.1.1.1 Linear Acceleration Method

Similar to the interpolation of input earthquake motion, a linear interpolation of the response acceleration is one of the relevant methods, which is shown in Fig. 11.1a by $\beta = 1/6$. This method is called a linear acceleration method as linear interpolation of the acceleration is used. Then displacement and velocity at time t+dt are obtained by integrating the acceleration from the response at time t as

$$\begin{aligned} u_{t+\mathrm{d}t} &= u_t + \mathrm{d}t\,\dot{u}_t + \frac{\mathrm{d}t^2}{3}\ddot{u}_t + \frac{\mathrm{d}t^2}{3!}\ddot{u}_{t+\mathrm{d}t} \\ \dot{u}_{t+\mathrm{d}t} &= \dot{u}_t + \frac{\mathrm{d}t}{2}\left(\ddot{u}_t + \ddot{u}_{t+\mathrm{d}t}\right) \end{aligned} \tag{11.8}$$

Substitution of this equation into Eq. (9.1) at time t+dt yields

$$\left(m + c\frac{\mathrm{d}t}{2} + k\frac{\mathrm{d}t^2}{6}\right)\ddot{u}_{t+\mathrm{d}t} = -m\ddot{u}_{g,t+\mathrm{d}t} - c\left(\dot{u}_t + \frac{\mathrm{d}t}{2}\ddot{u}_t\right) - k\left(u_t + \mathrm{d}t\,\dot{u}_t + \frac{\mathrm{d}t^2}{3}\ddot{u}_t\right) \tag{11.9}$$

This is a simultaneous equation for multi-degree-of-freedom system. Acceleration $\ddot{u}_{t+\mathrm{d}t}$ at time $t+\mathrm{d}t$ is obtained by solving this equation. Displacement and velocity at time $t+\mathrm{d}t$ are calculated from Eq. (11.8).

Another procedure may be used under the same assumption. Acceleration and velocity are first derived from Eq. (11.8). Then it is substituted into Eq. (9.1), which results in

$$\left(\frac{6m}{\mathrm{d}t^2}+\frac{3c}{\mathrm{d}t}+k\right)u_{t+\mathrm{d}t}=m\left(\frac{6u_t}{\mathrm{d}t^2}+\frac{6\dot{u}_t}{\mathrm{d}t}+2\ddot{u}_t\right)+c\left(\frac{\mathrm{d}t\,\ddot{u}_t}{2}+2\dot{u}_t+\frac{3u_t}{\mathrm{d}t}\right) - m\ddot{u}_{g,t+\mathrm{d}t} \qquad (11.10)$$

Equation (11.9) is frequently shown in many textbooks, but Eq. (11.10) is convenient in the nonlinear seismic ground response analysis because displacement is obtained at first. The stiffness k used in the analysis is usually a tangent modulus at the beginning of the interval, but it may change significantly when unload occurs. Some programs make one trial calculation to check whether unload occurs or not, and if unload is identified to occur, stiffness at unload is used as k instead of the tangent modulus, in which case the total amount of calculation becomes small if displacement is calculated at first. Equation (11.9) has another advantage that total amount of calculation is smaller as the stiffness matrix remains alone.

Incremental form is also used in the nonlinear analysis. Equation (11.9) is rewritten in an incremental form as

$$\left(m+c\frac{\mathrm{d}t}{2}+k\frac{\mathrm{d}t^2}{6}\right)\mathrm{d}\ddot{u}=-m\mathrm{d}\ddot{u}_g-\mathrm{d}t\,c\ddot{u}_t-k\left(\mathrm{d}t\,\dot{u}_t+\frac{\mathrm{d}t^2}{2}\ddot{u}_t\right) \qquad (11.11)$$

where $\mathrm{d}\ddot{u}$ and $\mathrm{d}\ddot{u}_g$ are acceleration and earthquake motion increment, respectively. In the same manner, Eq. (11.10) can also be written in the incremental form.

The linear acceleration method is introduced in almost all the textbooks on the numerical analysis, and it has good accuracy (Sato and Hino 1982), but it is hardly used in the practical calculation because of the stability problem explained in Sect. 11.1.2. Instead of the linear acceleration method, Newmark's β methods and Wilson's θ method become to be used.

11.1.1.2 Newmark's β Method

A more general formulation is proposed by Newmark (1959) instead of Eq. (11.8) as

$$\begin{aligned}\dot{u}_{t+\mathrm{d}t}&=\dot{u}_t+(1-\gamma)\,\mathrm{d}t\;\ddot{u}_t+\gamma\mathrm{d}t\;\ddot{u}_{t+\mathrm{d}t}\\ u_{t+\mathrm{d}t}&=u_t+\mathrm{d}t\;\dot{u}_t+(0.5-\beta)\,\mathrm{d}t^2\ddot{u}_t+\beta\mathrm{d}t^2\ddot{u}_{t+\mathrm{d}t}\end{aligned} \qquad (11.12)$$

where γ and β are adjusting parameters. The procedure to obtain the solution is the same as the linear acceleration method. In the original paper, 0.5 is recommended for γ because numerical damping is associated by other value, and 1/5 to 1/6 is recommended for β. When $\gamma = 0.5$ and $\beta = 1/6$, the equation completely coincides with the linear acceleration method. Therefore, Newmark's β method was another name of the linear acceleration method in the old day (Togawa 1975).

However, as described above, $\beta = 1/6$ (linear acceleration method) is not used because of the stability reason. Instead of it, as the reason shown in Sect. 11.1.2, $\beta = 1/4$ is frequently used, which is called an average acceleration method. Acceleration is assumed to be constant as average of the accelerations at the beginning and at the end of the time interval as shown in Fig. 11.1a. The combination of $\gamma = 0.6$ and $\beta = 0.3025$ is also said to be good for stability (Association for Disaster Prevention Research 1990).

11.1.1.3 Wilson's θ Method

Wilson's θ method is an extension of the linear acceleration method. As shown in Fig. 11.1b, response acceleration is assumed to be linear between t and $t + \theta \mathrm{d}t$ ($\theta > 1$) (Clough and Penzien 1975), whereas it is assumed to be linear between t and $t + \mathrm{d}t$ in the linear acceleration method. When acceleration at time $t + \theta \mathrm{d}t$ is obtained by the same procedure as the linear acceleration method, acceleration at $t + \mathrm{d}t$ is calculated as

$$\ddot{u}_{t+\mathrm{d}t} = \frac{(\theta - 1)\,\ddot{u}_t + \ddot{u}_{t+\theta \mathrm{d}t}}{\theta} \tag{11.13}$$

Then displacement and velocity are obtained from Eq. (11.8) (Togawa 1975; Clough and Penzien 1975).

High-frequency response is constrained by the setting longer linear range, which works to reduce or to remove instability problem described in Sect. 11.1.2. As the high-frequency response is not very important in the engineering practice, this method may be powerful.

The equilibrium at time $t + \mathrm{d}t$ is not guaranteed in this procedure, which means that unbalanced force remains even in the linear elastic problem. In order to solve this problem, displacement, velocity, or acceleration is to be recalculated by using the equation of motion at time $t + \mathrm{d}t$. In the nonlinear analysis, however, procedure above does not produce any problem as the effort to cancel unbalanced force is usually made.

Figure 11.2 shows response spectrum against the El Centro 1940 wave. The result of Nigam's method is the most accurate in this case. Compared with this method, Wilson's θ method shows larger acceleration in longer periods. If unbalanced force is canceled, however, the result (modified Wilson θ in the figure) agrees with that by Nigam's method. The linear acceleration method and Newmark's β method ($\beta = 1/4$) agree with Nigam's method. They are not shown in the figure as they cannot distinguish from Nigam's method in the figure.

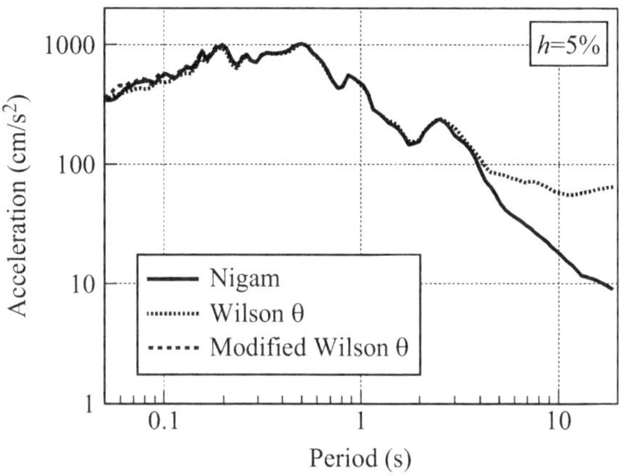

Fig. 11.2 Response spectrum for evaluating accuracy of Wilson's θ method

11.1.1.4 Central Difference Method

This method is somewhat different from the previous methods. If displacement is assumed to change by a quadratic expression (parabola) in two time intervals, acceleration and velocity can be expressed as

$$\dot{u}_t = \frac{u_{t+\mathrm{d}t} - u_{t-\mathrm{d}t}}{2\mathrm{d}t}$$
$$\ddot{u}_t = \frac{u_{t+\mathrm{d}t} - 2u_t + u_{t-\mathrm{d}t}}{\mathrm{d}t^2} \tag{11.14}$$

Substituting this equation into Eq. (9.1) results in

$$\left(\frac{m}{\mathrm{d}t^2} + \frac{c}{2\mathrm{d}t}\right) u_{t+\mathrm{d}t} = \left(\frac{2m}{\mathrm{d}t^2} - k\right) u_t + \left(-\frac{m}{\mathrm{d}t^2} + \frac{c}{2\mathrm{d}t}\right) u_{t-\mathrm{d}t} - m\ddot{u}_{g,t} \tag{11.15}$$

Displacement at time $t + \mathrm{d}t$ is obtained by solving this simultaneous equation. Acceleration and velocity are calculated from Eq. (11.14).

This procedure uses the response not only time t but also $t - \mathrm{d}t$. This method has two advantages. The one is that coefficient matrix of the simultaneous equation does not include the stiffness k; the coefficient matrix is constant. One of the typical methods to solve simultaneous equation is a sweep out method. This method is composed of two steps. At first, the coefficient matrix is converted to the upper triangular matrix, which is called a forward elimination. Then solution of the simultaneous equation is obtained by the backward substitution. The forward elimination takes huge time than the backward substitution. When coefficient matrix is constant, the computing time can be saved significantly as the forward elimination is carried out only once. Therefore the computing time can be saved significantly. The second advantage is that stiffness is not necessary, which can be recognized by replacing ku_t in Eq. (11.15) by a restoring force $f(u_t)$. Therefore, the constitutive models that cannot output tangent modulus can be used.

On the other hand, it has disadvantage on stability. As explained later, stability problem can be avoided by using a smaller time interval, but this process requires more computing time. If reduction of the computing time in the central difference method is larger than the increase of the computing time by using the small time interval, central difference method is effective. In this sense, it is applicable to the analysis of the ground, but is not practical when there are large stiffness elements such as piles and structures because required time interval becomes significantly short.

11.1.1.5 Other Methods

There are many differential methods in theory. Euler's method and the Runge–Kutta method, for example, frequently appear in the textbook. However, these methods do not used in the practice of seismic ground response analysis. The methods explained in the preceding are specially designed for the second-order differential equation such as an equation of motion.

11.1.1.6 Accuracy of Methods

As explained at the beginning of this section, the problem is not completely defined in the ordinary seismic ground response analysis in which the input earthquake motion is not defined as a continuum function, but it is given by digits with the small time interval. Therefore, infinite number of solutions can be obtained by using different interpolation functions in the small time interval. In this sense, all the method gives solutions of the equation of motion under given conditions; the difference comes from the different assumption in the small time interval.

On the other hand, it is also true that there are acceptable solutions and unacceptable solutions. For example, let us consider a couple of solution

$$u_n = \dot{u}_n = 0, \ddot{u}_n = \ddot{u}_{g,n} \quad \text{for all } n \tag{11.16}$$

This is a solution of the equation of motion because equation of motion holds all the time when the acceleration is specified. It is, of course, unacceptable solution because behavior within the time interval is against our implicit assumption that behavior within the time interval is simple.

In this sense, a linear interpolation of the input motion is acceptable and reasonable. A linear interpolation of the response acceleration used in the linear acceleration method is also a good assumption. One can expect good solution by using these interpolations. The average acceleration method or Wilson's θ method will show less accurate result because the interpolation is not natural. When the time increment is sufficiently small, however, error can be made sufficiently small as explained in Sect. 11.1.3.

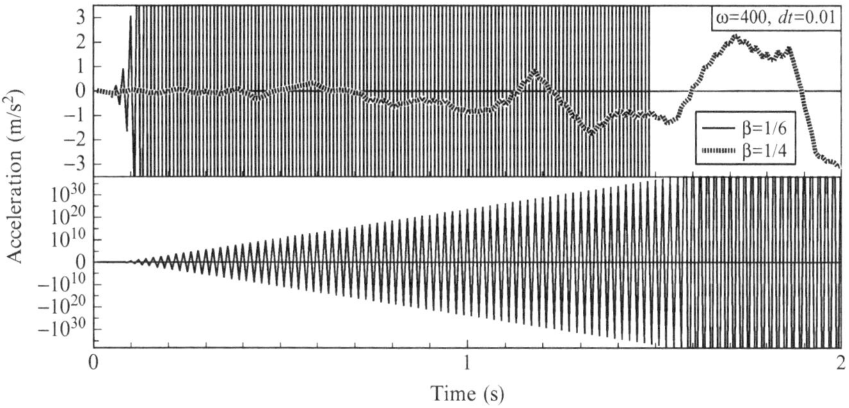

Fig. 11.3 Example of divergence

11.1.2 Stability of Numerical Integral

Figure 11.3 shows a result of the seismic ground response analysis of a one-degree-of-freedom system whose natural circular frequency $\omega = 400$ under the El Centro earthquake motion. Newmark's β method is used for time integration. The ordinate of the upper figure is the acceleration in the linear axis. The thick dotted line in the figure is a result with $\beta = 1/4$, which is a good solution in this problem. On the other hand, thin solid line is the result with $\beta = 1/6$ or linear acceleration method, in which acceleration rapidly becomes very large and its sign changes at every time increment. This is the unstable phenomenon that is discussed in this subsection.

Let us consider the reason why this kind of divergence occurs. In order to make the discussion simple, a free vibration of an undamped system is considered. Then equation of motion is

$$\ddot{u} + \omega_0^2 u = 0 \tag{11.17}$$

Elimination of acceleration term by substituting Eq. (11.17) into Eqs. (11.12) and (11.14) yields

$$\begin{bmatrix} 1 & \omega_0 \Delta t/2 \\ 0 & 1 + \omega_0^2 \beta \Delta t \end{bmatrix} \begin{Bmatrix} u_{n+1} \\ \dot{u}_{n+1} \end{Bmatrix} = \begin{bmatrix} 1 & -\omega_0^2 \Delta t/2 \\ dt & 1 - \omega_0^2 \left(1/2 - \beta\right) \Delta t^2 \end{bmatrix} \begin{Bmatrix} u_n \\ \dot{u}_n \end{Bmatrix}$$

$$\begin{Bmatrix} u_{n+1} \\ u_n \end{Bmatrix} = \begin{bmatrix} 2 - (\omega_0 \Delta t)^2 & -1 \\ 1 & 0 \end{bmatrix} \begin{Bmatrix} u_n \\ u_{n-1} \end{Bmatrix} \tag{11.18}$$

for Newmark's β and the central difference methods, respectively. The case of the central difference method is discussed here as it is simpler. Equation (11.18) can be written as

$$\begin{aligned}
\begin{Bmatrix} u_2 \\ u_1 \end{Bmatrix} &= [A] \begin{Bmatrix} u_1 \\ u_0 \end{Bmatrix} \\
\begin{Bmatrix} u_3 \\ u_2 \end{Bmatrix} &= [A] \begin{Bmatrix} u_2 \\ u_1 \end{Bmatrix} = [A]^2 \begin{Bmatrix} u_1 \\ u_0 \end{Bmatrix} \\
&\cdots\cdots \\
\begin{Bmatrix} u_m \\ u_{m-1} \end{Bmatrix} &= [A] \begin{Bmatrix} u_{m-1} \\ u_{m-2} \end{Bmatrix} = [A]^{m-1} \begin{Bmatrix} u_1 \\ u_0 \end{Bmatrix}
\end{aligned} \tag{11.19}$$

where

$$[A] = \begin{bmatrix} 2 - (\omega_0 \Delta t)^2 & -1 \\ 1 & 0 \end{bmatrix} \tag{11.20}$$

It indicates that matrix $[A]$ is multiplied many times. Therefore, if the magnitude of $[A]$ is too large, the solution diverges. The determinant of the eigenvalue of matrix $[A]$ is to be smaller than unity in order to avoid this divergence. The eigenvalue λ of matrix [A] is

$$\lambda = \frac{2 - (\omega_0 \Delta t)^2}{2} \pm \sqrt{\frac{\left\{2 - (\omega_0 \Delta t)^2\right\}^2}{4} - 1} \tag{11.21}$$

If the numerator in the square root term is negative, λ becomes complex number and this condition is satisfied. Therefore, stability criterion yields

$$\Delta t \leq \frac{2}{\omega_0} = \frac{T}{\pi} \tag{11.22}$$

where T denotes natural period of the system. In the multi-degree-of-freedom system, T is minimum natural period of the system.

The stability criteria for Newmark's β method is obtained by the same manner, yielding

$$(\omega_0 \Delta t)^2 (1 - 4\beta) \leq 4 \quad \text{or} \quad \Delta t \leq \frac{2}{\omega_0 \sqrt{1 - 4\beta}} \tag{11.23}$$

If $\beta \geq 1/4$, Eq. (11.23) holds regardless of the time increment; the system becomes unconditionally stable. On the other hand, error becomes larger as β increases. Considering this balance, $\beta = 1/4$ is usually used in practice.

The unstable condition can also be obtained in Wilson's θ method by setting $\theta \geq 1.366025$. However, accuracy decreases very rapidly as θ increases. In practice, therefore, $\theta = 1.37$ or 1.4 is usually used, and value larger than 1.4 is not suggested.

In order to see this unstable behavior in detail, acceleration is written in a log scale in the lower figure in Fig. 11.3, by neglecting the acceleration whose absolute value is less than unity. Acceleration increases linearly in log axis as a result that constant value larger than 1 is multiplied every time as the theory above indicates. The maximum value of the acceleration in this figure is 10^{38}, which is the maximum number that a computer can handle.

According to Eq. (11.23), $\Delta t \leq 2\sqrt{3}/\omega_0 \approx 0.0086$ is necessary at $\beta = 1/6$ for the case in Fig. 11.3. Time increment used in the analysis (0.01 s) is a little longer than this, but the solution diverged.

In a multi-degree-of-freedom system, the shortest natural period must satisfy the stability criteria in order to obtain a stable solution. In the old days, as the computer power is not strong, mesh size of the FE model is fairly large and minimum natural period is long. Mesh size becomes finer as the computer power increases, which caused natural period shorter and shorter so that stability criteria cannot be satisfied at the reasonable time increment. This is the reason why the linear acceleration method ($\beta = 1/6$ for Newmark's β method) is not used at present although it is more accurate than the other methods. The same discussion can be made for a central difference method. However, as aforementioned, computing time required for forward elimination is much shorter than other methods; it can be used if required time increment is as short as the order of 1/10,000 s.

Divergence may occur even if stability criterion above is satisfied in a nonlinear analysis. Divergences during a nonlinear analysis were sometimes reported, especially in Wilson's θ method (Souza and Garg 1984). The reason is not sure for the author. It is also noted that stable solution does not mean that it is a good solution. An extreme example is shown by Gudehus (1977).

Curious solution may be obtained in the practical nonlinear seismic ground response analysis. However, it is noted that all the curious solution does not always come from the stability problem discussed in this section. The reason of the curious response comes from other causes unless the behavior such as in Fig. 11.3 is not seen because a numerical integration is made relevant if the response does not diverge.

Response accelerations are interpolated in the time interval. It indicates that free response is constrained. As a result, artificial damping occurs, which is called a numerical damping. Because of this damping, response becomes gradually small even in the undamped system. Wilson's θ method shows large numerical damping (Gudehus 1977). This is a disadvantage from the accuracy viewpoint of the analysis. However, it is an advantage from the stability viewpoint in a practical analysis, because high-frequency response (short-wavelength behavior) is suppressed. An impulse is applied at the beginning of the increment as the load is applied by a step function, which sometimes appears as a pulse of the acceleration especially that stiffness is very small (e.g., see Sect. 15.4). This pulse is not actual behavior but appears as a result of numerical analysis. Therefore, it is relevant to erase them, and the numerical damping works to erase these pulses.

11.1.3 Choice of Numerical Integration Scheme

When a computer program has several options on the numerical integration, the user needs to select one of them. As shown in Fig. 11.2, there is no significant difference in the response spectrum. When looking at the figure in detail, however, some difference is observed in a short period region.

In order to see this difference clearer, similar calculations are made as shown in Fig. 11.4. Damping ratio is set 2 %, and input earthquake motion is the El

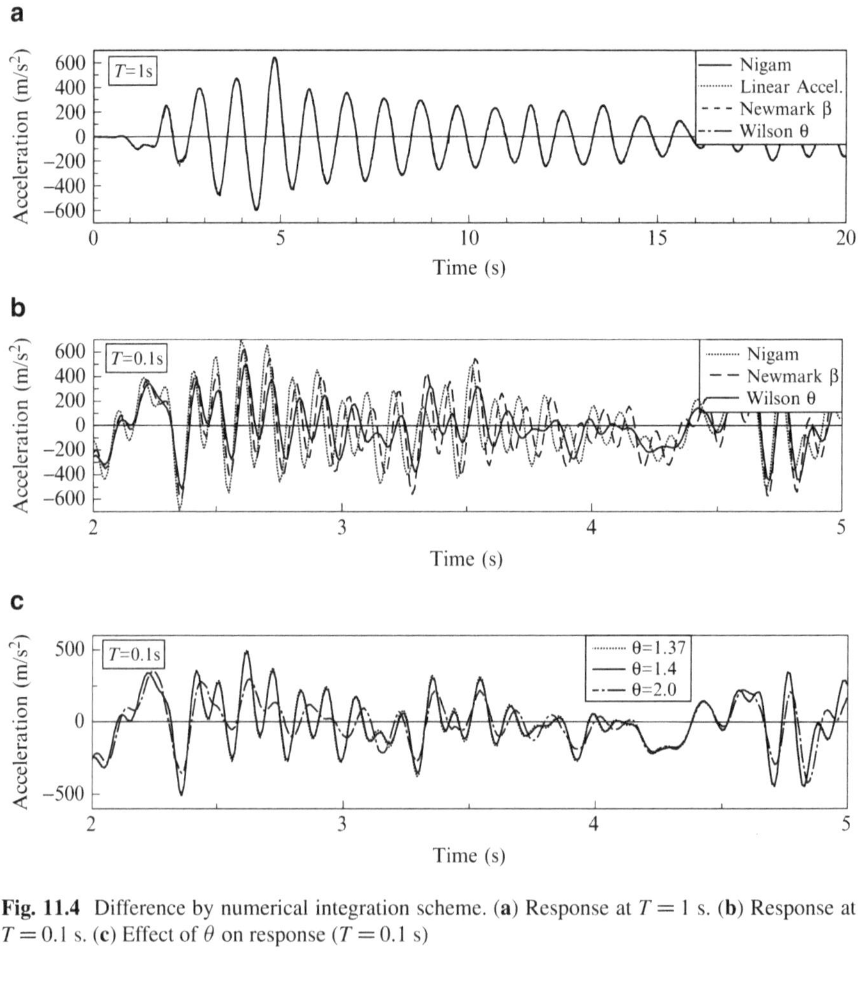

Fig. 11.4 Difference by numerical integration scheme. (**a**) Response at $T = 1$ s. (**b**) Response at $T = 0.1$ s. (**c**) Effect of θ on response ($T = 0.1$ s)

Centro wave with time increment 0.01 s. Four integration methods, Nigam, linear acceleration, Newmark's β ($\beta = 1/4$) method, and Wilson's θ ($\theta = 1.37$) method, are used. Figure 11.4a shows the results of a system with natural period $T = 1$ s. All the responses agree well and there is no difference, which is an identical observation as shown in Fig. 11.2.

Differences begin to appear in Fig. 11.4b, in which natural period T is 0.1 s. It is noted that abscissa is enlarged so that difference can be seen clearly. The result of the linear acceleration method is not shown because it diverged. Some phase difference is seen in Newmark's β method, and amplitude is similar with Nigam's method (exact solution). On the other hand, the difference both on phase and amplitude is seen in Wilson's θ method. Natural period of 0.1 s is just 10 times of time increment of the input earthquake motion. This frequency range is usually out of consideration in the engineering practice, but behavior in this frequency range cannot always be negligible. Extraordinary behavior in this frequency range may affect the overall behavior as the natural periods in the higher mode are generally smaller than this period.

Figure 11.4c shows the effect of the θ value in Wilson's θ method. Attenuation of response becomes larger as θ increases. This is the reason why $\theta = 1.37$ or 1.4 is usually used in the engineering practice.

Three factors, accuracy, stability, and computing time, are to be considered to choose numerical integration scheme. They are summarized in Table 11.1.

Newmark's β method ($\beta = 1/6$) or linear acceleration method is good at accuracy, but it is hardly used in the practice because of the stability problem. It is not realistic to use smaller time increment so that stability criterion is satisfied.

Both Newmark's β method ($\beta = 1/4$) and Wilson's θ method have advantage on stability, especially that unconditionally stable scheme can be used. Instead, error is largest. In addition, high-frequency behavior is suppressed especially by Wilson's θ method. However, they do not become problem in many practical calculations.

There is no unconditionally stable scheme in the central difference method. Therefore, time increment is to be very small. However, the computing time can be saved very much at the same time. Therefore, it may be used in the analysis of the ground, but cannot be used when structure is included.

Error caused by the numerical integration is one of the most important interests to the engineer. Research by Sakai et al. (2005) may help on this topic. Many numerical methods are compared in this research, and it is concluded that behavior in the interested frequency range for the ordinary civil or structural engineers is almost identical although difference appears at high-frequency range. In addition, solutions tend to be identical as the time increment becomes small.

Table 11.1 Choice of numerical integration scheme

	Newmark's β $\beta = 1/6$	Newmark's β $\beta = 1/4$	Wilson's θ	Central difference
Accuracy	○	△	△	○
Stability	×	○	○	×
Computing time	△	△	△	○

This observation can be used to the ensure error of the analysis by conducting an analysis with a half time increment. In many cases, results with regular time increment and a half of it are almost identical. If difference is not negligibly small in two analyses, there are some problems in using the program, such as choice of the numerical integration scheme, method to solve the nonlinear equation, etc. If possible, calculation with 1/5 or 1/10 time increment is better. Smaller time increment requires more time, but it is valuable work to ensure the numerical error of the analysis.

11.2 Frequency Domain Analysis

Partial differential equation with time and space is solved by a variable separation method. It is assumed that solution can be separated into time and space as

$$u(x,t) = U(x)\mathrm{e}^{i\omega t} \tag{11.24}$$

where U is a function of space only. Variable $\mathrm{e}^{i\omega t}$ indicates sinusoidal function from Euler's equation (Eq. 8.40). Therefore, if ω is constant, the system vibrates only one frequency, and it is obviously not an actual solution. Exactly speaking, therefore, Fourier series expansion is applied in Eq. (11.24), i.e.,

$$u(x,t) = U_0(x) + \sum_{n=1}^{N} U_n(x)\mathrm{e}^{i\omega_n t} \tag{11.25}$$

where ω_n is expressed by an integer n and duration of the earthquake T as

$$\omega_n = \frac{2\pi n}{T} \tag{11.26}$$

The mathematical meaning of this equation is schematically shown in Fig. 11.5. The original function defined in the period T is redefined as a periodic function with period T. The function is expressed by a trigonometric function with period T/n. As each trigonometric function is independent from the trigonometric functions with different period, it is sufficient to write for a representative period.

Substitution of Eq. (11.25) into equation of motion (Eq. 9.1) yields

$$\left(-m\omega_n^2 + ic\omega_n^2 + k\right)U_n = -mU_{n,g}\ (n = 1 - N) \tag{11.27}$$

where $U_{n,g}$ denotes a coefficient of the nth Fourier term of the input earthquake motion. Equation (11.27) is a differential equation with respect to space and does not include time. There are N terms and constant term, among which constant term is usually neglected. Then, the displacement starts from zero and ends to zero. It means that residual deformation does not remain.

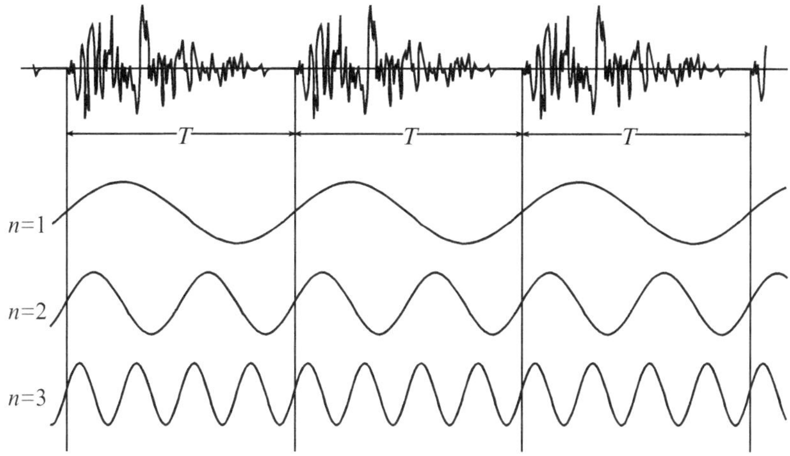

Fig. 11.5 Schematic figure showing the method of separation of variable or Fourier series expansion

Displacement is separated by Eq. (11.25) and solved frequency by frequency by Eq. (11.27). In other words, a variable separation method transfers the time domain analysis into the frequency domain analysis. Therefore, this method is frequently called a frequency domain analysis.

Complex representation is used in Eq. (11.24) because the stress–strain relationships are expressed by complex numbers in order to consider the hysteretic damping by the linear stress–strain relationships. Real part of complex corresponds to the requested solution.

Error can become small by increasing N. However, high-frequency response is not important in the engineering practice. Therefore calculation time can be saved by considering the interested frequency region, which is an advantage of this method. In addition, use of the fast Fourier transform (FFT) reduces computing time significantly when N is a power of 2. A subroutine for FFT is shown by Ohsaki (1994) in which there are only 30 lines.

There are many advantages in a frequency region analysis from the point to solve differential equations, but there is a big disadvantage. As shown in Eq. (11.25), solution in time domain is obtained by summing up the response in different frequencies, which is possible when the system is linear. In other words, stress–strain relationships must be linear. An equivalent linear technique described in Sect. 11.4 is used to transfer the nonlinear system into the linear system.

There are some notes to use the frequency domain analysis.

The highest frequency that can be considered in the analysis is 0.5/dt when time increment is dt, which frequency is called a Nyquist frequency. Frequency a little smaller than the Nyquist frequency is reproduced in the engineering sense. Since the Nyquist frequency is usually much larger than the natural frequency of the ground and the structure, the engineer need not consider it.

Another issue is duration of the analysis. It must be longer than the duration of the input motion because of the following reasons. As mentioned before, the computing time reduces significantly when N is a power of 2; computing time increases very much when N is not a power of 2. The computing time is, for example, 28, 77, and 460 times when N is 4,097, 4,098, and 4,099, respectively, compared with the computing time when N is 4,096 (Ohsaki 1994). In addition, N that is not a power of 2 cannot be accepted in many computer programs. Therefore, trivial zeroes are to be added at the end of the input earthquake motion in order to make N to be a power of 2, which is called as trailing zeroes. Then the most efficient N seems to be the power of 2 that just exceeds the number of input earthquake data, but it may be insufficient.

As the duration of the analyzed time is divided by N in the frequency domain analysis, there are $N+1$ points in the analyzed region. It indicates that $N+1$-th point is the end point in the considered region, but it is, at the same time, the starting point in the next cycle as schematically shown in Fig. 11.5; behavior in duration T is not independent, but interacts by the region before and after the region.

A ground is analyzed to see the effect of the interaction in Fig. 11.6, in which thickness of the ground $H = 5$ m, S-wave velocity $V_s = 100$ m/s, predominant natural period $T_0 = 0.2$, and damping ratio is 2 %. The first 120 data (1.2 s) of the El Centro wave is used with adding trailing zeroes. Figure 11.6a is the result when $N = 1{,}024$ (duration $T = 10.24$ s). Free vibration nearly stops and the system is almost quiet at the end of the analysis. In the same manner, analysis is made by $N = 128$, 256, and 512. Responses at the first 2 s are compared in Fig. 11.6b; there

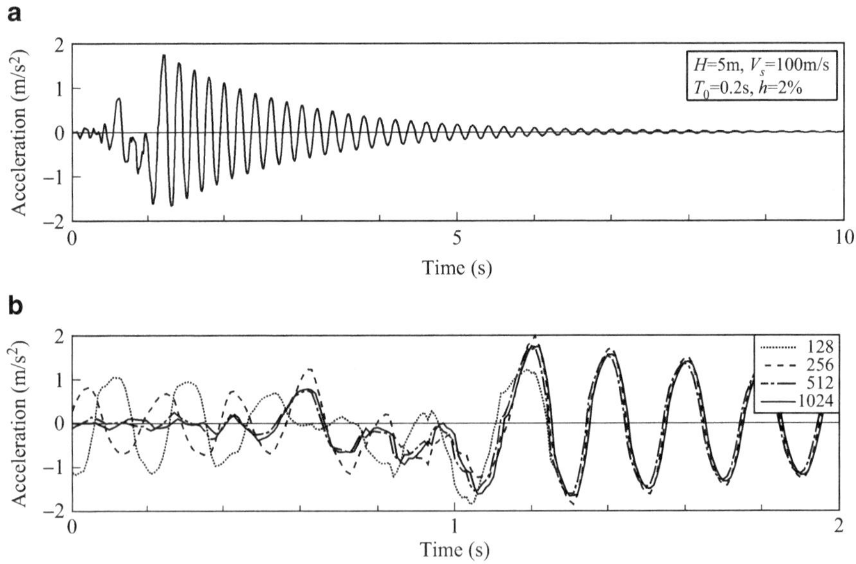

Fig. 11.6 Effect of trailing zeroes. (**a**) $N = 1{,}024$. (**b**) Effect of time increment

are significant differences between the results. As free vibration does not terminate when N is less than 1,024, free vibration response comes into the next analyzed region, which is the reason why significant difference appears at the beginning of the analysis.

As shown in this example, the length of the trailing zeroes is to be determined so that free vibration stops. This length is not very large in many seismic response analyses as input earthquake motion gradually tends to zero at the end of the earthquake. Termination of the free vibration cannot be confirmed unless time history is drawn, which may be additional job. Instead of it, check of the response at time $t=0$ may be useful, which is usually zero. If it is not the negligible order, therefore, duration may be short.

The last note may not be important. Response at time $t=0$ is usually assumed to be zero in the step-by-step time integration method. Therefore, the first data of the input earthquake motion is at $t=\mathrm{d}t$, and output of the response starts from $t=\mathrm{d}t$, although some computer programs output the response at $t=0$ to be zero. On the other hand, the first data corresponds to $t=0$ in the frequency domain analysis. Therefore, there are one time interval differences between the time domain and the frequency domain analyses. However, computer program may use different assumptions. For example, instead of quiet state at $t=0$, $t=-\mathrm{d}t$ can be used in the time domain analysis. Zero input motion can be added at the beginning of the input motion in the frequency domain analysis. Depending on the differences the output time may differ by one time increment. Difference of the time difference of $\mathrm{d}t$ is not a big issue in the practical problem. However, if this difference becomes an important issue depending on the problem, it is suggested to check the theory or practical procedure used in the computer program.

11.3 Multiple Reflection Theory

This method is used only in the frequency domain analysis of the one-dimensional ground. Equilibrium of an infinitesimally small element is already shown in Fig. 1.15a, and equilibrium condition is derived in Eq. (1.8), which was

$$\frac{\partial \tau}{\partial z} = \rho \frac{\partial^2 u}{\partial t^2} \tag{1.8}$$

This equation is called a wave equation as explained before. Substitution of the stress–strain relationships in the complex representation (Eq. 8.43) and the strain–displacement relationship

$$\gamma = \frac{\partial u}{\partial z} \tag{11.28}$$

result in the following equation:

$$\frac{\partial^2 u}{\partial t^2} - \frac{\overline{G}}{\rho}\frac{\partial^2 u}{\partial z^2} = 0 \tag{11.29}$$

Here subscript 0 in Eq. 8.43 is omitted because there is no confusion. Equation (11.29) is an equation of motion with respect to displacement.

This equation is solved by the variable separation method. Procedure is the same as before. Then the differential equation in space yields

$$\frac{\mathrm{d}^2 U}{\mathrm{d}z^2} + \frac{\rho\omega^2}{\overline{G}}U = 0 \tag{11.30}$$

This equation is solved in a small region in the FEM, but it is solved analytically in the multiple reflection theory. General solution of this equation is expressed as

$$U(z) = E\mathrm{e}^{ikz} + F\mathrm{e}^{-ikz} \tag{11.31}$$

where E and F are integral constants, and

$$k = \omega\sqrt{\frac{\rho}{\overline{G}}} \tag{11.32}$$

is called a complex wave number, which is the number of waves in a unit length. General solution of the original differential equation is obtained by multiplying the time term as

$$u(z,t) = E\mathrm{e}^{i(kz+\omega t)} + F\mathrm{e}^{-i(kz-\omega t)} \tag{11.33}$$

Let us consider a term including integral constant E to see the nature of this solution. The location of the wave is discussed when time advances (t increases). In order to keep the exponent constant, z is to become smaller when t increases. As the positive z direction is downward, the location of the wave is upward. In other words, the term including E is propagating upward, or it is called an incident wave. In the same manner, terms including the integral constant F propagate downward, or it is a reflected wave. The name of the multiple reflection comes from the method of analysis in which waves multiply reflected at the layer boundaries are considered (Sezawa 1932).

Integral constants E and F are amplitude of the incident and reflected waves, respectively. Since they are integral constant, any other symbol can be used instead of E and F. In the engineering practice, however, E and F are commonly used. Moreover, E and F are frequently used to indicate incident and reflected waves, respectively, such as a common noun. For example, $E+F$ is used for composite wave and $2E$ for outcrop motion (double of the incident wave). As described in Sect. 10.4.2, $E+F$ is specified as input motion into the rigid base and $2E$ is specified as input motion at the elastic base.

The displacement and the shear stress are to be continuous at the boundary between the layers. Among them, continuity condition of the displacement is expressed as

$$u_m\,(z = h_m, t) = u_{m+1}\,(z = 0, t) \tag{11.34}$$

where coordinate z starts at the top of each layer. Substitution of this equation into Eq. (11.33), one obtains

$$E_m \mathrm{e}^{ikh_m} + F_m \mathrm{e}^{-ikh_m} = E_{m+1} + F_{m+1} \tag{11.35}$$

In the same manner, continuity condition of the shear stress result in

$$G_m^* k_m \left(E\mathrm{e}^{ikh_m} - F\mathrm{e}^{-ikh_m}\right) = G_{m+1}^* k_{m+1}\,(E_{m+1} - F_{m+1}) \tag{11.36}$$

Then one obtains the continuity condition as

$$\begin{aligned} E_{m+1} &= \tfrac{1}{2} E_m\,(1 + \alpha_m)\,\mathrm{e}^{ik_m h_m} + \tfrac{1}{2} F_m\,(1 - \alpha_m)\,\mathrm{e}^{-ik_m h_m} \\ F_{m+1} &= \tfrac{1}{2} E_m\,(1 - \alpha_m)\,\mathrm{e}^{ik_m h_m} + \tfrac{1}{2} F_m\,(1 + \alpha_m)\,\mathrm{e}^{-ik_m h_m} \end{aligned} \tag{11.37}$$

where

$$\alpha_m = \frac{k_m G_m^*}{k_{m+1} G_{m+1}^*} \tag{11.38}$$

is an impedance ratio between two layers.

When the number of the layers above the base is set N, there are $N + 1$ layers including the semi-infinite region beneath the base. There are $2(N + 1)$ unknowns as each layer has incident and reflected wave. On the other hand, displacement and stress continue at the boundary between the layers, which gives $2N$ equations. In addition, $E = F$ at the ground surface. Then, there are $2N + 1$ equations, and one condition is short for $2(N + 1)$ unknowns. The last condition is given as magnitude (amplitude) of the wave. It can be E, $E + F$, or F at any layer. Therefore, for example, it is possible to specify the earthquake motion observed at the ground surface or in the underground in order to obtain the incident wave into the base. This process is called a deconvolution, whereas ordinary analysis in which earthquake motion is specified at the base is called convolution. The multiple reflection theory is a useful tool in the engineering practice because a deconvolution analysis is possible with ease.

The multiple reflection theory is employed in SHAKE (Schnabel et al. 1972), DYNEQ (Yoshida et al. 2002), and FDEL (Sugito et al. 1994), which will be explained in the next section.

The idea how to divide a layer in the step-by-step time integration method is shown in Sect. 10.3, but it is quite different in the multiple reflection theory because

	Model/method	12	22	44
Accel. (m/s^2)	Hyperbolic	0.747	0.615	0.632
	Ramberg-Osgood	0.204	0.151	0.192
	Equivalent linear	0.367	0.367	0.367
Vel. (m/s)	Hyperbolic	0.670	0.521	0.535
	Ramberg-Osgood	0.120	0.116	0.112
	Equivalent linear	0.307	0.307	0.307
Disp. (cm)	Hyperbolic	1.828	1.394	1.418
	Ramberg-Osgood	0.253	0.249	0.247
	Equivalent linear	0.401	0.401	0.400

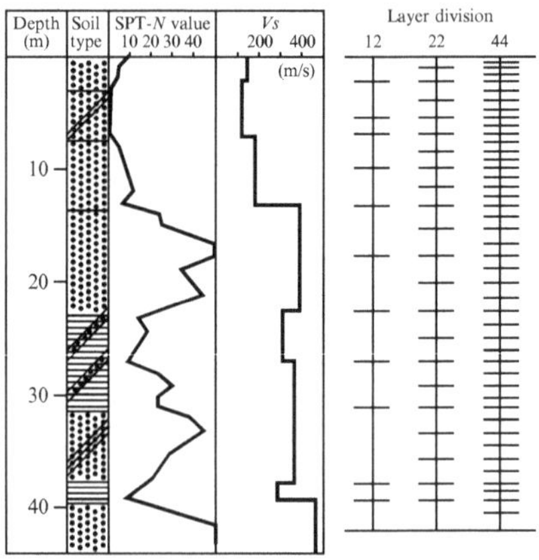

Fig. 11.7 Effect of layer subdivision

analytical solution is obtained. Therefore subdivision of a layer is not necessary in the elastic analysis even if thickness of the layer is 1 km, for example. In the nonlinear analysis, however, subdivision is required from different background.

Material property changes when nonlinear behavior takes place. Then material property becomes to depend on depth even in the same layer; mesh subdivision becomes to be required. Here, the thickness of the sublayer is to be determined so that the property at the middle of the sublayer is representative properties of the sublayer. Thickness of the sublayer based on this requirement can be larger than that required in the step-by-step time integration method. Figure 11.7, for example, compares the effect of the layer subdivision by different methods (Kokusho et al. 1995). Here, the hyperbolic and the Ramberg–Osgood are names of the stress–strain model in the step-by-step time integration method, and the result depends on the number of layers significantly. On the other hand, the result of the equivalent linear method using the multiple reflection theory is not affected by the layer subdivision. It is noted that there are significant differences among three methods, which come from the difference of the stress–strain model, but are not discussed here.

11.4 Equivalent Linear Method

The stress–strain relationships must be linear in the frequency domain analysis. On the other hand, the soil shows nonlinear behavior from the small strains. The equivalent method is a tool to obtain linear stress–strain relationships from the nonlinear ones. Therefore, it is not "equivalent" in the mathematical sense that two methods show exactly the same results, but is just an approximate method. Because of this reason, the author does not like this name, but this name already becomes like a common noun.

Mechanical properties (stiffness and damping) are kept constant in the whole duration of the earthquake. As the Fourier series expansion is used in the equivalent linear method, there is no residual deformation. Therefore, the applicability of this method decreases when material property changes monotonically or changes significantly. The liquefaction problem fits this condition, in which stiffness and strength reduce as the decrease of the effective confining stress changes significantly to nearly zero when liquefaction occurs. Therefore, it is natural to consider that the equivalent linear method cannot be used in the liquefaction method. However, probably because it is an easy handling method as explained later, there are not a few reports indicating that equivalent linear analysis can be applied or has a possibility to liquefaction problems. Sometimes, it uses extraordinary mechanical properties (Yoshida 1994). Sometimes, data is adjusted as the result is known prior to the analysis. As far as the author knows, there is no method that shows the procedure to evaluate material property used in the analysis by building up from fundamental data. Since this book does not deal with effective stress analysis, no more discussion is made on the applicability to the liquefaction problem.

The equivalent linear method has two big advantages in the engineering practice. The first one is the stability of the numerical integration, and the other is the preparation of the input data on the mechanical property. As there is no stability problem that the step-by-step time integration has as described in Sect. 11.1.2, the engineer need not consider divergence. There are cases when it diverges (see Sect. 11.4.1) because of different reasons, but it occurs very few. Mechanical properties (strain-dependent modulus and damping ratio) can be input as table type, whereas the procedure to evaluate the model parameters from the table-type strain-dependent modulus and damping ratio is necessary in the nonlinear analysis. Because of these advantages, the so-called engineering judgment is hardly required, which makes the equivalent linear method an easy handling method. However, as shown in Sect. 15.3 in which significant differences exist in the analysis by using SHAKE in the blind test, the engineering judgment is important for building the model the same as the nonlinear method.

11.4.1 Method in SHAKE

SHAKE (Schnabel et al. 1972) is one of the most popular computer programs based on the equivalent method. The name of SHAKE is so famous that it is frequently used as common noun although it is the name of a particular program. For example, the statement "the analysis is made by using SHAKE" frequently means that calculation is made by a computer program different from SHAKE, but the method is the same with SHAKE. Actually, there are many computer codes that use the same procedure with SHAKE or have a function to simulate SHAKE as one function of the program. Many engineers already did not see the original SHAKE. This book stands for the same standpoint; the name of SHAKE is used as a common noun of the method that is shown in this subsection.

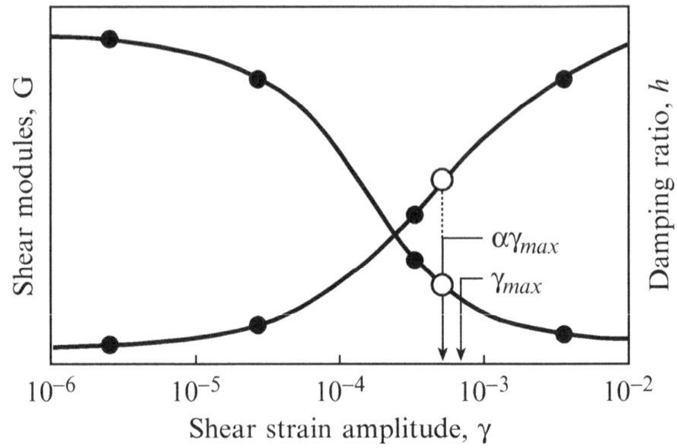

Fig. 11.8 Method to evaluate effective strain and strain-dependent properties

As described in Sect. 8.5 and in the preceding section, only *G* and *h* are necessary as the mechanical properties in the equivalent linear method. They are evaluated by the effective shear strain γ_{eff}, which is calculated from the maximum shear strain $\gamma_{\max}$ as

$$\gamma_{\text{eff}} = \alpha\gamma_{\max} \tag{11.39}$$

where α is a correction factor. There are several researches on the relevant value of α (see Sect. 11.4.3), but 0.65 is usually used in the engineering practice. In this sense, α is treated as a constant.

The procedure is schematically shown in Fig. 11.8. As the maximum strain is not known at the beginning of the analysis, relevant value such as values at small strain or specified strain is assumed in the first analysis. Both *G* and *h* are calculated at the end of the analysis by using the effective strain obtained at the analysis. The job terminates when difference between the used and new *G* and *h* is within the allowable error. If error is large, γ_{eff} is reevaluated from Eq. (11.39), and the same procedure is repeated. Difference of 5 % is frequently used as an allowable error.

As this procedure hardly diverges according to the author's experience, it is a good method. However, there are cases when convergence is not obtained, mainly because of the incorrect setting of the ground condition and mechanical property. One of the reasons comes from the limitation of SHAKE and will be explained later. Another case occurs in the deconvolution analysis in which acceleration at or near the ground surface is given as input motion. Specification of the acceleration is equivalent with the specification of the inertia force. If shear strength of a layer is smaller than the inertia force, displacement increases as iteration and analysis do not converge. The last case occurs when there are thick layers whose S-wave velocity is very small.

It may be difficult to recognize why the last case causes divergence unless actual procedure in SHAKE is understood. In Sect. 1.3, it is shown that the earthquake motion attenuates at the ground surface when there are layers with small shear strength. Because of the same reason, the earthquake motion attenuates to the ground surface when the soft soil deposited. From the different sayings, earthquake motion amplifies into the underground. When solving the equation of motion in the frequency domain analysis, $E = F = 1$ is assumed at the ground surface at first. Then E and F in the lower layer are obtained from the continuity condition (Eq. 11.37). When this recurrent procedure reaches to the semi-infinite layer, the amplification ratios in the all layers are obtained. Then a factor is multiplied so that response at the layer where input earthquake motion is specified becomes relevant. In the first procedure, amplitudes E and F become larger and larger when there exist thick soft soil layers as earthquake motion amplified as depth. This calculation diverges when amplitude becomes larger than the number that the computer can handle (about 10^{38} in the ordinary case). According to the author's experience, this problem occurred in the risk analysis where the ground models are built mechanically and automatically, but never happened in the analysis of the individual ground because this kind of extraordinary ground is not modeled.

11.4.2 Limitation of SHAKE

There are two big limitations in SHAKE, which is an overestimation of the maximum acceleration and is an underestimation of the amplification at high frequency.

11.4.2.1 Overestimation of the Maximum Acceleration (Finn et al. 1978)

Figure 11.9 shows the mechanism of the overestimation schematically. Here the solid line is the specified stress–strain curve. The effective strain γ_{eff} is evaluated from the converged $\gamma_{\max}$. The linear stress–strain curve used in the analysis is

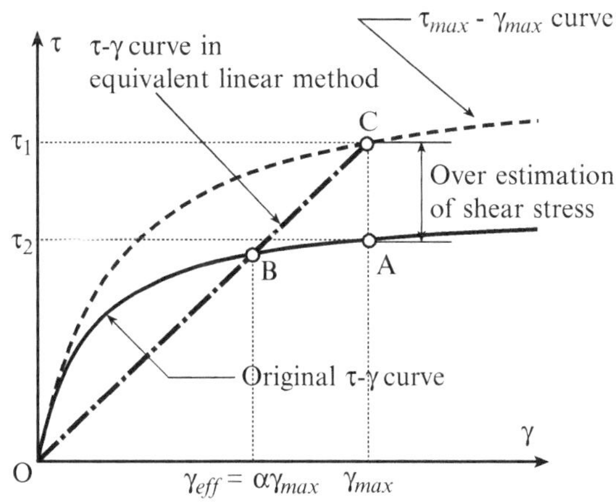

Fig. 11.9 Mechanism of overestimation of shear stress

defined to pass the stress at $\gamma = \gamma_{eff}$, which is the line OB in the figure (chained line). Therefore, maximum stress at $\gamma = \gamma_{max}$ is τ_1 at point C. On the other hand, relevant shear stress must be on the stress–strain curve at $\gamma = \gamma_{max}$, which is τ_2 at point A. It indicates that maximum stress is overestimated. The maximum stress–maximum strain curve is shown in Fig. 11.9 by the dotted line. The maximum overestimation occurs when stress–strain curve is perfectly plastic state; the overestimation is $1/0.65 = 1.54$ under $\alpha = 0.65$.

The maximum acceleration at the ground surface is shown to be controlled by the shear strength of the weakest layer in Sect. 1.3. This indicates that overestimation of the shear stress result in the overestimation of the acceleration above this layer. As shown in Sect. 14.1, this overestimation is clearly observed when the maximum strain is about 0.08 %.

The fact that SHAKE overestimates maximum acceleration is known long time ago. Figure 11.10, for example, shows the comparison of three computer programs (Iwasaki et al. 1980). Here, DESRA and CHARSOIL are computer programs that use step-by-step time integration scheme in which the hyperbolic and the Ramberg–Osgood models are used, respectively. Model parameters are determined so that stress–strain curve becomes nearly the same. Figure 11.10 shows that the maximum shear stress is much larger by SHAKE than those by other two programs. Acceleration response is larger in period between 0.3 and 0.5 s in SHAKE, too. It was explained that resonance occurs in the particular frequency range because the equivalent linear analysis is a linear analysis. The same opinion is also suggested by the Japanese researchers (Kokusho 1982; Masuda et al. 2001). As shown in Sect. 14.1, however, this opinion is not correct and the mechanism shown in Fig. 11.9 is reasonable.

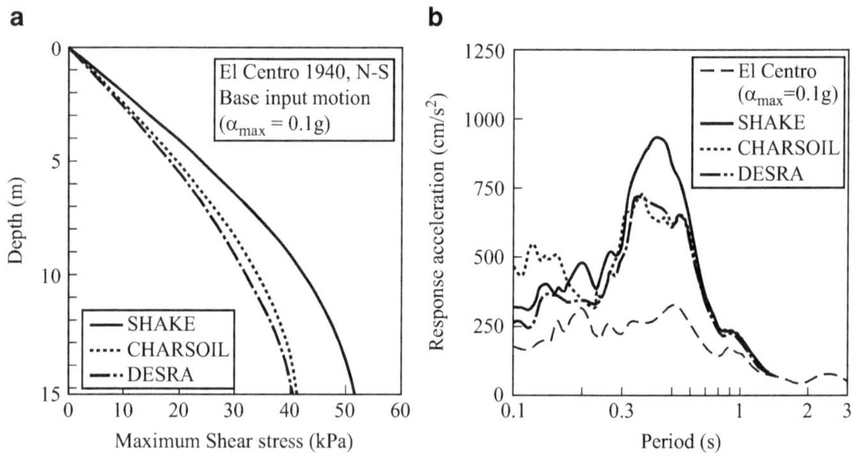

Fig. 11.10 Comparison between equivalent linear and truly nonlinear methods. (**a**) Maximum shear stress. (**b**) Response spectrum

The reason why the maximum stress is larger than the stress at the maximum strain seems to be simple; stress–strain curve is not drawn. The shear stress was calculated as a product of shear strain and converged shear modulus in the original SHAKE. Obviously, the stress–strain curve is linear and there is no interesting place. On the other hand, shear modulus and damping ratio seems correct as it is evaluated from the effective strain. Therefore, correct stress–strain curve was believed to be used.

Relationships between the real part of the complex stress and the real part of the complex strain are to be drawn in order to obtain the elliptic shape stress–strain curve such as in Figs. 8.20, 14.5, 14.12, and 14.16.

11.4.2.2 Underestimation of Amplification

Another limitation of SHAKE is seen in Fig. 11.11, in which vertical array record during the 1987 Chibaken toho-oki earthquake at the fill of the Tokyo bay area is analyzed (Masuda et al. 2001). Amplification ratio by SHAKE is underestimated at frequencies higher than several Hz. This underestimation of amplification at high frequency is the second limitation.

Frequency higher than several Hz is not important in many engineering practices. Therefore, this limitation seems not to be important in the practice. It is true in the ordinary convolution analysis, but it becomes a significant problem in the deconvolution analysis. A typical example is seen in Fig. 11.12 in which vertical array record at Lotung, Taiwan, is analyzed. Two earthquake records, Nos. 11 and 16, were analyzed by the deconvolution. Good agreement is obtained in the No. 11 earthquake, but acceleration at deep depths is significantly overestimated in the No. 16 earthquake. Strain is much larger in the No. 16 earthquake than No. 12 earthquake, and nonlinear behavior is expected.

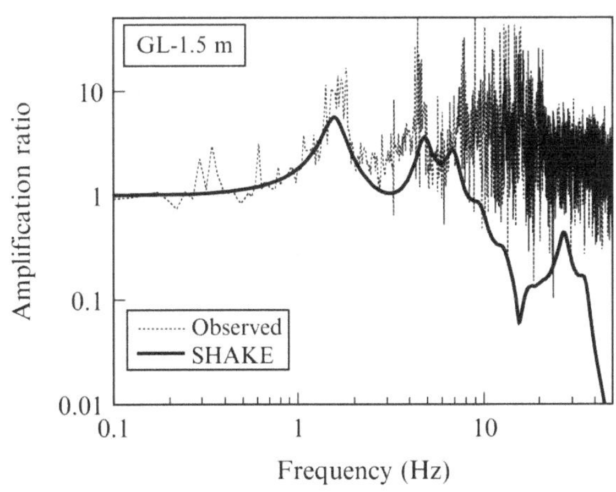

Fig. 11.11 Comparison of amplification ratio

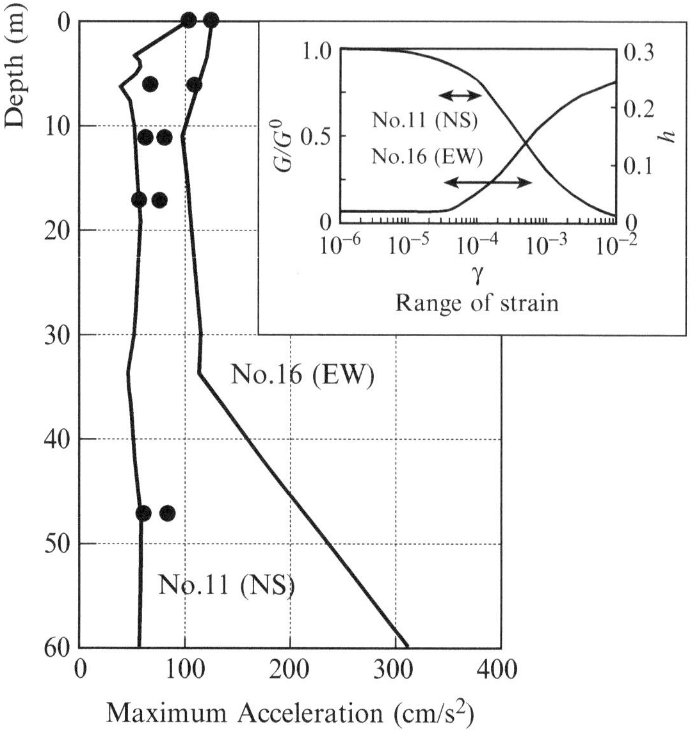

Fig. 11.12 Example of extraordinary deconvolution analysis

Similar behavior was known empirically. For example, a deconvolution analysis of the Hachinohe wave (see Chap. 3), which is one of the frequently used earthquake records in the earthquake resistant design in Japan, diverges when frequency up to the Nyquist frequency is considered. Frequencies higher than 10–15 Hz are to be neglected in order to obtain the converged solution.

These two behaviors are caused by the same mechanism. Amplification by SHAKE in Fig. 11.12 is less than unity at high frequencies. It indicates that the amplitude of the acceleration becomes larger in the underground. If this inverse amplification of acceleration becomes very large, it affects overall behavior causing larger acceleration at deep depths. Divergence may occur when this overestimation becomes too large as described at the last of Sect. 11.4.1.

11.4.3 Improvement of SHAKE

It is clear that the limitation of SHAKE comes from the method to evaluate the effective strain or the correction factor α. There are some researches in which relevant value of α is evaluated from the simulation of the vertical array records. The back analyzed α was 0.3 (Nozawa et al. 1987) and 0.4 (Tazoh et al. 1987), which are fairly smaller than the frequently used value 0.65 in the engineering practice. It is clear that this small α value results because earthquake motions used in the back analyses are small, in which case the contribution of high-frequency component to the acceleration becomes large. SHAKE, however, underestimates amplification of the high-frequency component significantly as the second limitation; α should be smaller to make high-frequency acceleration large.

On the other hand, α must be close to unity in order to avoid the overestimation of the maximum acceleration under large earthquakes. However, acceleration is underestimated when $\alpha = 1$ because of the underestimation of amplification at high frequency. Therefore, relevant value is usually a little smaller than 1, which is seen in Sect. 14.1.

As can be seen from these two extremes, α is to be smaller under the small or medium earthquakes, whereas it must be larger under the large earthquakes. Therefore, there is no relevant unique value. It may depend on the maximum strain, but there is no such research. The relevant α may be different between the convolution and deconvolution analyses even in the same site in the same earthquake.

Alternate methods are introduced in this section.

11.4.3.1 Effective Strain as a Function of Frequency

Kiyota et al. (Hagiwara and Kiyota 1992) pointed out the reason why SHAKE works well for stiff ground such as rock, but it does not work well in the soft ground that comes from the constant α value in all frequency region, and proposed a method to evaluate the effective strain; this method is named GRAMER (ground response analysis by modal equivalent elastic ratio). The system is separated into each mode, and effective strain is evaluated mode by mode. Since modal analysis was used, this method does not fit the frequency region analysis; it had not become popular. In addition, since effective strain is evaluated by the method the same with SHAKE, overestimation of the maximum acceleration remains as a limitation.

On the other hand, Sugito et al. (1994) expressed the effective strain by using the Fourier spectrum of the strain $F(\omega)$ and its maximum value $F_{\max}$ as

$$\gamma_{\text{eff}} = \alpha \gamma_{\max} \frac{F(\omega)}{F_{\max}} \tag{11.40}$$

where ω denotes a circular frequency. Since $F(\omega)$ is smaller at high frequency, effective strain also becomes small. This method is named FDEL (frequency-dependent equi-linearized technique).

Figure 11.13 shows a result of the site similar in Fig. 11.11 (Masuda 2001). Amplification at high frequency becomes large and the limitation of SHAKE is improved. In practice, the use of Eq. (11.40) may cause divergence of the calculation because $F(\omega)$ is not a smooth function but changes significantly as a function of ω

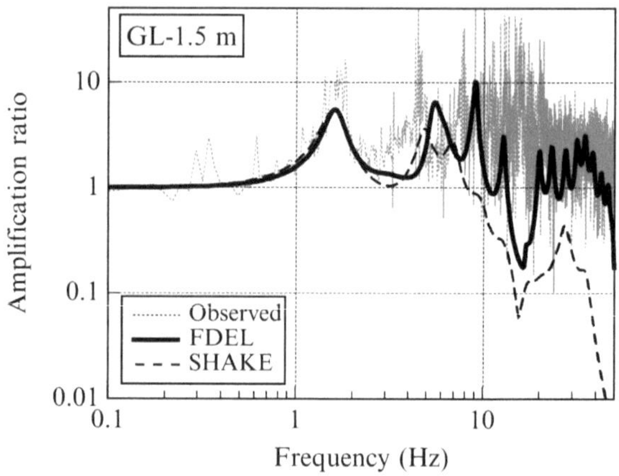

Fig. 11.13 Improvement by FDEL (Yoshida 1995)

and it changes very much at every iteration, too. This problem can be solved by smoothing the $F(\omega)$, and the triangular window was used in the original paper.

They carried a parametric study by changing α from 0.6 to 1.0 and pointed out that the disagreement of amplification at the predominant natural period becomes large when α is larger. Then they suggest that the value of α is relevant to be between 0.65 and 0.7. Therefore, the first limitation of SHAKE, overestimation of the maximum acceleration, is not solved, but accelerated because the acceleration increases more than SHAKE. In addition, problem shown in 14.3 may appear under the large earthquakes.

It can be understood that one of the limitations in SHAKE can be solved by introducing the frequency-dependent effective strain. However, another limitation of SHAKE, overestimation of the maximum acceleration, is not solved yet. In addition, if effective strain depends on frequency, S-wave velocity also depends on frequency. This indicates that the S-wave velocity has dispersibility, which is against a simple theory that the S-wave velocity is constant in all frequencies.[1]

It is well known that the damping term has frequency-dependent characteristics from the earthquake observation (see Sect. 12.2). Then, if the second limitation of SHAKE is solved by considering the frequency-dependent nature of the damping term, the problem of velocity dispersibility does not occur. Unfortunately, this trial did not work well.

Figure 11.14 shows amplification characteristics of the analysis at the site explained in Sect. 14.4 (Suetomi and Yoshida 1996). Here, three analyses were compared. They are SHAKE, FDEL, and FDEL-G in which stiffness is evaluated the same with SHAKE and damping is evaluated the same with FDEL. It is clearly

[1] S-wave velocity of the saturated soil depends on frequency because of the interaction between the soil particle and porewater in the exact sense (Miura et al. 2001), but it can be considered constant in the engineering practice as translation occurs at very high frequency. This is a quite different nature discussed here.

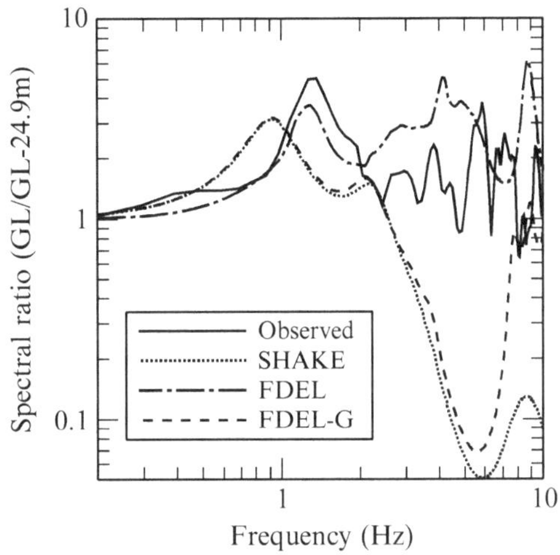

Fig. 11.14 Comparison with amplification ratio (Suetomi and Yoshida 1996)

seen that FDEL-G cannot improve the limitation of SHAKE to underestimate the amplification at high frequency. Therefore, frequency-dependent characteristics of the shear modulus are necessary to improve SHAKE.

11.4.3.2 Method by DYNEQ

DYNEQ (Yoshida 1995) (*dyn*amic response analysis of level ground by *eq*uivalent linear method) is an open source program developed by the author. The program has various functions the same with SHAKE, FDEL, etc., in which method by Yoshida et al. (2002) is introduced here. In order to distinguish this method from other methods, it is called DYNEQ in this book. It is noted that all analyses in the multiple reflection theory is made by the computer program DYNEQ in this book. Namely, SHAKE, FDEL, and DYNEQ are the name of the method but not the name of the computer programs.

It is clear that frequency dependency on both the stiffness and the damping must be considered in order to improve SHAKE. However, frequency-dependent stiffness is difficult to explain mechanically. This problem can be solved by considering the nature of the stress–strain relationships.

Figure 11.15 shows hysteresis loops at large and small strain amplitude schematically. The stiffness is large and the damping is small under the small strain amplitude, whereas the stiffness is small and the damping is large under large strain amplitude. Period under the small amplitude vibration is generally short, whereas it is long under the large amplitude. In order to see this explanation well, period of each half cycle of the strain waveform is evaluated by the zero-cross and double amplitude methods shown in Fig. 11.16a. As can be seen in Fig. 11.16b, which is the result of SHAKE shown in Sect.14.1, the shear strain amplitude becomes small as the frequency becomes large in general. Although the stiffness and the damping are

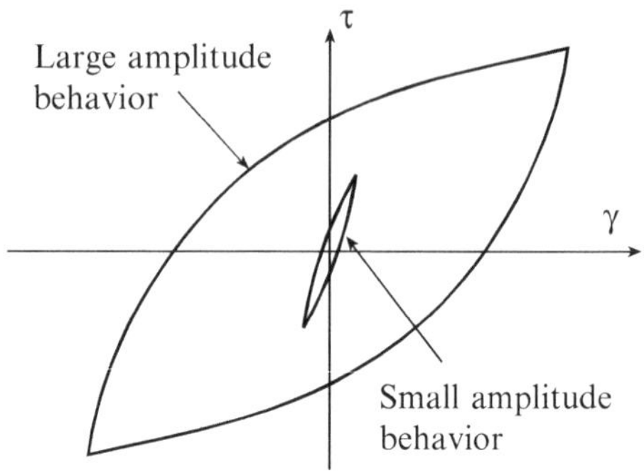

Fig. 11.15 Hysteresis loop at small and large strain amplitudes.

not frequency dependent, the shear strain has frequency-dependent characteristics as shown in this figure. Then, if these relationships are kept valid in the frequency region, SHAKE may be improved. In other words, strain-dependent characteristics appear as the result of mapping of the frequency-dependent strain characteristics in the time domain into the frequency domain. This is the basic concept in DYNEQ, which can explain the dispersibility of S-wave velocity.

Each method to evaluate the period of the wave shown in Fig. 11.16a has advantages and disadvantages. The zero-cross method is easy to understand, but it cannot catch the period well when strain drifts in one direction. On the other hand, large amplitude waves may not be caught when a cycle of strain reversal and reload with small amplitude occurs in the large strain amplitude. As there is no residual strain in the equivalent linear method, zero-cross method is used in the following calculations.

Figure 11.17 shows frequency-dependent characteristics by different earthquakes that will be explained as examples of the analysis in Sects. 14.1, 14.2, and 14.3. They all have similar tendency. By the way, the effect of the nonlinear behavior becomes prone at strains larger than 10^{-5}–10^{-4}; corresponding frequency is 5–15 Hz. Then effective strains are set zero at high frequency from the practical view point, and it is defined as

$$\begin{cases} \gamma_{\mathrm{eff}} = \gamma_{\max} & f_p > f \\ \gamma_{\mathrm{eff}} = \gamma_{\max} - \left(\frac{\log\ f - \log\ f_p}{\log\ f_e - \log\ f_p}\right)^m & f_p \le f \le f_e \\ \gamma_{\mathrm{eff}} = 0 & f > f_e \end{cases} \tag{11.41}$$

As schematically shown in Fig. 11.18, it is very simple. The effective strain γ_{eff} is constant at low-frequency region ($f < f_p$), and it is zero (elastic behavior) at high-frequency region ($f > f_e$). In the middle frequency region ($f_p \le f \le f_e$), a parabolic equation is used in logarithm expression (practically $m = 2$). Here f_p is a frequency at which shear strain amplitude is largest. On the other hand, f_e is a key frequency in this method. As mentioned before, if shear strains between 10^{-5} and 10^{-4} are boundary to the nonlinear behavior, f_e lies between 5 and 15 Hz. According to the past experience, frequency about 10 Hz seems to work well.

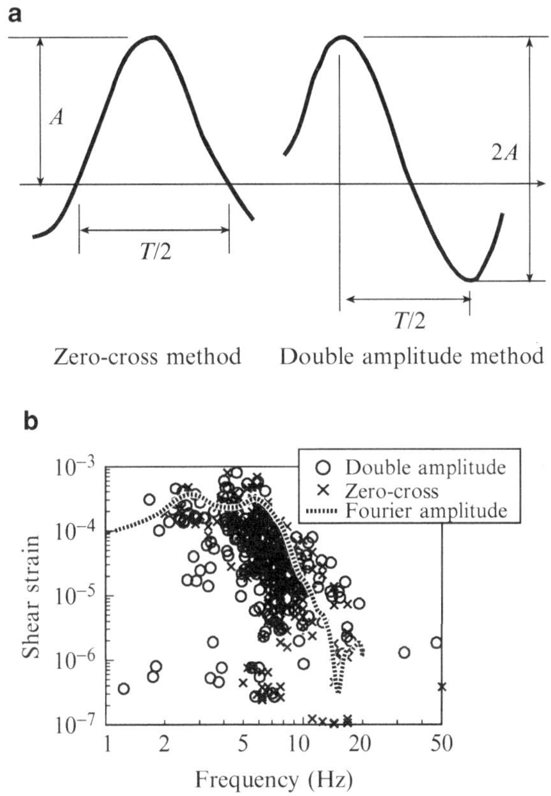

Fig. 11.16 Frequency dependency of shear strain amplitude (**a**) Evaluation of period. (**b**) Strain amplitude vs. frequency

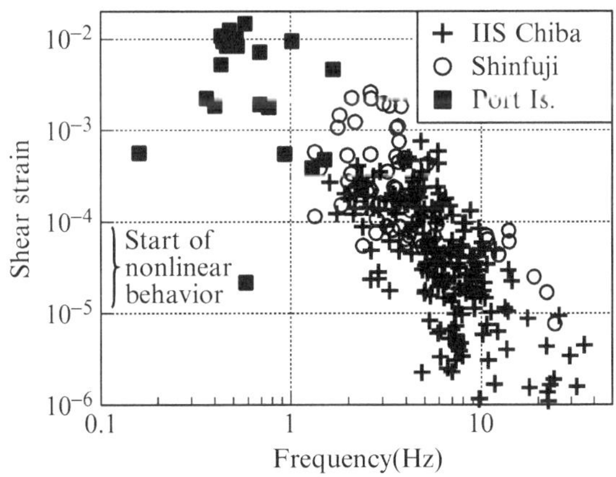

Fig. 11.17 Frequency-dependent characteristics under different earthquakes/sites

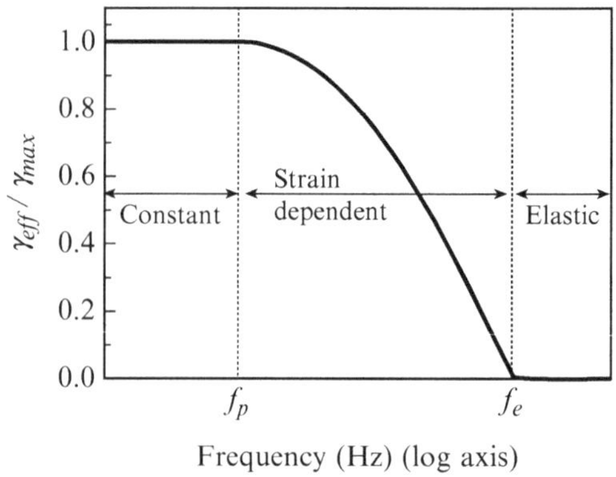

Fig. 11.18 Frequency-dependent effective strain

11.4.4 Equivalent Linear Analysis in Time Domain

The idea of the equivalent linear method can be applied for both the frequency domain and the time domain, but it is limited to the frequency domain analysis by means of the complex modulus in Japan. There are computer programs that use the equivalent linear method in the time domain analysis such as Hudson et al. (1994). This program is hardly used in Japan, but is sometimes used outside Japan as the source list is opened to public.

Technical difficulty may remain in evaluating the damping matrix so as to express the hysteretic damping in the velocity proportional damping. Three methods may be possible to build the damping matrix under the specified damping constant. It is explained in Chap. 12 in detail, and the referent section is shown here.

The first method is to use the Rayleigh damping in Sect. 12.2.1, in which damping ratio is determined based on the natural frequencies. The second method is also to use the Rayleigh damping, but the coefficients are determined from the interested frequency range, which is also described in the middle of Sect. 12.2.1. The coefficient of the damping matrix is specified element by element. These methods had a disadvantage that the damping ratio is frequency dependent. It can be solved by employing mode proportional damping explained in Sect. 12.2.2. This method, however, has a disadvantage that it requires huge computing time as the damping matrix is a full matrix.

11.4.5 Nonlinear Method and Equivalent Linear Method

When soil shows nonlinear behavior, mechanical property changes time by time. Therefore, an analysis that considers this instantaneous change is theoretically the most accurate method. This method is sometimes called a truly nonlinear method, but it is simply called a nonlinear analysis or a nonlinear method. The equivalent linear analysis or the equivalent linear method is also an analysis to consider the nonlinear behavior, but it is not called a nonlinear analysis in this book.

The nonlinear method has the following characteristics compared with the equivalent linear method:

1. Fast computing time

It is sometimes explained that the equivalent linear method is a fast method because it is a linear method and calculation time can be reduced by means of the fast Fourier transform (FFT). In addition, calculation time can be reduced by neglecting the high-frequency response. However, this is not true. Generally, the nonlinear method is much faster than the equivalent linear method. It comes from the total amount of output. All the output is obtained in one run in the nonlinear analysis. On the other hand, only Fourier amplitudes of the strain and the displacement are obtained at the end of the analysis in the equivalent linear method. Then if one wants peak acceleration, it is calculated from the displacement Fourier amplitude by differentiating with respect to time twice, which increases the amount of calculation.

In the old day when computer power is significantly less than present, engineer needs to satisfy himself by the limited number of output. The location of the layer where maximum output is required is to be specified through input data in the original SHAKE (Schnabel et al. 1972), for example, and there is no way to obtain displacement. The latter may come from another reason that original developers may feel that the equivalent linear method is not so accurate to predict the displacement. At present, however, not only the maximum accelerations but also other outputs such as maximum velocities, displacements, shear stresses, and shear strains are automatically output. Time histories of them may also be required. These outputs require huge computing time, which is the reason why the nonlinear analysis is faster than the equivalent linear analysis.

2. Difficult usage

The preparation of the input data for SHAKE is very simple. Stress strain curve can be input as table type; test data can be input as it is. Other data is also easy and simple. Therefore, there is no difficult engineering sense. In addition, there is no problem such as divergence.

On the other hand, there are many places where the engineering judgment is required in the nonlinear analysis. As stress–strain relationships are usually expressed by a mathematical equation (see Sects. 8.2, 8.3, and 8.4), engineers need to evaluate the value of parameters. It is necessary to evaluate the coefficient of the velocity proportional term (see Chap. 12 in detail) and to determine the numerical integral scheme (Sect. 11.1.1). The result of the analysis may depend on these parameters.

3. Theoretically accurate

The equivalent linear analysis should use the approximation of the stress–strain relationships developed by the equivalent linearization. On the other hand, the nonlinear analysis can consider mechanical properties that change momentarily.

It is, however, noted that nonlinear analysis is more accurate only when relevant stress–strain model is used. Therefore, the use of the nonlinear analysis does not imply more accurate result than the equivalent linear method.

4. Difficulty to express frequency-dependent characteristics

Frequency-dependent characteristics are difficult to handle in the nonlinear analysis because the analysis is made based on the instantaneous state.

Generally speaking, the applicability of the equivalent linear method is as follows. Case studies of the equivalent linear analysis will be shown in Chap. 14:

1. The applicability of SHAKE is frequently discussed by means of the strain, and it is a little larger than 0.01 % for a deconvolution analysis. The limitation of SHAKE becomes apparent at strains about 0.1 %.
2. The applicability of the improved equivalent linear method DYNEQ extends to strain larger than 1 %, which is nearly the same applicable strain of the nonlinear method whose limitation mainly comes from the production of the stress–strain behavior in test.
3. It is suggested to set $\alpha = 1$ to improve equivalent linear method FDEL. However, predominant frequency is not predicted well when $\alpha = 1.0$. If smaller value is used, overestimation of the maximum acceleration appears.
4. The applicability of the improved method for the convolution and deconvolution analyses is nearly the same.

References

Association for Disaster Prevention Research (1990) Report on investigation and research on development of two-dimensional liquefaction analysis computer program (in Japanese)
Clough RW, Penzien J (1975) Dynamics of structures. McGraw-Hill Kogakusha, Tokyo, 634 pp
Finn WDL, Martin GR, Lee MKW (1978) Comparison of dynamic analyses for saturated sands. In: Earthquake Engineering and Soil Dynamics, ASCE, GT special conference, vol 1, pp 472–491
Gudehus G (ed) (1977) Finite elements in geomechanics. Wiley, London, 673 pp
Hagiwara T, Kiyota Y (1992) Seismic response analysis based on equivalent linearization of soil constants in modal region Part I theory and example of inverse analysis, summaries of the technical papers of annual meeting of AIJ (Hokuriku), pp 487–488
Hudson M, Idriss IM, Beikae M (1994) User's manual for QUAD4M, a computer program to evaluate the seismic response of soil structures using finite element procedures and incorporating a compliant base. Center for Geotechnical Modeling, Department of Civil & Environmental Engineering, University of California, Davis
Iwasaki T, Kawashima K, Tatsuoka F (1980) Effect of strain dependence of soil properties on seismic response of subsurface grounds. Civ Eng J PWRI 22(12):27–32 (in Japanese)
Kokusho T (1982) Dynamic deformation characteristics of soil and nonlinear response of ground, Report no. 301, Central Research Institute of Electric Power Industry (in Japanese)
Kokusho T, Satoh K, Matsumoto M (1995) Nonlinear dynamic response of soil ground during 1995 Hyogo-ken nanbu earthquake. Tsuchi-To-Kiso 43(9):39–43 (in Japanese)
Masuda T (2001) Study on damage to electric facilities due to earthquake and remedial measures against it. Theses submitted to the University of Tokyo (in Japanese)

Masuda T, Yasuda S, Yoshida N, Sato M (2001) Field investigations and laboratory soil tests on heterogeneous nature of alluvial deposits. Soils Found 41(4):1–16
Miura K, Yoshida N, Kim Y (2001) Frequency-dependent property of waves in saturated soil. Soils Found 42(2):1–19
Newmark NM (1959) A method of computation for structural dynamics. Proc ASCE J Eng Mech Div 85:67–94
Nigam NC, Jennings PC (1964) Calculation of response spectra from strong motion earthquake records. BSSA 59(2):909–922
Nozawa T, Ohki H, Annaka T (1987) Analysis of amplification characteristics of ground by one-dimensional wave propagation theory considering frequency dependent characteristics of damping. In: Proceedings of the 19th JSCE earthquake engineering symposium 1987, JSCE, Tokyo, pp 109–112 (in Japanese)
Ohsaki Y (1994) Introduction of spectrum analysis of ground motion, new edn. Kajima Institute Publishing, Tokyo (in Japanese)
Sakai H, Yoshida N, Sawada S (2005) Numerical errors in non-linear earthquake response analyses caused by time integration procedures. J Struct Mech Earthq Eng Proc JSCE No. 794/I-72:291–300 (in Japanese)
Schnabel PB, Lysmer J, Seed HB (eds) (1972) SHAKE a computer program for earthquake response analysis of horizontally layered sites. University of California, Berkeley, Report No. EERC72-12
Sezawa K (1932) Theory of vibration, Iwanami Shoten, 799 pp; facsimile edition (1971), Gendai Kogaku-sha, Tokyo
Shibata A (2010) Dynamic analysis of earthquake resistant structures. Tohoku University Press, Sendai, 369 pp
Souza AFD, Garg VK (1984) Advanced dynamics, modeling and analysis. Prentice-Hill, Englewood Cliffs, pp 213–227
Suetomi I, Yoshida N (1996) A comparison of frequency-dependent damping in equivalent linear analysis. In: Proceedings of the 31st symposium of geotechnical engineering, JGS, Kitami, pp 1119–1120 (in Japanese)
Sugito M, Goda H, Masuda T (1994) Frequency dependent equi-linearized technique for seismic response analysis of multi-layered ground. J Geotech Eng Proc JSCE 493/III-27:49–58 (in Japanese)
Sato T, Hino T (1982) Error measure for time integration schemes in structural dynamics. In: Proceedings of the 6th Japan symposium of earthquake engineering, Tokyo, pp 1049–1056
Tazoh T, Sato M, Shimizu K, Koyama K, Watanabe O (1987) Effect of layer-subdivision and effective strain coefficient in nonlinear seismic response analysis. In: The 22nd Japan national conference of soil mechanics and foundation engineering, pp 655–658 (in Japanese)
Togawa H (1975) Vibration analysis by finite element method. Science-sha (in Japanese)
Yoshida N (1994) Applicability of conventional computer code SHAKE to nonlinear problem. In: Proceeding symposium on amplification of ground shaking in soft ground, JSSMFE, Tokyo, pp 14–31 (in Japanese)
Yoshida N (1995) DYNEQ a computer program for DYNamic response analysis of level ground by EQuivalent linear method, version 3.25 (September 2004), http://www.civil.tohoku-gakuin.ac.jp/yoshida-j.html/ [2010]
Yoshida N, Kobayashi S, Suetomi I, Miura K (2002) Equivalent linear method considering frequency dependent characteristics of stiffness and damping. Soil Dyn Earthq Eng 22(3):205–222, Elsevier

Chapter 12
Evaluation of Damping

The equation of motion is introduced in Chap. 9 as

$$m\ddot{u} + c\dot{u} + ku = -m\ddot{u}_g, \tag{9.1}$$

or

$$\ddot{u} + 2h\omega_0\dot{u} + \omega_0^2 u = -\ddot{u}_g \tag{9.2}$$

The damping term is recognized as the velocity proportional term in general. There is, however, more damping in the practice, which are listed in the following and will be explained in this chapter:

1. Hysteresis damping
2. Viscous damping (velocity proportional damping)
3. Damping caused by wave scattering (scattering attenuation)
4. Radiation damping
5. Numerical damping
6. Damping as alternative

12.1 Hysteresis Damping

The hysteresis damping is caused by the nonlinear characteristics of the stress–strain behavior. When the hysteresis loop has area, the mechanical energy is lost and the vibration becomes small. The lost energy is transferred into heat, which can be recognized when a wire's temperature increases as it is subjected to the cyclic bend. This damping is already explained in Sects. 7.2 and 8.2.2.

N. Yoshida, *Seismic Ground Response Analysis*, Geotechnical, Geological and Earthquake Engineering 36, DOI 10.1007/978-94-017-9460-2_12

12.2 Velocity Proportional Damping

The velocity proportional damping or the viscous damping is caused when the material with viscosity vibrates. There are two types of damping, which are named an external damping and an internal damping. The external damping is caused from the interaction between the considered body and outside of the system, which can be understood by considering the vibration in the water. In the actual situation, however, they are very small as the external system is air. On the other hand, the internal damping is caused by the difference of the internal velocity, which can be imagined by a dashpot. The damping of this mechanism is not well known at present although it is observed in the practical material. Even if it exists, it is difficult to distinguish it from the damping caused by the wave scattering. From this point, it can be discussed with the damping caused by the wave scattering as one thing, or both the internal damping and the damping caused by the wave scattering. This book stands for the former standpoint.

Both the hysteretic damping and the viscous damping are caused by the material property. By the way, the damping measured in situ and in the laboratory is compared in Fig. 12.1 (Kokusho 1986, 1992), but they do not agree. This indicates that there is damping that cannot be measured in the laboratory.

Two methods are known to build velocity proportional damping. Theoretically, it may be preferable to use the damping coefficient because it is related to the actual damping. However, discussions are made by means of the damping constant as it is very difficult to discuss using the damping coefficient.

12.2.1 Rayleigh Damping

The Rayleigh damping is the most frequently used damping because it is very easy to handle. The damping coefficient c is expressed by

$$c = \alpha m + \beta k \tag{12.1}$$

where m denotes the mass (matrix), k denotes the stiffness (matrix), and α and β are coefficients. Substitution of Eq. (12.1) into Eq. (9.2) yields

$$\ddot{u} + \left(\alpha + \beta\omega_0^2\right)\dot{u} + \omega_0^2 \dot{u} = 0 \tag{12.2}$$

Comparison of this equation with Eq. (9.2) leads to

$$h = \frac{\alpha}{2\omega_0} + \frac{\beta\omega_0}{2} \tag{12.3}$$

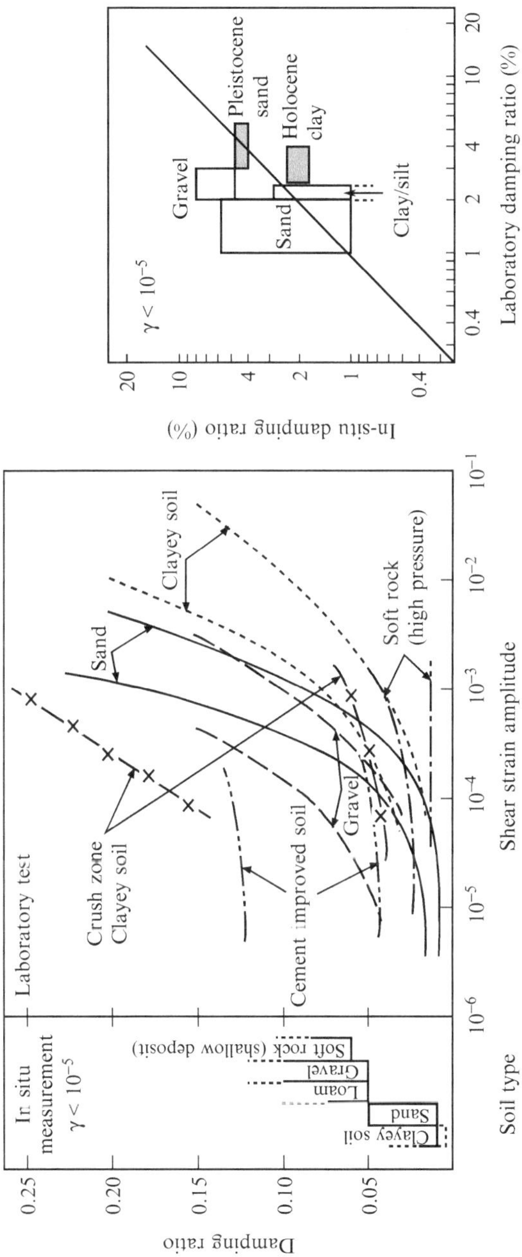

Fig. 12.1 Comparison of damping measured in situ and in laboratory

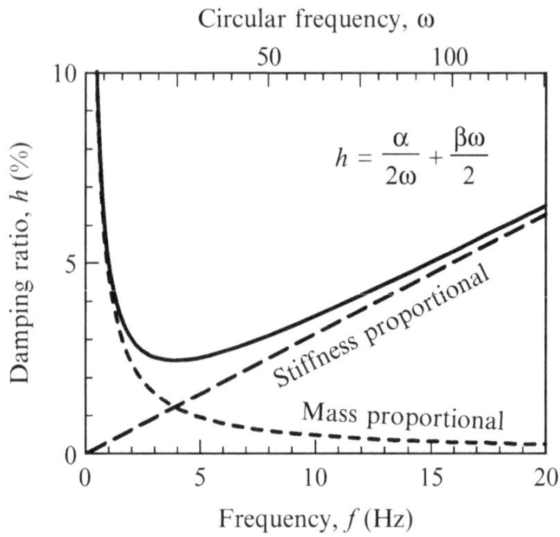

Fig. 12.2 Schematic figure showing image of Rayleigh damping

The multi-degree-of-freedom system can be separated to the group of the single-degree-of-freedom system by the eigenvalue analysis. Therefore, the following equation is obtained for multi-degree-of-freedom systems

$$h_i = \frac{\alpha}{2\omega_i} + \frac{\beta\omega_i}{2} \tag{12.4}$$

where subscript i denote ith mode quantities and ω_i denotes natural circular frequency of the ith mode. This relationship can be schematically shown in Fig. 12.2. Although ω_i are scattered quantities, they are written as continuum quantities in the figure. Each term of the Rayleigh damping is sometimes called a mass proportional or a stiffness proportional damping,

As the mass and stiffness proportional damping has characteristics inverse proportional and proportional to the (circular) frequency, addition of them results in concave curve as schematically shown in Fig. 12.2.

As described in the preceding, almost all the damping is to be expressed by the velocity proportional damping. Therefore, parameters α and β are to be evaluated so that they can express the actual damping. On the other hand, there are only two parameters that can be handled in the Rayleigh damping. Therefore, they are determined to focus in the frequency that is important in the engineering practice.

Deformations important in the engineering practice occur by the low-frequency mode. Since there are only two parameters in the Rayleigh damping, only two conditions can be satisfied at maximum. When the damping ratios at the first and second modes are specified, two parameters are calculated as

$$\alpha = 2\omega_1\omega_2\frac{h_1\omega_2 - h_2\omega_1}{\omega_2^2 - \omega_1^2}, \qquad \beta = 2\left(\frac{h_2\omega_2 - h_1\omega_1}{\omega_2^2 - \omega_1^2}\right) \tag{12.5}$$

This method is called method-1 hereafter. As explained in Sect. 15.4, the stiffness proportional damping is more important than the mass proportional damping. Therefore, β is frequently evaluated from the damping at the predominant period keeping $\alpha = 0$.

This method intends to reproduce the actual damping. One condition must be, however, satisfied so that this method works well; frequency where the damping ratio is specified is within the important frequency range. If this condition is not satisfied, this method does not work well, which can be seen in the case study in Sect. 15.6. In addition, eigenvalue analysis is to be made in order to obtain the natural period, although approximate method exists for the first natural period of the one-dimensional soil column (see Sect. 1.2.4).

The important frequency range is limited in the engineering practice. For example, JMA instrumental seismic intensity is developed based on the acceleration between 0.5 and 8 Hz, and spectral intensity uses pseudo velocity response spectrum in frequency 0.4–10 Hz. Since these indices are related to the earthquake damage, this frequency range can be applied in many structures.

The risk analysis by the Central Disaster Prevention Council against the Nankai and Tonankai earthquake (Central Disaster Prevention Council and Cabinet Office of Japan 2005), for example, the Q-value (see Sect. 12.3) of the surface layer is set 35, which corresponds to the damping ratio of 1.43 %. As described in Sect. 13.2, important frequency range in the engineering practice is between 0.5 and 6 Hz. Then the damping ratios at 0.5 Hz and 6 Hz are set 1.43 %, which leads to $\alpha = 0.083$ and $\beta = 0.00070$. However, the damping ratios between 0.5 and 6 Hz become less than the specified value; minimum damping ratio is 0.76 %. This may underestimate the damping ratio. Finally, parameters are set $\alpha = 0.109$ and $\beta = 0.00092$ by setting the minimum damping ratio to be 1 %, by which average damping ratio in this frequency range is about 1.4 %. This method is called method-2 hereafter. The advantage of the method-2 is that the same parameters can be used regardless of ground, whereas eigenvalue problem is to be solved site by site in the method-1.

Here, let us consider the mechanical meaning of the Rayleigh damping. Let mass matrix be a lumped mass (see Sect. 10.5.2); then the mass proportional damping works to be proportional to the absolute value of the velocity, or it corresponds to the external damping. On the other hand, as the stiffness proportional damping works by the relative velocities, it works similar to the internal damping. Therefore, Rayleigh damping has some mechanical meaning (Togawa 1975).

As the stiffness proportional damping works as an internal damping, it is theoretically to be constant such as stiffness; it need not be modified under the change of the analytical conditions such as change of the depth of the base and the movement from the one-dimensional to the two-dimensional analysis. However, it will change when the natural period changes by method-1. This contradiction is to be accepted because the parameters are determined through the natural periods.

Considering the problem explained in Sect. 15.6, in addition, method-2 looks more practical than the method-1. From the different point of view, however, the Rayleigh damping has an important role for the stability of the numerical integration. Therefore, it need not be an actual damping constant.

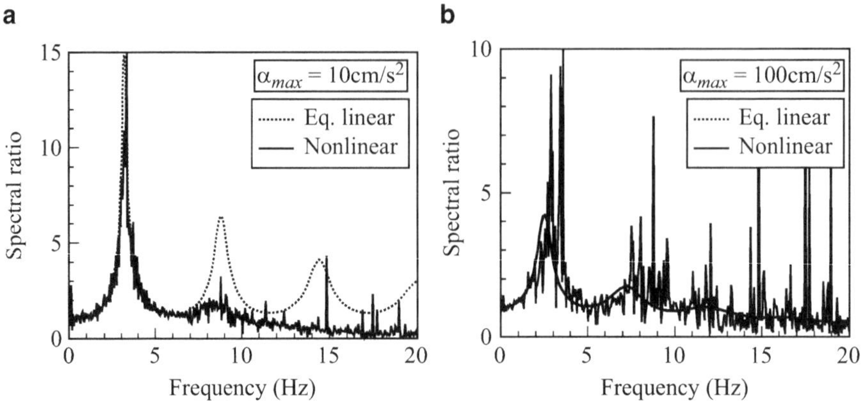

Fig. 12.3 Effect of Rayleigh damping. (**a**) Close to linear. (**b**) Nonlinear

Effect of the Rayleigh damping on the seismic response can be seen in Fig. 12.3 where model ground with 15 m thickness used by Finn et al. (1978) is analyzed. $\alpha_{\max}$ denotes the maximum acceleration of the input motion. Amplification ratio or spectral ratio of the Fourier amplitude at the ground surface to that of base is compared by two methods (Yoshida et al. 1997). One is SHAKE and the other is nonlinear analysis in which stiffness proportional damping of 3 % at the first mode is used. Damping ratio used in SHAKE is the sum 3 % and the hysteretic damping. Input earthquake motion is very small in Fig. 12.3a so that responses are close to elastic. The amplification ratio by the nonlinear analysis (but close to the elastic state) decreases at high frequency compared with the amplification by SHAKE, which clearly indicates the role of Rayleigh damping in Eq. (12.4) that the damping increases as frequency. On the other hand, difference between two analyses is not seen clearly in Fig. 12.3b where nonlinear behavior is expected as input earthquake motion is large. This case study indicates that the Rayleigh damping can be used against stability problem under the large earthquake motion and it need not be the actual damping coefficient because it hardly affects the result of the nonlinear seismic ground response analysis under the large earthquake motion.

12.2.2 Mode Proportional Damping

Choice of the damping ratio as a function of frequency is limited in the Rayleigh damping because there are only two parameters. A mode proportional damping may be used for more freedom (Ohsaki 1980; Yoshida et al. 2006).

Matrix and vector are expressed by the bracket and the brace, respectively, in this subsection for easy understanding. Then equation of motion is expressed by

$$[M]\{\ddot{u}\} + [C]\{\dot{u}\} + [K]\{u\} = \{F\} \tag{12.6}$$

where $\{u\}$ denotes displacement vector and $\{F\}$ denotes external load vector. $[M]$, $[C]$, and $[K]$ are mass, damping, and stiffness matrices, respectively. The mass matrix is assumed to be the lumped mass, or it is a diagonal matrix. It is also assumed that the eigenvalue analysis is possible, or the system can completely be uncoupled into each mode. Then displacement $\{u\}$ can be expressed as a superposition of the eigenvectors $\{X_i\}$ as

$$\{u\} = [\{X_1\}\{X_2\}\{X_3\}\cdots\{X_n\}]\{q\} = [\xi]\{q\} \tag{12.7}$$

where $\{q\}$ denotes a modal displacement and $[\xi]$ denotes a matrix whose columns are eigenvector $\{X_i\}$. Substitute this equation into Eq. (12.6) and multiply $[\xi]^{\mathrm{T}}$ from the left; we obtain

$$[\xi]^{\mathrm{T}}[M][\xi]\{\ddot{q}\} + [\xi]^{\mathrm{T}}[C][\xi]\{\dot{q}\} + [\xi]^{\mathrm{T}}[K][\xi]\{q\} = [\xi]^{\mathrm{T}}\{F\} \tag{12.8}$$

From the orthogonal condition of the eigenfunctions, both $[\xi]^{\mathrm{T}}[M][\xi]$ and $[\xi]^{\mathrm{T}}[K][\xi]$ are diagonal matrices. They are expressed here as

$$[\xi]^{\mathrm{T}}[M][\xi] = [m_j], \qquad [\xi]^{\mathrm{T}}[K][\xi] = [k_j] \tag{12.9}$$

where $[m_j]$ and $[k_j]$ indicate diagonal matrix whose diagonal components are m_j and k_j, respectively. The diagonal terms are expressed by

$$m_j = \{X_j\}^{\mathrm{T}}[M]\{X_j\}, \qquad k_j = \{X_j\}^{\mathrm{T}}[K]\{X_j\} \tag{12.10}$$

Equation (12.6) can be uncoupled, if the second term of Eq. (12.8) is a diagonal matrix or

$$[\xi]^{\mathrm{T}}[C][\xi] = [c_j] \tag{12.11}$$

Then the damping constant c_j in the jth mode is calculated from the separated equation of motion (Eq. 12.8) into each mode as

$$c_j = 2\omega_j h_j m_j \tag{12.12}$$

where h_i denotes damping constant at jth mode. Substitution of this equation into Eq. (12.11) yields

$$[C] = [M][\xi][2h_j\omega_j][1/m_j][\xi]^{\mathrm{T}}[M] \tag{12.13}$$

where $[1/m_j] = [m_j]^{-1}$. The damping matrix can be developed by specifying the damping ratio in each mode.

As the Rayleigh damping is obtained by setting the damping constants to be proportional or inverse proportional to the frequency, it is a special case of the mode proportional damping.

In practice, the damping ratio is desired to be specified material by material. A strain energy proportional or a kinematic energy proportional relationship (Ohsaki 1980; Muto and Kobayashi 1977) can be used to evaluate the modal damping. They are

$$h_j = \frac{\sum_{n=1}^{N} h_n \{X_j\}^{\mathrm{T}} [k]_n \{X_j\}}{\{X_j\}^{\mathrm{T}} [K] \{X_j\}}, \text{ or } h_j = \frac{\sum_{n=1}^{N} h_n \{X_j\}^{\mathrm{T}} [m]_n \{X_j\}}{\{X_j\}^{\mathrm{T}} [M] \{X_j\}} \tag{12.14}$$

where $[k]_n$ and $[m]_n$ are element stiffness and mass matrix written in the global coordinate and summation is made to all elements.

The mode proportional damping looks as if frequency-dependent damping ratio is specified at a glance, but it is not true; the eigen mode-dependent damping ratio is specified. However, earthquake motion amplifies near the predominant period. In this sense, it can be used instead of the frequency-dependent damping. Actually it is reported to work as the frequency-dependent damping ratio (Yoshida et al. 2006).

The mode proportional damping is a powerful tool theoretically because arbitrary frequency-dependent damping can be specified. However, it has disadvantage and there are notes for the practical use.

The biggest disadvantage is that the damping matrix is a full matrix. In other words huge computing time is required as all the components are nonzero. Therefore it is not practical to use it in the two- or three-dimensional analysis. In addition, if damping ratio at high frequency is small, pulse appears in the acceleration. Therefore, time increment is to be much smaller than the case when the stiffness proportional damping is used.

The modal damping can be made by considering the arbitrary number of modes, in which case $[\xi]$ in Eq. (12.7) includes only considered eigenvectors. The amount of calculation can be reduced by using this nature, but it is not recommended because an amplification of the mode that is not considered may amplify very much. It is strongly suggested to use all modes to develop the damping matrix.

12.3 Wave Scattering

As explained in Sect. 6.7.4, the damping ratio at small strains is nearly zero in the careful laboratory test. On the other hand, the back analysis of the vertical array records shows the existence of the damping as seen in Fig. 12.1. This damping, therefore, is the internal damping, the radiation damping, or the scattering attenuation. Since internal damping and scattering attenuation cannot be separated as explained in the preceding, they are treated as the scattering attenuation.

When a ground is modeled, the same layer is considered to be homogeneous as shown in Fig. 12.4a. Then a wave that incident into this layer at point A will arrive at B which is located just above the point A; the path is a linear line. On the other hand, if there is a region that is different from other place as shown in Fig. 12.4b,

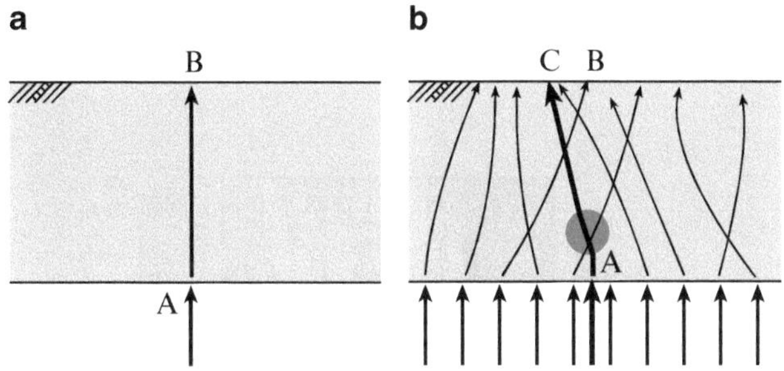

Fig. 12.4 Mechanism of scattering attenuation. (**a**) Homogeneous ground. (**b**) Inhomogeneous ground

the wave refracts at the inhomogeneous zone and arrives at C. Since the ground is not perfectly homogeneous but mechanical property scatters, all the incident waves do not arrive at the point just above the incident point, but locations apart from it. It indicates that arrival time is late as the wave moves longer distance.

On the one hand, late arrival time indicates the existence of the damping when the ground is modeled into homogeneous media. Although this phenomenon occurs because of the wave scattering, it appears as if there is damping or attenuation. This is the reason why it is called the scattering attenuation. It is noted that the energy that is stored in the ground does not change by the scattering attenuation, but the maximum acceleration becomes small because of several reasons. Internal damping becomes larger as the travel path increases. As the effect of the scattering depends on the wavelength or frequency, phase lag may appear near the maximum acceleration.

The damping ratio is used in the engineering field in order to show the attenuation characteristics, but a Q-value is used in the field of the seismology to express the attenuation along the propagating path. Relationships between the Q-value and the damping ratio h are expressed as

$$Q = \frac{1}{2h} \tag{12.15}$$

Scattering attenuation can be seen by the numerical analysis, too. A 32 m wide and 32 m deep ground with the average S-wave velocity of 200 m/s is shown in Fig. 12.5 (Suetomi and Yoshida 1998). Properties of the ground are shown in Fig. 12.5a, in which impact distance is assumed to be 4 m in the horizontal direction and 1 m in the vertical direction. No damping is considered. An example of the wave velocity distribution before the earthquake is shown in Fig. 12.5b and that after the earthquake is shown in Fig. 12.5c. Amplification ratio at the center of the ground surface against the base is shown in Fig. 12.6 by the thin dotted line, and average of them is shown by the thick solid line. On the other hand, spectral ratio in one-dimensional analysis is shown by a dashed line. Difference between the one-dimensional and the two-dimensional analyses corresponds to the scattering attenuation.

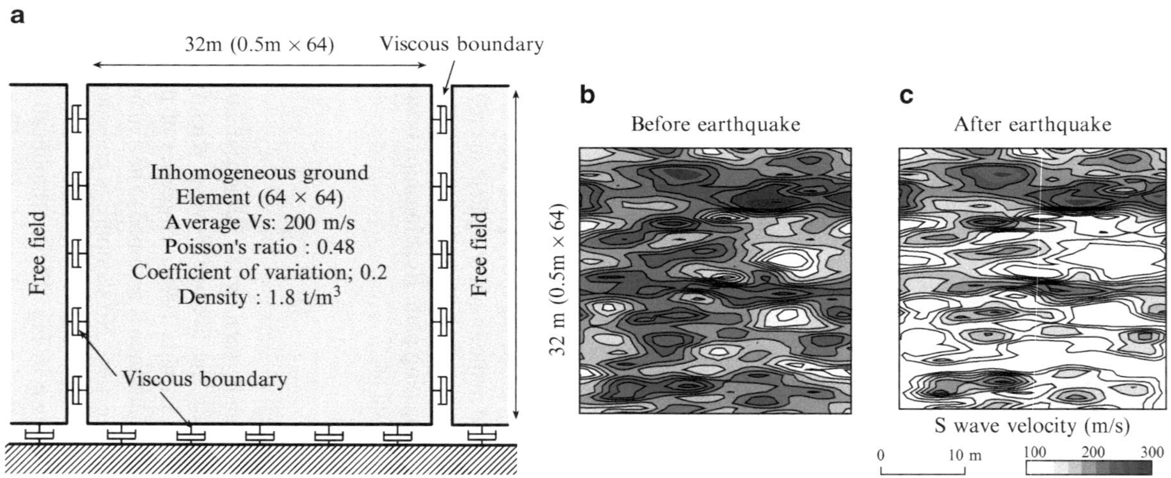

Fig. 12.5 Analyzed ground and distribution of V_s. (**a**) Analyzed ground. (**b**) V_s before Eq. (**c**) V_s after equation

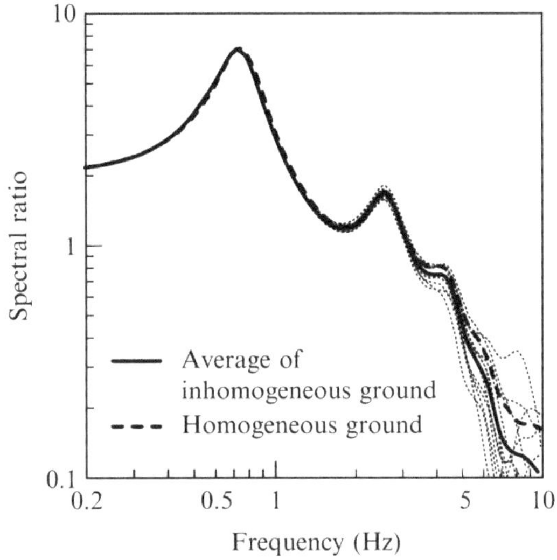

Fig. 12.6 Comparison of spectral ratio between homogeneous and inhomogeneous ground

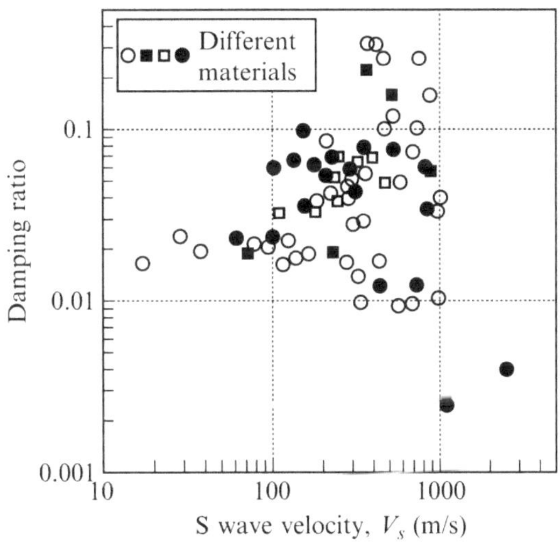

Fig. 12.7 S-wave velocity vs. damping ratio

There are researches on the damping characteristics at small strain as the earthquake records under the small earthquakes are easy to obtain. Back analyzed damping ratio can be classified into two types, a constant damping and a frequency-dependent damping.

An example of the constant damping is already shown in Fig. 12.1. Figure 12.7 shows relationships between the S-wave velocity and the damping ratio (Ishida et al. 1984). Damping ratio spreads widely, but many data lie between 1 and 10 %. There is no correlation between the damping ratios and the S-wave velocities in general. However, different features can be seen if the soil is classified. The damping ratio is

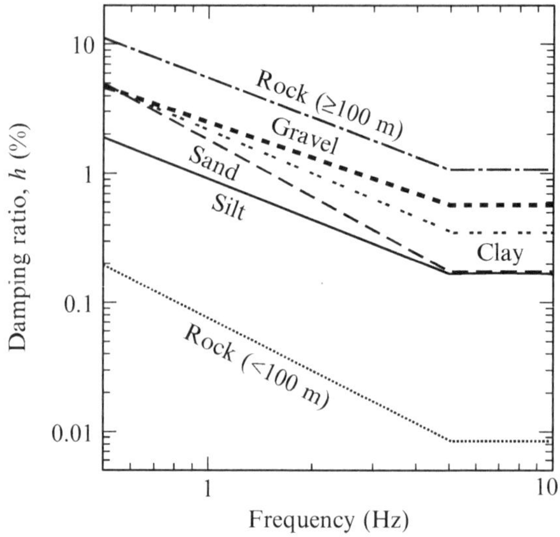

Fig. 12.8 Frequency dependency of scattering attenuation

nearly constant ($Q = 20, h = 0.025$) for clay in wide range of V_s. On the other hand, the damping ratio scatters such that $Q = 7 - 20$ ($h = 0.07 - 0.025$) for sand.

Both Figs. 12.1 and 12.7 indicate that the damping ratio obtained in situ is in order of several percent.

Since the scattering attenuation comes from the inhomogeneous nature of the soil, it is natural to consider it as frequency dependent. Actually, it is reported that the frequency-dependent characteristics are not observed in the laboratory test, but the damping ratio observed by the S-wave logging shows frequency-dependent nature; the damping ratio is larger in the low-frequency region than that in the high-frequency region (Hara and Kiyota 1977). An example of the frequency-dependent damping is shown in Fig. 12.8 (Fukushima and Midorikawa 1994), which is obtained by compiling the past researches. Other researches also showed the similar tendency that the damping ratio is larger at low frequency. Empirical relation is frequently expressed by

$$h = af^{-b} \tag{12.16}$$

where f denotes frequency and a and b are positive numbers. It is noted that this equation does not hold for all frequency, but the applicable range is 0.5 5 Hz in Fig. 12.8, for example. The damping ratio is fairly large at low frequency, but it is generally less than 0.1 (10 %). This damping ratio agrees with the constant damping ratio that does not consider frequency dependency.

According to Kiyota and Hagiwara (1993), the damping ratio is 2–4 % when the frequency-dependent characteristics are not considered and the following equation is proposed for the frequency-dependent damping:

$$h = (0.03 \sim 0.05)\, f^{-0.5} \tag{12.17}$$

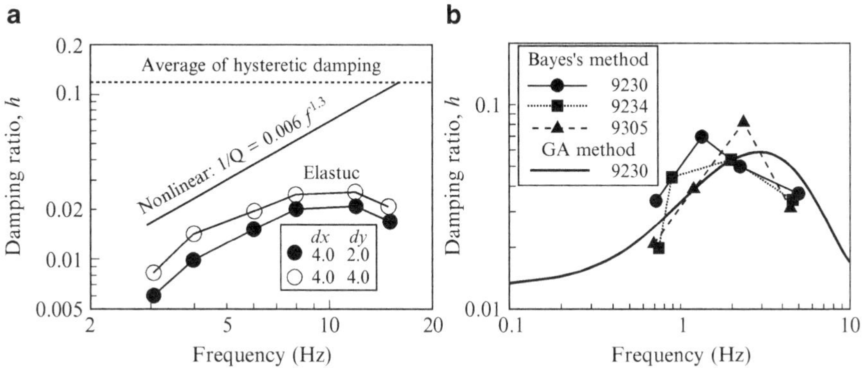

Fig. 12.9 Back analyzed scattering attenuation. (**a**) Elastic vs. nonlinear behaviors. (**b**) Back analyzed damping ratio (Modified from Suetomi 2000 and Nakamura et al. 2005)

On the other hand, back analysis of Fig. 12.6 leads to

$$\frac{1}{Q} = 0.006f^{1.3} \quad \left(h = 0.003f^{1.3}\right) \tag{12.18}$$

Equation (12.18) shows different characteristics from Eqs. (12.16) and (12.17); the damping ratio increases as frequency in the former case, whereas it decreases in the latter case. Both equations are, however, correct. Back analyzed scattering attenuation is shown in Fig. 12.9 (Suetomi 2000; Nakamura et al. 2005). Damping ratio becomes maximum in the middle and decreases to both high and low frequencies. Therefore, power can take both signs depending on the interested frequency region.

Figure 12.9a compares two researches under nearly the same conditions. The damping attenuation by the nonlinear analysis (Suetomi 2000) is larger than that by the elastic analysis (Kurita et al. 1966). This is explained as the scattering becomes large under the nonlinear analysis because deformations have a tendency to concentrate on the relatively weak portions. In addition, the scattering attenuation is smaller than the hysteretic damping. These observations agree with Fig. 12.3.

The scattering attenuation was hardly considered in the past seismic response although it is in order of several percent. However, conventional analysis by using SHAKE, for example, has succeeded to explain the earthquake observation. The following reason is considered to this confliction (Yoshida 1994).

As explained in Sect. 6.7.3, conventional cyclic shear deformation characteristics test has shown 2–4 % damping ratio at small strains. It is also explained that a careful recent test shows almost no damping at small strains. Therefore, small damping observed in the conventional test comes from test apparatus or test method. It indicates that the analysis has been carried by considering the damping that does not exist in the material. On the other hand, the scattering attenuation is also on the same order. Therefore, the scattering attenuation is automatically considered in the analysis as low strain damping that does not exist.

Other reasons are also considered. One of the reasons is that the low strain damping does not affect the result of the equivalent linear analysis if the effective strain is larger than this strain range and the low strain behavior is not used in the analysis. Another reason is that the value of the damping ratio itself is insensitive to the result of the seismic ground response analysis.

The scattering attenuation at low strain level is discussed here, but the effect at large strain level is not known well. Figure 12.5c, for example, shows scattering of the S-wave velocity, and inhomogeneous nature of the ground seems to increase more than that before the earthquake. This is because the deformation has a tendency to concentrate at weaker places. If so, scattering attenuation increases under the large earthquakes.

However, it may be a different issue that this increase affects the analysis. The effect of the hysteretic damping is predominant at large strains, and the role of other dampings becomes small as seen in Fig. 12.3. In addition, there is no relevant method to consider this damping in the nonlinear analysis. Therefore, engineers need not worry about this damping as there is no way to consider relevantly.

12.4 Radiation Damping

Waves that incident from the base reflect in the analyzed area and go out of it. Therefore, the energy stored in the analyzed region decreases, which result in the attenuation of the vibration in the analyzed region. This mechanism is called a radiation damping. The radiation damping occurs on both the lateral boundary and the base.

These phenomena are automatically considered if the semi-infinite region is added in the analyzed region. It is, however, not a simple work. A dashpot and the spring introduced in Sect. 10.4.2 are used as the alternate method. It is, however, noted that these alternate methods are difficult to express frequency-dependent characteristics that the actual boundary has.

In the old days, the base is modeled into the rigid boundary, in which all the wave reflects at the boundary; the radiation damping cannot be considered. Internal damping was used in order to consider the radiation damping, which will be described in Sect. 10.4.

The dashpot or the viscous damping is frequently used at the base, which is already explained in Sect. 10.4.2.

12.5 Numerical Damping

This damping appears when the equation of motion is solved by the step-by-step time integration scheme. As explained in Sect. 11.1.1, the response acceleration is interpolated in the small time increment. It is, for example, assumed to be linear in the linear acceleration method. These assumptions constrain the free response of the system, by which response is suppressed. This damping is called the numerical damping.

Intensity of the numerical damping depends on the interpolation function. It becomes larger as the suppression from the free response becomes large. The linear acceleration method, for example, shows smaller damping than the Wilson's θ method as the latter method assumes linear response longer than the former. The effect is already discussed in Fig. 11.4.

It is theoretically clear that the numerical damping is not a welcome nature. It, however, cannot be said to be always bad from the point of view of the engineering practice.

The step-by-step time integration scheme is based on the impulse response by which a step function is applied as the inertia force at the beginning of the increment. If, therefore, there is no damping, this sudden increase of the load is carried by the inertia term, resulting in rapid change of the acceleration or the pulse. It can be, of course, made negligibly small by taking the time increment sufficiently small, but it may not always be possible in the engineering practice because the computing time increases. If, however, there is damping, it absorbs this pulse and the acceleration response becomes smooth or natural. This issue is discussed in Sect. 15.4.

The appearance of the pulse is natural from the theoretical point of view, but people will not believe the acceleration in Fig. 15.6 although it is theoretically correct. In this sense, some damping is necessary and it is by introducing the velocity proportional damping and/or numerical damping. Especially, the numerical damping works well at high frequency where pulse appears. The stiffness proportional damping also works at high frequency.

It sometimes becomes practically important by making the acceleration response smooth. Generally, pulse wave attenuates as a damped free vibration as discussed in Sect. 15.4 or as wave travels to the neighboring elements even if it appears. It indicates that pulse wave may affect the behavior in the neighboring elements. Here, if the stiffness of the neighboring element is nearly zero under the strong nonlinear state, extraordinary behavior may occur even if the disturbance is very small. If element that shows extraordinary behavior appears, then its neighboring element becomes the same situation. The propagation of the extraordinary behavior may yield incorrect result or divergence of the calculation. It is easier to occur in the effective stress analysis where the stiffness becomes almost zero. In these situations, a function that can constrain this kind of pulse is variable.

12.6 Damping as Alternative

The external damping as an actual mechanism is hardly considered in the engineering practice because it is too small. The internal damping is not specified element by element, but given as global one, which indicates that actual internal damping is not used. Then viscous damping seems not to be necessary, but it is used to represent the various other dampings.

There are various dampings that are to be considered in the analysis. However, there is little method to consider them in the analysis with several exceptions such as the dashpot boundary to consider the radiation damping, and others should be taken into account as the velocity proportional damping. In this sense, the damping treated in this subsection is quite artificial. They are used instead of the following two purposes:

1. Alternate to actual damping
   As the radiation damping cannot be considered by the rigid base, for example, another damping that compensates it is to be considered in order to consider the radiation damping, and the use of the velocity proportional damping is one of the possible methods. The scattering attenuation can also be considered in the same manner as explained in Sect. 12.3.
2. Adjustment to actual behavior
   The equation of motion is composed of three terms, the inertia, the viscosity, and the restoring forces. Among them, the inertia force is related to the mass, and the restoring force is related to the stiffness or the stress-strain relationships. They can be evaluated fairly accurately. On the other hand, the viscous term is difficult to evaluate relevantly as explained in the preceding. Then, if an engineer is simulating the test result, for example, and the result is not satisfactory, he/she can use the damping term to obtain better agreement with the test result. Large change cannot be made by the adjustment of the damping term, but it may be sufficient in some case.

References

Central Disaster Prevention Council, Cabinet Office of Japan (2005) Report of expert committee on Tonankai and Tokai earthquakes (in Japanese)

Finn WDL, Martin GR, Lee MKW (1978) Comparison of dynamic analyses for saturated sands, earthquake engineering and soil dynamics, ASCE, GT special conference, vol 1, pp 472–491

Fukushima Y, Midorikawa S (1994) Evaluation of site amplification factors base on average characteristics of frequency dependent Q-1 of sedimentary strata. J Struct Constr Eng AIJ 460:37–46 (in Japanese)

Hara A, Kiyota Y (1977) Study on dynamic property of soil for the earthquake response analysis of ground -elastic shear modulus. In: Proceedings of the 14th Japan national conference of soil mechanics and foundation engineering, JSSMFE, Okinawa, pp 533–536 (in Japanese)

Ishida K et al (1984) On damping of ground -estimation of Q value by optimization-, programme and abstracts, the seismological society of Japan, p 149 (in Japanese)

Kiyota Y, Hagiwara T (1993) Earthquake response analysis based on equivalent linearization of soil constants in modal region part III estimation of damping property during very small ground motion, summaries of the technical papers of annual meeting of AIJ (Kanto), Tokyo, pp 217–218 (in Japanese)

Kokusho T (1986) Damping characteristics of soil, dynamic characteristics of soil for seismic response analysis. JSSMFE, pp 45–51 (in Japanese)

Kokusho T (1992) Dynamic characteristics of ground, lecture: analytical method of the interaction between ground and structure. Tsuchi-to-Kiso 40(4):76–84 (in Japanese)
Kurita T, Tsuzuki T, Annaka T, Shimada M, Fujitani M (1966) Scattering attenuation of seismic waves in inhomogeneous media. In: Proceedings of the 11th WCEE, Mexico
Muto K, Kobayashi T (1977) Comparative study of various damping theory frequently used in aseismic design of nuclear power plant. J Struct Constr Eng AIJ 255:35–45 (in Japanese)
Nakamura S, Sawada S, Yoshida N (2005) Modeling of surface ground for nonlinear analysis, Part 2. In: Proceedings of the 3rd symposium on strong motion prediction master plan for earthquake disaster reduction, pp 101–104 (in Japanese)
Ohsaki Y (1980) Vibration theory. Shokoku-sha, 302 pp (in Japanese)
Suetomi I (2000) A study on real time estimation of earthquake motion distribution during strong earthquake, theses submitted to Kyoto University for partial fulfillment of doctor's degree (in Japanese)
Suetomi I, Yoshida N (1998) Damping characteristics of soil deposits during strong ground motions. In: Proceedings of the 2nd international symposium on the effect of surface geology on seismic motion, Yokosuka, Japan, vol 2, pp 765–772
Togawa H (1975) Vibration analysis by finite element method. Science-sha (in Japanese)
Yoshida N (1994) Applicability of conventional computer code SHAKE to nonlinear problem. In: Proceedings of symposium on amplification of ground shaking in soft ground, JSSMFE, pp 14–31 (in Japanese)
Yoshida N, Suetomi I, Kato N, Miura K (1997) Parametric study on effectiveness of dynamic response analysis by equivalent linear method. In: Proceedings of the 32nd Japan national conference of geotechnical engineering, pp 877–878 (in Japanese)
Yoshida N, Sawada S, Nakamura S (2006) Effect of damping on earthquake response of ground and its accuracy. J JAEE 4(4):55–73 (in Japanese)

Chapter 13
Evaluation of Accuracy and Earthquake Motion Indices

It is impossible to simulate the earthquake motion perfectly in the seismic ground response analysis. In other words, seismic ground response analysis includes error because of the reasons in the following:

1. Soil profile and its modeling

   It is impossible to grasp the soil profile, distribution of mass, etc., perfectly. Even if they are grasped, it is almost impossible to model it perfectly. Analyzed region is divided into several blocks, and mechanical properties in each block are assumed to be the same. The distribution of the mass is also not perfectly obtained. In addition inhomogeneous nature of the ground is also impossible to measure perfectly. What we can obtain is somewhat average nature of the ground. Therefore, response in the predominant frequency range may be well examined, but higher-frequency response is difficult to explain in detail.
2. Mechanical property and its modeling

   As the in situ and the laboratory tests are carried by a routine procedure, all the behavior cannot be grasped. If the mathematical equation is used to represent them, in addition, error is included because the mathematical equation cannot express the actual behavior perfectly.
3. Numerical calculation

   The viscous damping includes various mechanisms and some of them are difficult to evaluate quantitatively. The method of the numerical integration also affects the result of the analysis.

The order of error can be grasped through the simulation of the vertical array records, the shake table tests, and centrifugal tests. It is important to have experience of the analysis of these records in order to get the engineering sense and to recognize the error that the methods have.

N. Yoshida, *Seismic Ground Response Analysis*, Geotechnical, Geological and Earthquake Engineering 36, DOI 10.1007/978-94-017-9460-2_13

It is also important to grasp the general tendency that the method of the analysis has and the effect of the modeling on the numerical analysis. Several case studies are introduced in this chapter in order to give some hint to recognize the numerical analysis.

It is noted that calculation in this section is made by the computer programs DYNEQ (Yoshida 1995a), DYNES3D (Yoshida 2004), and STADAS (Yoshida 1993). Although the names of SHAKE and FDEL are used in the equivalent linear analysis, they just indicate that the procedures the same as SHAKE and FDEL are used as common noun.

The accuracy and the applicability of the analysis are related to each other. One of the frequent questions on equivalent linear analysis is the applicability of this method such as the maximum strain. The author, however, does not prepare the answer, because it cannot be answered without knowing how the result will be used. Error of 50 % may be allowed and 5 % may not be allowed depending on the situation. The answer, therefore, must be prepared by the engineer who is going to make the analysis. This book sometimes discusses on the applicability, which comes from the author's sense based on his expense.

Five earthquake motion indices, the maximum acceleration, the maximum velocity, the maximum displacement, the spectral intensity (SI value), and the seismic intensity scale, are explained in the following as representative earthquake motion indices. The accuracy of the analysis depends on the choice of the earthquake motion index. In the engineering practice, the accuracy of the analysis is frequently discussed by means of the maximum acceleration, but it is not so simple to discuss the error by only one index. The maximum acceleration is the most sensitive index among five indices. Therefore, it can be a good index to discuss the accuracy, but it may emphasize the differences of the analyses too sensitive.

Figure 13.1 shows a result of a questionnaire on the error of the analysis (Standardization committee on test method of dynamic deformation characteristics of soil 1994), which may help to recognize the order of the error although it is a little old data. The error between the observed and the calculated maximum accelerations and the acceptable error are shown in the figure. There are about two thirds of data that shows error more than 8 %. It is also shown that acceptable error is much smaller.

According to the author's experience, however, error between the observation and the analysis seems conservative. There is not many vertical array data that show nonlinear behavior (see Chap. 14). As shown there, error is fairly large in SHAKE. In addition, as shown in Sect. 15.3, blind test also has shown much larger error. Therefore, Fig. 13.1 seems to be based on fairly small earthquake record, which indicates that this data cannot be applied in the ground where nonlinear behavior becomes prone.

13.1 Acceleration, Velocity, and Displacement

The acceleration, the velocity, and the displacement are representative earthquake motion indices that are output in almost all the computer programs. Here, the acceleration indicates an absolute acceleration and the displacement indicates a

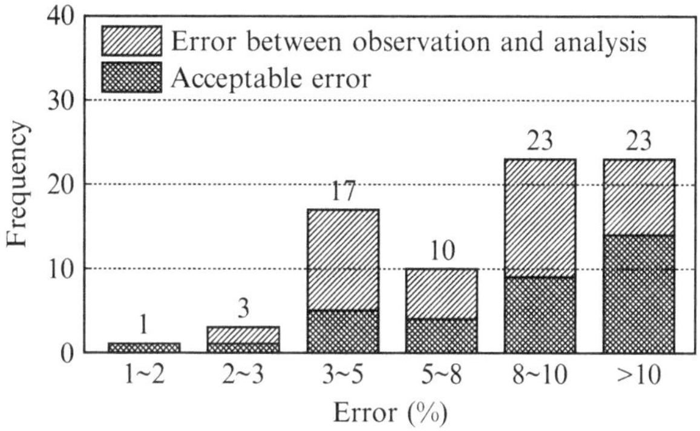

Fig. 13.1 Expected and experienced error on seismic ground response analysis

relative displacement. Velocity may be absolute or relative depending on the computer program.

Since the inertia force is a product of the mass and the acceleration, the absolute acceleration is a good index for rigid structures. However, when a natural period of the structure becomes longer, it cannot be a good index and the absolute velocity becomes more important. On the other hand, the displacement is important in the underground lineal structure such as piles and pipe lines. However, since the measurement of the displacement is very difficult, the evaluation of the error is a hard work.

The inertia force is important in the earthquake resistant design. The seismic coefficient, a ratio of the earthquake force to the weight of the structure, is frequently used in order to evaluate the earthquake load in the static analysis. The product of the seismic coefficient and the acceleration of gravity yields response acceleration. Confusion of this acceleration and the maximum response acceleration is sometimes seen in the author's experience as both accelerations are used to evaluate the inertia force acting to the structure. They are quite different quantities because design force works statically, whereas the maximum acceleration is just an instantaneous value and energy transmitted to the structure is small. An example is shown in Fig. 13.2, in which the ordinate is the seismic coefficient and the abscissa is the maximum acceleration at the ground surface (Noda et al. 1976). Data on damaged gravity quay walls are plotted in the figure, where • denotes upper or lower limit of the seismic intensity and ↑ and ↓ denote that the seismic intensity is greater than or less than these symbols, respectively. It is seen that seismic coefficient does not become greater than 0.25. They also suggest the upper boundary of the seismic coefficient as

$$k = \frac{1}{3}\left(\frac{\alpha_{\max}}{g}\right)^{1/3} \tag{13.1}$$

where $\alpha_{\max}$ denotes maximum acceleration at the ground surface and g is the acceleration of gravity based on the data in which upper boundary of the seismic coefficient is sure.

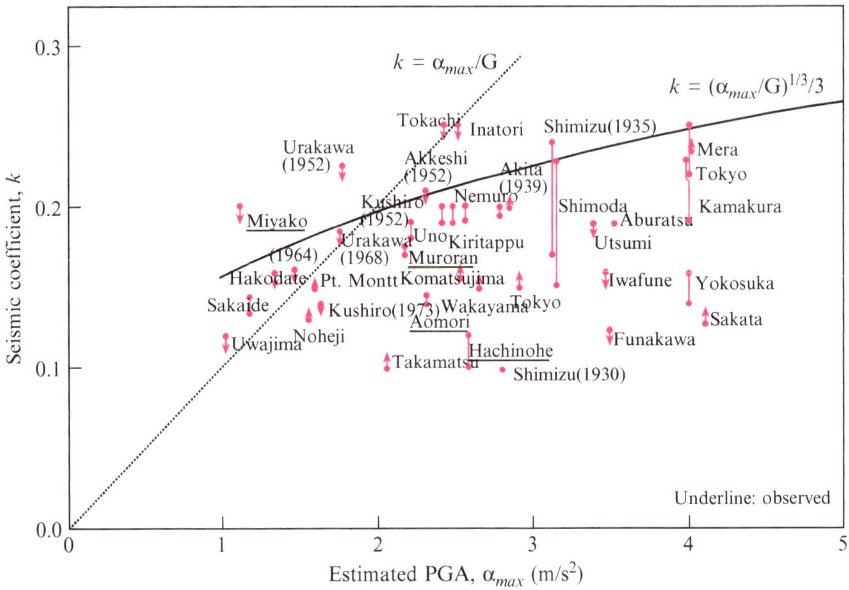

Fig. 13.2 Relationships between seismic coefficient and maximum acceleration at ground surface (Modified from Noda et al. 1976)

It is noted that this equation is proposed on the gravity quay wall. If structural type changes, for example, the equation may be different although it seems to be applicable to similar structure such as retaining walls. Unfortunately, there are few researches of this type.

The observed maximum accelerations become large in recent earthquake observations, and maximum acceleration more than 1 g is not extraordinary. There may be at least two reasons. Many seismographs have been installed in Japan, especially after the 1995 Kobe earthquake. Earthquake records near the fault or just above the fault were obtained, which sometimes shows very large acceleration. The maximum vertical acceleration is sometimes larger than the maximum horizontal acceleration in these records. The second reason is improvement of the seismographs. Earthquake motions were drawn on the paper in the old days and damping ratio of the seismograph is large, in which case high-frequency response is not well recorded. An example is a SMAC-B2-type seismograph, which was a representative seismograph used in Japan known to have less sensibility at high frequency (Kawashima et al. 1982). Anyway, observed maximum acceleration will increase if higher-frequency response will be recorded. In many engineering practices, however, higher-frequency behavior more than 10 Hz does not affect the behavior of the structure. In this meaning, discussion on accuracy based on the acceleration is to be done with care.

13.2 Seismic Intensity Scale

The seismic intensity scale (the seismic intensity hereafter) has been used to express the intensity of the ground shaking or degree of damage. As it is defined based on the facts that occurred during the earthquake, agreement with the earthquake damage is considered to be good. In the historic earthquakes, seismic intensity is analyzed from the damage distribution (Usami 2003). In Japan, an office stuff of the meteorological observatory determined seismic intensity based on his cenesthesia and damage near the observatory. It indicates that it takes time to define the seismic intensity. However, quick announcement becomes important recently because of the rescue purpose. Japan Metrological Agency felt it (Kinoshita and Ohtake 2000), and a new method, called JMA instrumental seismic intensity, was developed (Notification of JMA 1996; Explanation of JMA seismic intensity and Metrological Bureau 2009) after the 1995 Kobe earthquake. As the new seismic intensity is developed so as to agree with the conventional seismic intensity, it has been usually called a seismic intensity the same as the former name.

As the new seismic intensity can be evaluated from the acceleration time history, it can be calculated simultaneously when the earthquake record is obtained. The calculated seismic intensity is round off at the second digit after the decimal point; output is made at one place of decimals. Then it is made integer and used for general use. There were 8 classes from 0 to 7 in the old seismic intensity. On the other hand, seismic intensities outside Japan usually have more classes. Considering them, intensities 5 and 6 are divided into 2 by putting + and –, resulting in 10 classes. It is noted that the maximum seismic intensity was 6 in the old day. However, significant damage was observed during the 1948 Fukui earthquake where almost all houses were collapsed. Considering these damages, intensity 7 was added after the earthquake. The intensity 7 first appeared during the 1995 Kobe earthquake in the so-called damage belt zone (Editorial Committee for the Repost on the Hanshin-Awaji Earthquake Disaster 2000).

The method of the calculation is described in detail by Kinoshita and Ohtake (2000) and JMA (1996). Roughly speaking, it is calculated by using three components of the earthquake records. Accelerations are filtered so as to emphasize the frequency range 0.1–10 Hz. It is explained that seismic intensity is a quantity between the acceleration and the velocity, and the applicability is an earthquake motion with period less than 2 s (Kinoshita and Ohtake 2000).

One directional analysis is common in the seismic ground response analysis in the engineering practice, whereas three components are used in the calculation of the seismic intensity. However, calculation is possible by using only one component.

Various equations have been proposed on the seismic intensity and the maximum acceleration (Kinoshita and Ohtake 2000). Historically, empirical equation by Kawasumi (1943) is famous. One of the recently proposed equations is (Cabinet Office 2001)

$$\begin{aligned} I &= 2.02 \log V_{\max} + 2.4 \\ A_{\max} &= 10^{0.908} V_{\max}^{1.13} \end{aligned} \tag{13.2}$$

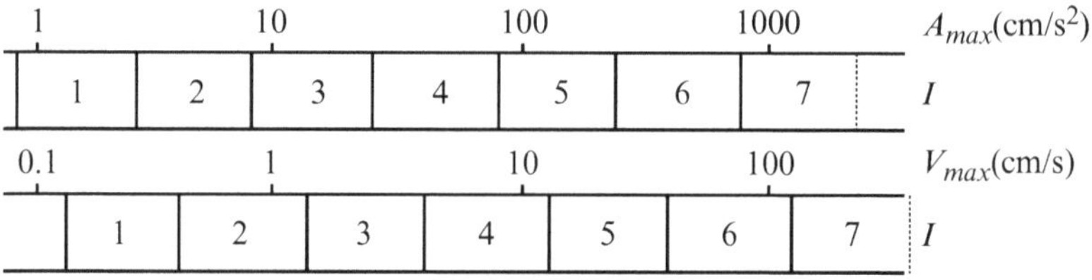

Fig. 13.3 Seismic intensity vs. maximum acceleration and velocity (Modified from Muramatsu 1967)

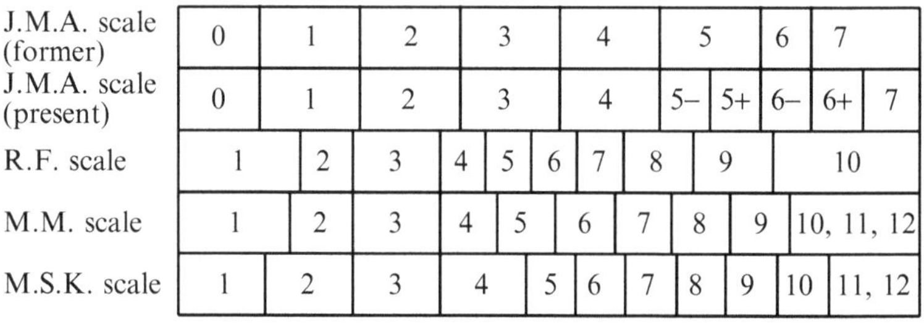

Fig. 13.4 Various seismic intensities (Modified from Kinoshita and Ohtake 2000)

where I denotes a seismic intensity and $A_{\max}$ and $V_{\max}$ denote the maximum acceleration and the maximum velocity in unit cm/s^2 and cm/s, respectively. Relationships by means of conventional (former) seismic intensity are shown in Fig. 13.3. Reference (Okada 2001) also discusses the seismic intensity.

Internationally, a variety of seismic intensities is used. Relationships between them are shown in Fig. 13.4.

13.3 Spectral Intensity

The spectral intensity (SI) is proposed by Housner (1965) at first as a new earthquake motion index because the maximum acceleration is not a relevant index to evaluate the damage to the structures. According to his original proposal, it is obtained by integrating the pseudo velocity response spectrum as

$$\mathrm{SI} = \int_{0.1}^{2.5} S_{v,20} \mathrm{d}t \tag{13.3}$$

where $S_{v,20}$ denotes pseudo velocity response spectrum with 20 % damping ratio. It is integrated between 0.1 and 2.5 s. Natural periods of many structures are included in this period range.

Dimension of the spectral intensity is displacement as it is obtained by integrating the velocity. It is, however, more convenient to make the dimension the same with the velocity. Therefore, in Japan, it is divided by the integral time interval (2.4 s) as

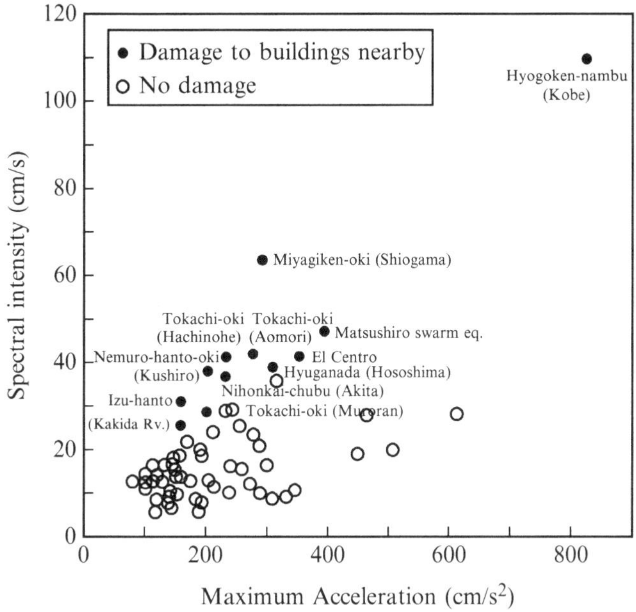

Fig. 13.5 Damage vs. *SI* value and maximum acceleration (Modified from Shimizu 2001)

$$SI = \frac{1}{2.4}\int_{0.1}^{2.5} S_{v,20}\mathrm{d}t \tag{13.4}$$

and is used instead of the original definition.

Housner pointed out that correlation between the SI value and the damage is good. Figure 13.5, for example, shows relationships between the damages, SI value and the maximum acceleration. Buildings may not be damaged at large acceleration, but is damaged when the SI value is larger than about 30. Of course, there are data where correlation is not good.

The SI value is also used to evaluate the displacement of the ground (Towhata et al. 1996) and the earthquake motion index for the liquefaction analysis (Yasuda et al. 1993). Pulse wave in the acceleration does not always have an effect to the damage to structures, and the use of the SI value compensates the disadvantage of the evaluation based on the acceleration.

13.4 Spectrum

The effect of wide frequency ranges is considered in the seismic intensity and SI value. Therefore, these indices can be applied to many structures, but the effect to the structure with a particular frequency cannot be grasped. It is also difficult to see

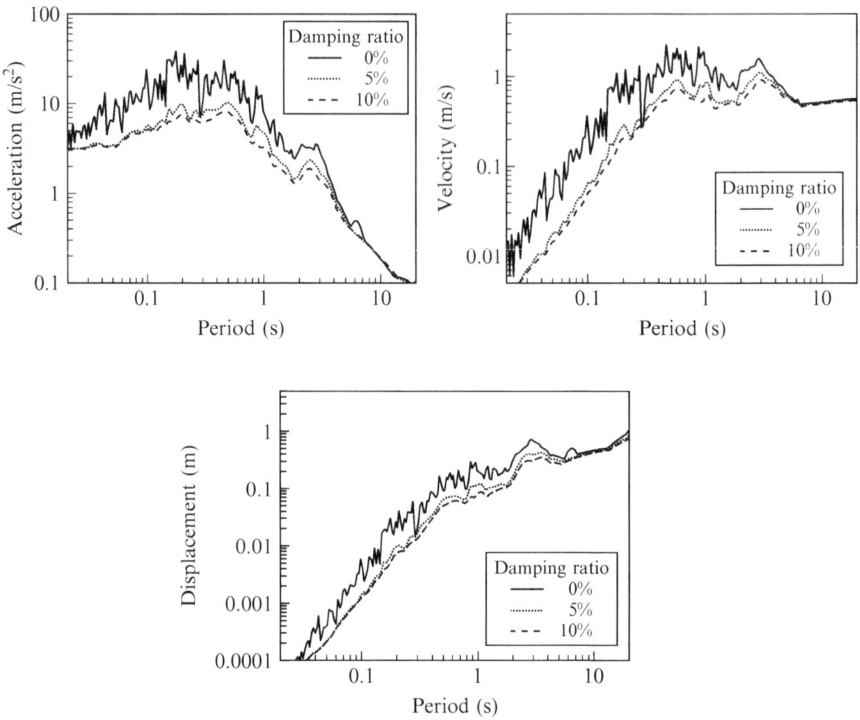

Fig. 13.6 Example of response spectrum against El Centro 1940 NS wave

the effect of the high-frequency component. Therefore, spectrum is usually used to see the effect to structures with different natural frequencies.

The most frequently used spectrum is a response spectrum, which shows maximum response of the one-degree-of-freedom system against natural period (or frequency). As shown in Fig. 13.6 as an example, there are three kinds of response spectrum, which are the acceleration response spectrum S_a, the velocity response spectrum S_v, and the displacement response spectrum S_d. Here, absolute value is used for the acceleration, whereas relative value is used for the velocity and displacement.

The following equations hold approximately for three response spectra:

$$\begin{aligned} S_a &\approx \omega S_v \approx \omega^2 S_d \\ S_v &\approx \omega S_d \end{aligned} \tag{13.5}$$

Then, all response spectra can be obtained from the displacement response spectrum (JSCE 1989). Among them, the velocity spectrum obtained from Eq. (13.5) is somewhat different from the original spectrum and is called a pseudo velocity spectrum. In the same manner, the acceleration response spectrum obtained from Eq. (13.5) is

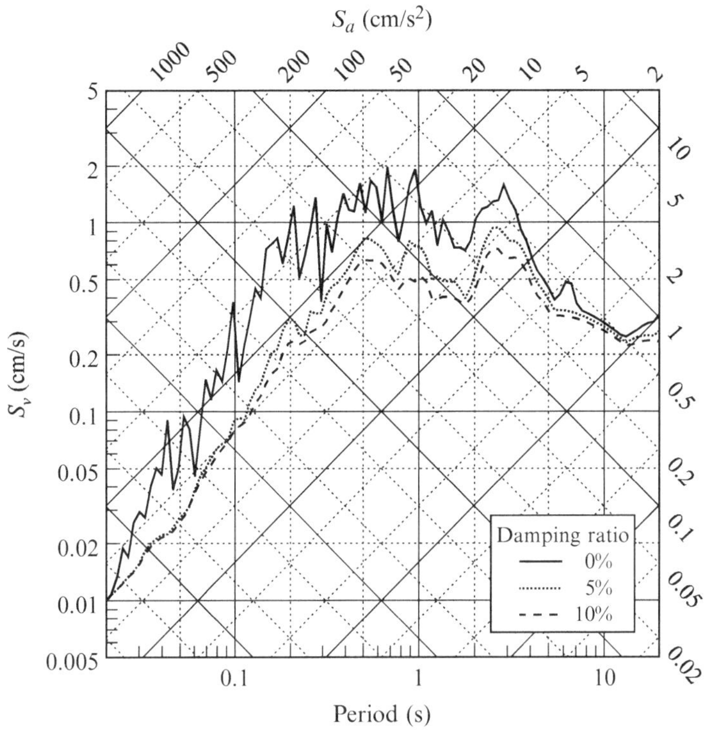

Fig. 13.7 Tripartite expression of response spectrum

called pseudo acceleration. The acceleration response spectrum obtained from Eq. (13.5) is, however, almost identical to the original response acceleration spectrum because both the acceleration and the displacement become maximum nearly the same time where the velocity is zero. The term "pseudo acceleration" is sometimes used to emphasize that it is obtained from Eq. (13.5).

The (pseudo) acceleration, pseudo velocity, and displacement response spectrum can be drawn in a figure such as in Fig. 13.7 by using the relationships in Eq. (13.5). The acceleration and the displacement can be read by the tilted axes. This expression is called a tripartite response spectrum.

The Fourier spectrum is also used to see the nature of the earthquake motion as it shows frequency component of the earthquake wave. The power spectrum is also used in the same purpose.

The amplification ratio spectrum is sometimes used to show the amplification characteristics of the ground in each frequency. It is a ratio of Fourier amplitude at the particular point (usually at the ground surface) to that at the base (or incident wave).

13.5 Other Quantities

The evaluation of the accuracy of the analysis is a difficult issue in the seismic ground response analysis. The maximum acceleration has been frequently used in the past researches. As described in the preceding, however, it is not a best index. Other quantities such as the similarity of the waveform, the agreement of the predominant period, etc., have also been used. Since there are many earthquake motion indices, comparison between observations and analyses can be done by means of many earthquake motion indices, but indices other than the above ones were not usually used. They were used in a limited way. For example, J-value defined as

$$J = \int (\ddot{u}_{\mathrm{obs}} - \ddot{u}_{\mathrm{cal}})^2 \mathrm{d}t \tag{13.6}$$

was used in order to evaluate the accuracy in the whole waveform, where subscripts obs and cal denote observation and analysis, respectively. The J-value becomes zero when both accelerations agree and the analysis is perfect. Accuracy becomes less as J-value increases. This method has a limitation that J-value increases if there is a time lag between two waves even if simulation is perfect. In order to avoid this limitation, calculations are made several times by drifting one wave and minimum value is taken as J-value. It is not used in the recent researches partly because this index cannot evaluate the effect of the earthquake motion to the structure and partly because the observed earthquake record is necessary.

The Arias intensity I_A (Hwang et al. 2004) is also sometimes used, which is defined as

$$I_A = \frac{\pi}{2g} \int \ddot{u}(t)^2 \mathrm{d}t \tag{13.7}$$

where g denotes an acceleration of gravity and the integral is made in whole duration. Two independently calculated intensities are added in order to evaluate two directional components of the earthquake motion in the original paper (Hwang et al. 2004). However, intensity changes when coordinate axis rotates. The use of vector summation seems better than the simple summation.

From the stand point of the comparison between the observation and the analysis, another aspect is possible. Figure 13.8, for example, compares a vertical array record and the analysis (Yoshida 1995b). The calculated waves cross the zero axis earlier than the observation at places shown by arrows, for example. These indicate that the wave velocity is faster in the analysis. On the other hand, if the arrival time of the analysis is late, the wave velocity is smaller than in situ. These differences occur because of two reasons. One is an elastic modulus and the other is a nonlinear behavior. If, for example, the wave velocity is larger or the strength is larger, apparent wave velocity is overestimated. The effect of the elastic modulus

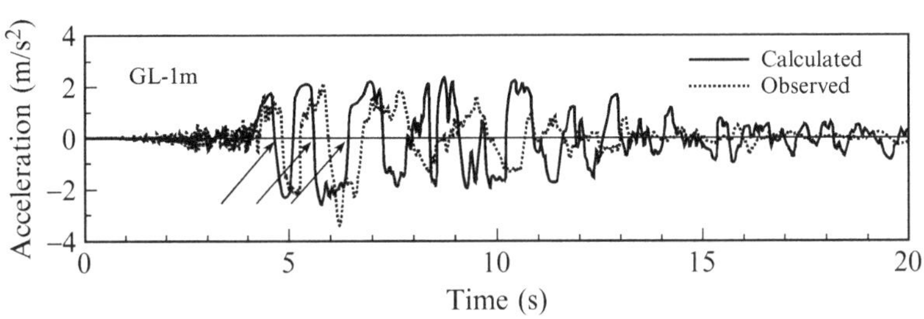

Fig. 13.8 Example of phase lag between observation and analysis

is investigated at the beginning of the earthquake where nonlinear behavior is not prone. If it agrees, difference at the large acceleration can be considered to be caused by the nonlinear behavior,

References

Cabinet Office (2001) Manual to support earthquake damage estimation (in Japanese)

Editorial Committee for the Repost on the Hanshin-Awaji Earthquake Disaster (2000) Report on the Hanshin-Awaji earthquake disaster, general issue volume 1, compendium of earthquake disaster including stochastics description, Maruzen, 549 pp (in Japanese)

Explanation of JMA seismic intensity, Metrological Bureau (2009) http://www.kishou.go.jp/know/jishin.html [2010]

Housner GW (1965) Intensity of earthquake ground shaking near the causative fault. In: Proceedings of the 3WCEE, New zealand, vol I, pp III-94-III-115

Hwang H, Lin CK, Yeh YT, Cheng SN, Chen KC (2004) Attenuation relations of Arias intensity based on the Chi-Chi Taiwan earthquake data. Soil Dyn Earthq Eng 24(9):509–517

JMA (ed) (1996) Study on seismic intensity, Gyosei, 238 pp

JSCE (1989) Seismic response analysis and earthquake resistant design, vol 1, Earthquake motion and dynamic property, Gihodo Shuppan, 170 pp (in Japanese)

Kawashima K, Takagi Y, Iazawa K (1982) Procedure of instrument correction and displacement calculation for SMAC-B2 accelerograph records with considering accuracy of digitization. Proc JSCE 325:35–44

Kawasumi H (1943) Earthquake intensity and intensity scale. Earthquake 15(1):6–12 (in Japanese)

Kinoshita S, Ohtake M (eds) (2000) Fundamentals of strong motion, National Research Institute for Earthquake Science and Disaster Prevention. http://www.k-net.bosai.go.jp/k-net/gk/publication/ [2010]

Muramatsu I (1967) Seismic risk, overviews of Japanese seismology. Jishin-2 20–4:281–290 (in Japanesc)

Noda S, Uwabe T, Chiba T (1976) Relation between seismic coefficient and ground acceleration for gravity quaywalls. Rep Port Harb Res Inst 14(4):67–111 (in Japanese)

Notification of JMA, No. 4, February 15, 1996

Okada N (2001) On earthquake motion index necessary for disaster prevention research. In: Discussion in the 26th JSCE earthquake engineering symposium-2001, Tokyo, pp 13–19

Shimizu Y (2001) A study on development of real-time earthquake disaster mitigation system for city gas network and utilization. A Thesis in Partial Fulfillment of the Requirements for the Degree of Doctor of Engineering, University of Tokyo (in Japanese)

Standardization committee on test method of dynamic deformation characteristics of soil (1994) State of art report on test equipment and test method for dynamic deformation characteristics (in Japan). In: Proc., Symp. dynamic deformation characteristics in dynamic problem of ground and soil structures -test and investigation methods and its application, JGS, pp 1–22 (in Japanese)

Towhata I, Park JK, Orense RP, Kano H (1996) Use of spectrum intensity for immediate detection of subsoil liquefaction. Soils Found 36(2):29–44

Usami T (2003) Materials for comprehensive list of destructive earthquakes in Japan, [416]-2001, Latest edition, University of Tokyo Press, 605 pp (in Japanese)

Yasuda S, Yoshikawa Y, Ushijima K, Ishikawa T (1993) A prediction method for liquefaction using spectral intensity. In: The 28th Japan national conference of soil mechanics and foundation engineering, Kobe, pp 1325–1328 (in Japanese)

Yoshida N (1993) STADAS, a computer program for static and dynamic analysis of ground and soil-structure interaction problems. Report, Soil Dynamics Group. The University of British Columbia, Vancouver, Canada

Yoshida N (1995a) DYNEQ, a computer program for DYNamic response analysis of level ground by EQuivalent linear method, Version 3.25 (September, 2004). http://www.civil.tohoku-gakuin.ac.jp/yoshida-j.html [2010]

Yoshida N (1995b) Earthquake response analysis at Port Island during the 1995 Hyogoken-nanbu earthquake. Tsuchi-to-Kiso 43(10):49–54 (in Japanese)

Yoshida N (2004) DYNES3D, a computer program for dynamic response analysis of level ground by effective stress-nonlinear method, Version 2.6. http://www.civil.tohoku-gakuin.ac.jp/yoshida-j.html [2010]

Chapter 14
Simulation of Vertical Arrays

Earthquake records both on the ground surface and in the underground are at least to be obtained in the vertical array earthquake observation; several records are sometimes obtained at different depths. Simulation of these data is a good tool to grasp the accuracy of the seismic ground response analysis and is also a good tool to cumulate experience on the seismic ground response analysis. Recently, a number of vertical array observatories increase such as Kiban-Kyoshin Network (KiK-net) (2010) which covers all Japan, although the number of the vertical array stations is limited in the old days. In order to examine the accuracy of the analysis, it is necessary that a strong ground motion record is obtained. In addition, ground conditions such as soil profile, velocity structure, nonlinear property, etc., are also well known. In this sense, KiK-net is not so useful because detailed soil profile and material properties are not known. As far as the author knows, data that can be used to investigate the accuracy of the nonlinear seismic ground response analysis are not sufficient. They are shown in Table 14.1.

14.1 Applicability of Equivalent Linear Method

A vertical array record was obtained during the 1987 Chibaken toho-oki earthquake at the Chiba test site, Institute of Industrial Science, University of Tokyo (Katayama et al. 1990; Katayama and Yamazaki 1990). Both the horizontal and vertical array observations are made in this site, among which vertical array record is analyzed. Soil profiles and material property are based on Lu et al. (1989).

The soil profiles of this site are shown in Fig. 14.1. Seismographs are installed at GL-1, GL-5, GL-10, GL-20, and GL-40 m, which are shown by a solid circle •. Acceleration waveforms at GL-1 and GL-40 m are shown in Fig. 14.2. According to Lu et al. (1989), stress–strain relationships are modeled into the Ramberg–Osgood

N. Yoshida, *Seismic Ground Response Analysis*, Geotechnical, Geological and Earthquake Engineering 36, DOI 10.1007/978-94-017-9460-2_14

Table 14.1 Vertical arrays where large ground shaking is recorded

Site	Earthquake	γ_{max}	Comments
IIS Chiba, University of Tokyo	1987 Chibaken toho-oki earthquake (Katayama et al. 1990)	0.09 %	Loam
Shin-Fuji substation, TEPCO	1983 Kanagawa and Yamanashi prefectures boundary earthquake (Yoshida et al. 1995)	0.4 %	Scoria, andosol
Lotung test site, Taiwan	1986.5.20 (Chang et al. 1991)	(0.1 %)	Silt
Research Institute, Kansai Electric Power Company	1995 Kobe earthquake (Soeda et al. 1996)	1 %	Fine sand
Port Island	1995 Kobe earthquake (Yoshida 1995)	6 %	Ma13 (Holocene clay)
Oh-i reclaimed land	1980.9.25 (Ishihara et al. 1981)	0.07 %	Sand
Shin-Ohta reclaimed land	1987 Chibaken toho-oki earthquake (Masuda et al. 2001)	0.4 %	Sand

Maximum strain γ_{max} is the result of numerical analysis or estimation by the author

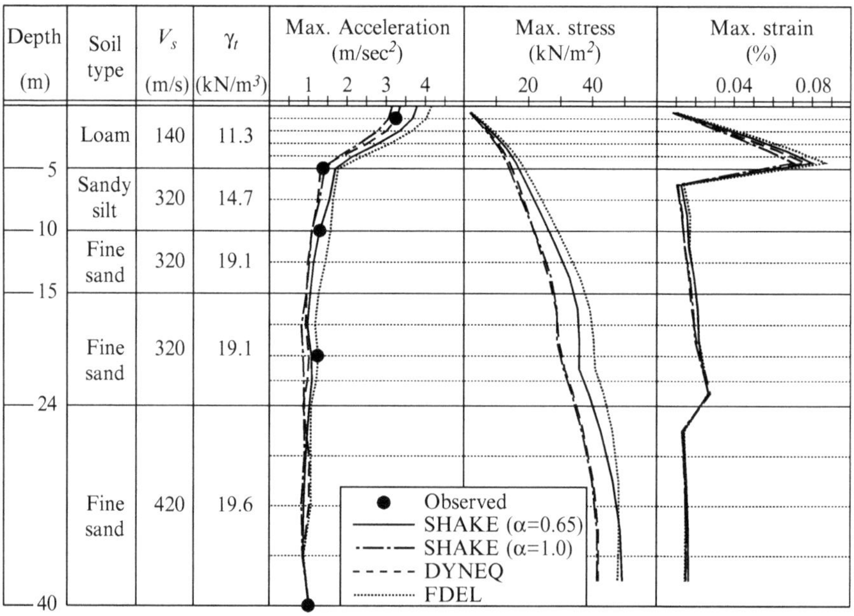

Fig. 14.1 Soil profiles and maximum responses

model. Reference strains at layers above and below GL-10 m are 3×10^{-3} and 5×10^{-4}, respectively. Damping characteristics are set as

$$h = h_{max}(1 - G/G_0) + h_0 \tag{14.1}$$

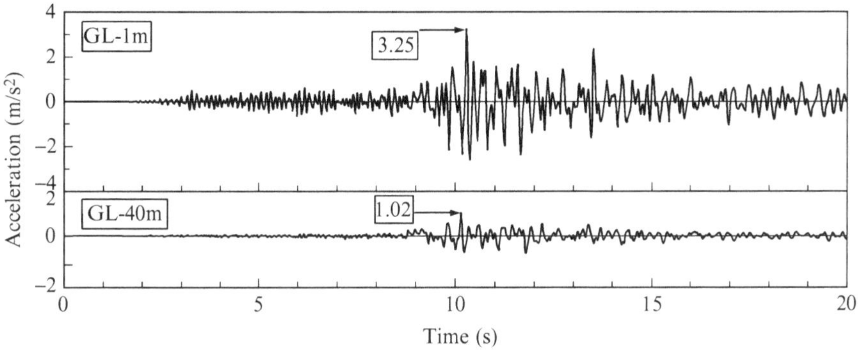

Fig. 14.2 Observed acceleration waveforms

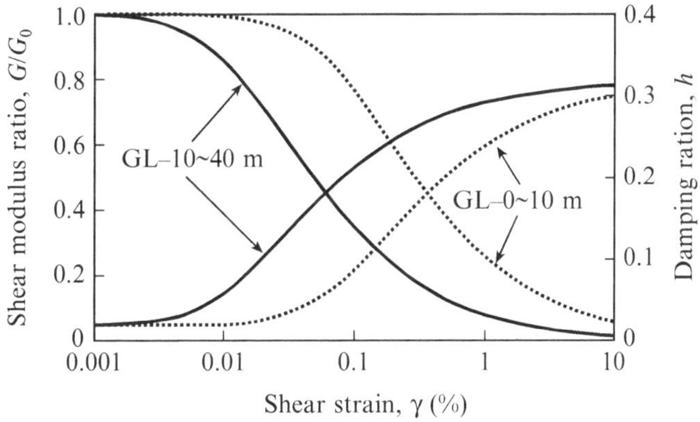

Fig. 14.3 Cyclic shear deformation characteristics

where the first term in the right-hand site comes from the Ramberg–Osgood model (Eqs. (8.10) and (8.11)) and h_0 denotes minimum damping ratio; $h_{max} = 0.3$ and $h_0 = 0.02$ are used. Parameters β of the Ramberg–Osgood can be calculated from h_{max}, and α is calculated from Eq. (8.13). The cyclic shear deformation characteristics of these models are shown in Fig. 14.3. The maximum damping ratio is 0.32, which is greater than 0.3 at strains larger than 10 %. This large damping ratio is hardly observed under the conventional test (undrained test) as explained in Sect. 6.7.2, but is frequently observed in the drained test as in Iwasaki et al. (1980).

Three equivalent linear analyses, SHAKE, DYNEQ, and FDEL, are used. In order to examine the reason of the overestimation of the maximum acceleration in SHAKE (see Sect. 11.4.2), correction factor α to obtain γ_{eff} from γ_{max} in Eq. (11.39) is set 0.65 and 1.0 in the SHAKE analysis. Therefore, four analyses are made in total.

The soil profiles and the maximum responses are shown in Fig. 14.1. The maximum acceleration is amplified a little below GL-5 m, but it amplifies above GL-5 m significantly, which is one of the proofs that amplification occurs at soft soil.

Maximum shear strain is about 0.08 %, which is less than the applicable range of the equivalent linear method, 0.1 % (Ishihara 1982).

Result by SHAKE is shown by the solid line. It looks to agree with the observed record; amplification characteristics below and above GL-5 m are well expressed. However, looking at the agreement in detail, acceleration at GL-1 m is a little larger than the observed one. The acceleration waveform is compared in Fig. 14.4 at GL-1 m. Arrows in the figure are maximum acceleration; it is 3.25 m/s^2 in the observed record, whereas it is 3.68 m/s^2 by SHAKE. Waveform agrees well in general, but difference appears near the maximum acceleration. Similar overestimations are also seen in local maximum accelerations, too.

The maximum acceleration becomes 3.06 m/s^2 in the analysis $\alpha = 1.0$, which is closer to the observed one than the case $\alpha = 0.65$. Acceleration waveform is quite similar with the case $\alpha = 0.65$. If the reason of the overestimation of the maximum acceleration is resonance (see Sect. 11.4.2), overestimation must appear in the $\alpha = 1.0$ analysis, too, but it does not. Therefore, the mechanism of the

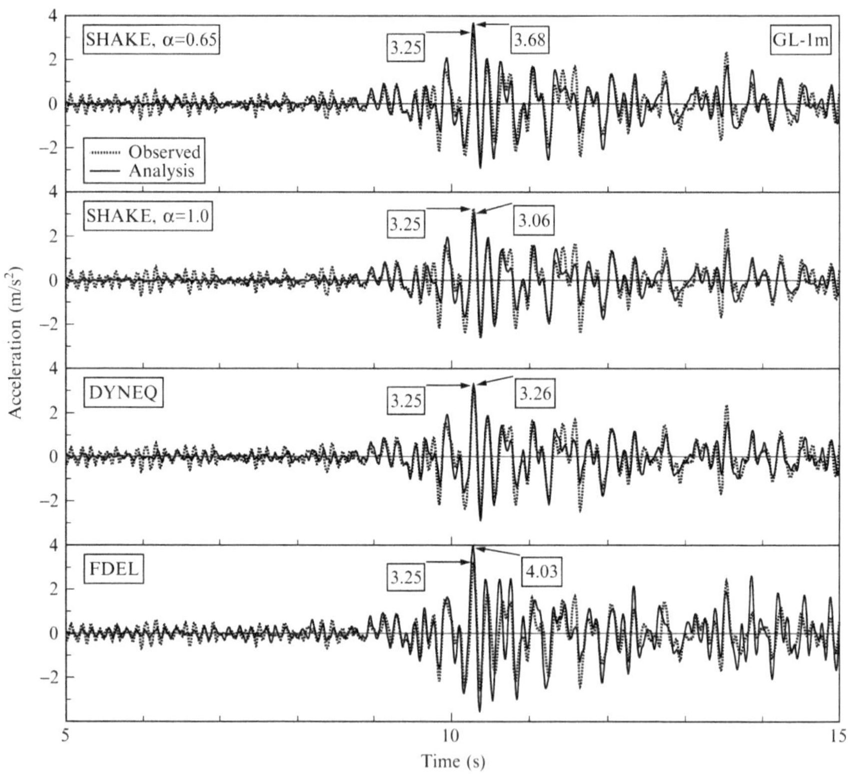

Fig. 14.4 Comparison with observed record (GL-1 m)

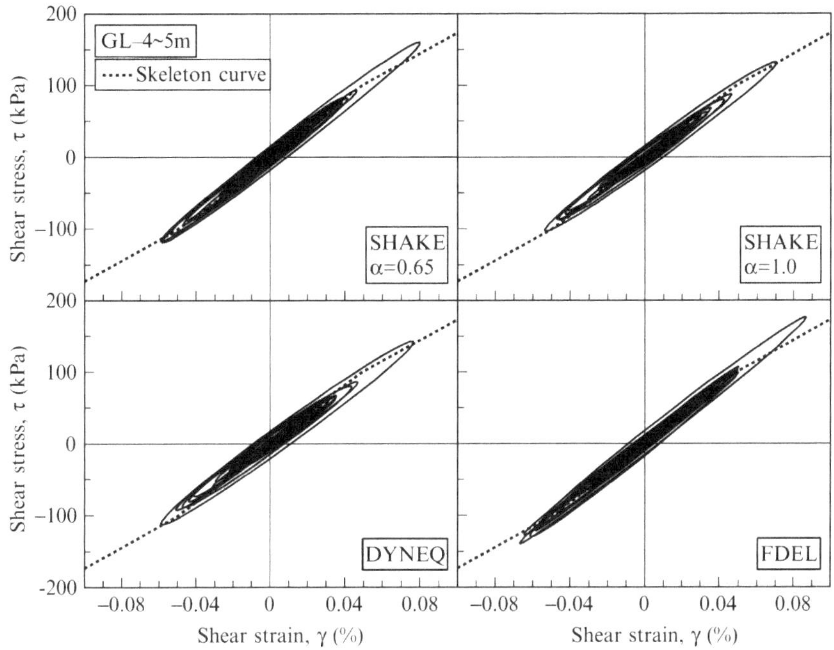

Fig. 14.5 Stress–strain curves

overestimation of the maximum acceleration is overestimation of the shear stress, but not a resonance. This can be clearly seen in Fig. 14.5 where stress–strain curve is compared with input skeleton curve; FDEL and SHAKE ($\alpha = 0.65$) overestimate the maximum stress, whereas DYNEQ and SHAKE ($\alpha = 1.0$) agree with the input data. The reason of the underestimation in SHAKE ($\alpha = 1.0$) comes from another limitation of SHAKE, i.e., underestimation of the amplification at high frequency.

Both the amplification ratio and the response spectrum are compared in Fig. 14.6. The result by FDEL is larger in both cases. The response acceleration by FDEL is very large because it overestimates the amplification around 5 Hz (0.2 s). Both DYNEQ and SHAKE ($\alpha = 1.0$) show similar behavior at low frequency, which is natural considering the formulation of DYNEQ.

Finally, the strain-dependent stiffness, the damping ratio, and the effective strain are compared in Fig. 14.7 at GL-4–GL-5 m where maximum strain is largest. They are constant in SHAKE. Effective strain by SHAKE ($\alpha = 0.65$) is smaller than that by DYNEQ at low strain because it is multiplied by 0.65 in SHAKE, whereas maximum strain is used as effective strain in DYNEQ. The effective strain of FDEL is very small around 5 Hz, which is the reason why FDEL shows amplification much larger than other case as seen in Fig. 14.6. In this meaning, the use of the Fourier spectrum in evaluating the effective strain may not be a good way.

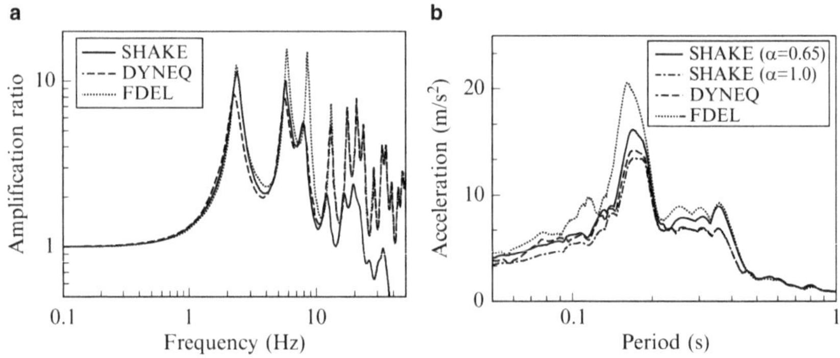

Fig. 14.6 Amplification ratio and acceleration response spectrum

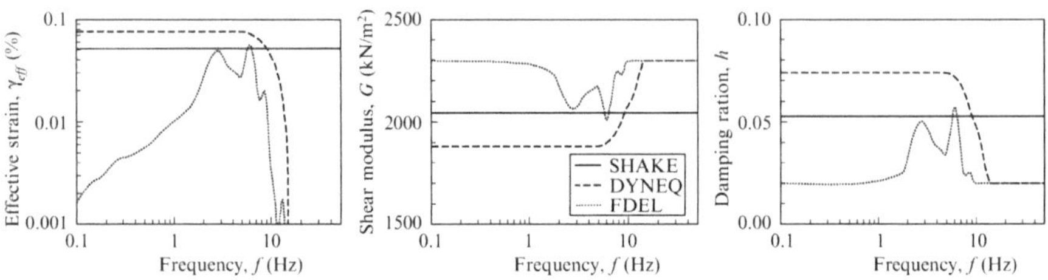

Fig. 14.7 Frequency-dependent characteristics

This example clearly shows that the limitation of SHAKE (overestimate maximum strain) appears even at strain range as small as 0.08 % or less.

14.2 Response at Medium Strain

Earthquake record was obtained at the Shin-Fuji substation, Tokyo Electric Power Company (TEPCO), during the 1983 Kanagawa and Yamanashi prefectures boundary earthquake with magnitude $M = 6.0$. Since the site is located only 12 km from the epicenter, a strong ground motion record is obtained although the earthquake is not big.

Soil profiles at this site are shown in Fig. 14.8. Earthquake records were obtained at the ground surface and at GL-28 m.

Sato et al. (1997) back calculated the frequency-dependent damping characteristics by setting

$$h = h_0 f^{-a} \tag{14.2}$$

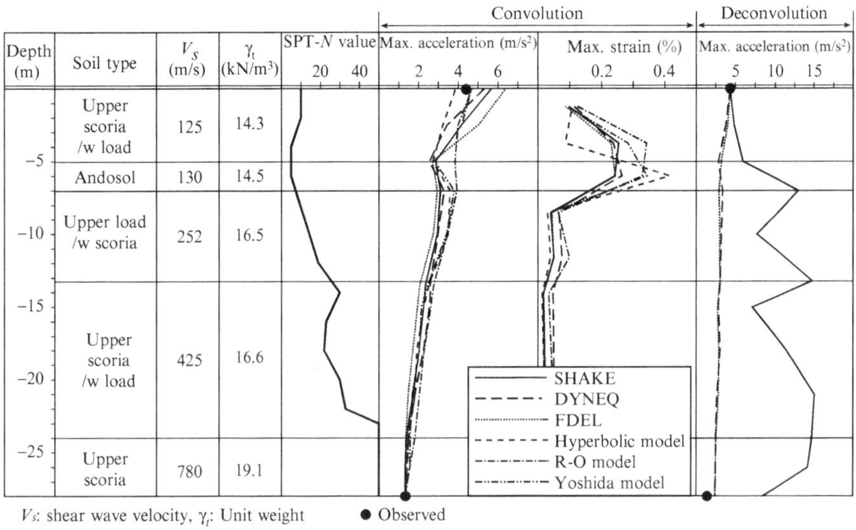

Fig. 14.8 Soil profile and maximum response

Fig. 14.9 Comparison of laboratory test and back analyzed characteristics

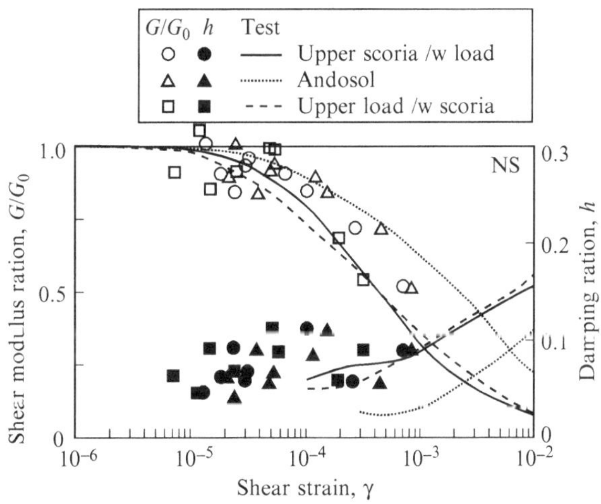

by SHAKE with frequency-dependent damping ratio under the four strong earthquake motions, which is shown in Fig. 14.9 with test data. Obtained damping characteristics agree with the laboratory test result in general. These types of comparison between the laboratory test and the back analyzed characteristics have been made at other sites by Chang et al. (1991), (Suzuki et al. 2009), etc., and the agreement is good in general.

The result of the six analyses are compared, among which three analyses are equivalent methods (SHAKE, DYNEQ, and FDEL) and the rest three are nonlinear methods. The hyperbolic model (Sect. 8.2.3), the Ramberg–Osgood model

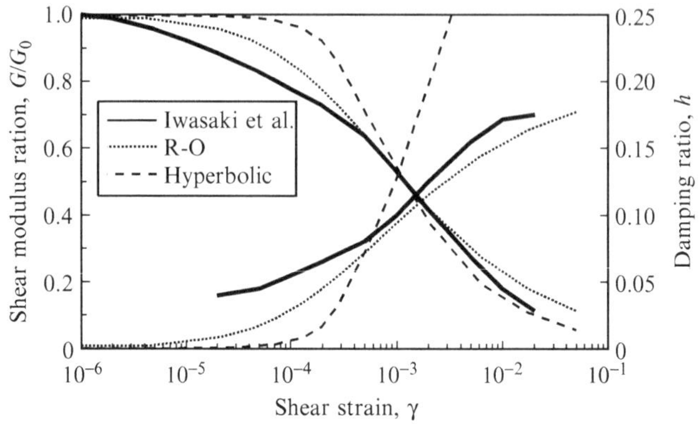

Fig. 14.10 Dynamic shear deformation characteristics used in the analysis (GL-5–7 m)

(Sect. 8.2.4), and the Yoshida model (Sect. 8.2.5) that can express given cyclic shear deformation characteristics perfectly are used as the stress–strain models.

The cyclic shear deformation characteristics for the Holocene clay (Iwasaki et al. 1979) (see Sect. 7.2.1) are used following Tazoh et al. (1987); they are compared with the hyperbolic and the Ramberg–Osgood models in Fig. 14.10. Those by the equivalent analyses and the Yoshida model are not shown in the figure because they accept table-type data and they can express given dynamic deformation characteristics perfectly.

The maximum responses are shown in Fig. 14.8 by the convolution analysis. The observed and the calculated accelerations are compared in Fig. 14.11. In addition, stress–strain curves of the andosol layer where strains are largest are shown in Fig. 14.12.

This case is a boundary or out of the applicable range of the equivalent linear analysis as the maximum strain is about 0.4 %. An agreement with the observed one is good in all analyses in general as seen in Fig. 14.11. Detailed investigation shows, however, that difference appears at the maximum or local maximum places. In other words, there is no significant difference between six analyses except the maximum and the local maximum responses. Therefore, discussion is made by focusing on the maximum acceleration. Three equivalent linear analyses overestimate the maximum acceleration at the ground surface. Overestimation is the largest by FDEL and then SHAKE and DYNEQ. The mechanism of the overestimation by SHAKE and FDEL is already explained. On the other hand, the Yoshida model and the Ramberg–Osgood model show a little larger acceleration than the observed one, and the hyperbolic model shows fairly smaller acceleration.

The shear stress is overestimated in SHAKE and FDEL as seen in Fig. 14.12. On the other hand, maximum stress is smaller than the input skeleton curve in the hyperbolic model, which can also be seen in Fig. 14.10; the hyperbolic model shows smaller shear modulus. It is also seen in the acceleration waveform that the acceleration seems to show upper bound acceleration. On the contrary, both the

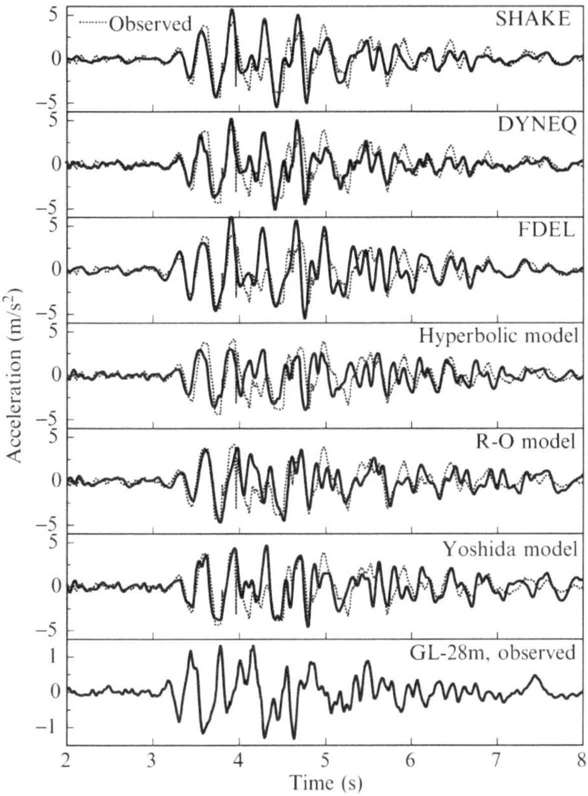

Fig. 14.11 Observed acceleration and comparison of acceleration at the ground surface

Yoshida model and the Ramberg–Osgood model show good agreement with input data at this strain range, which is the reason why they show good agreement with the observation.

As can be seen, difference between the analyses clearly appears near the maximum or the local maximum places, and the rest seems similar. It is also noted that the nonlinear analysis is theoretically the most accurate method, but actual accuracy depends on input data, especially on the modeling of the stress–strain relationships. In this case, accuracy becomes less than the equivalent linear method if stress–strain model is not relevant. This is true in many cases; the use of the relevant stress–strain model is one of the key issues to get accurate result.

The result of the deconvolution analysis is shown at the right side of Fig. 14.8, in which acceleration at the ground surface is used as input motion. It is noted that the scale of the abscissa is different from the convolution analysis. Both DYNEQ and FDEL show maximum acceleration distribution that agrees with the observed one. On the other hand, that by SHAKE seems extraordinary, which comes from the same reason in Fig. 11.12 in Sect. 11.4.2.

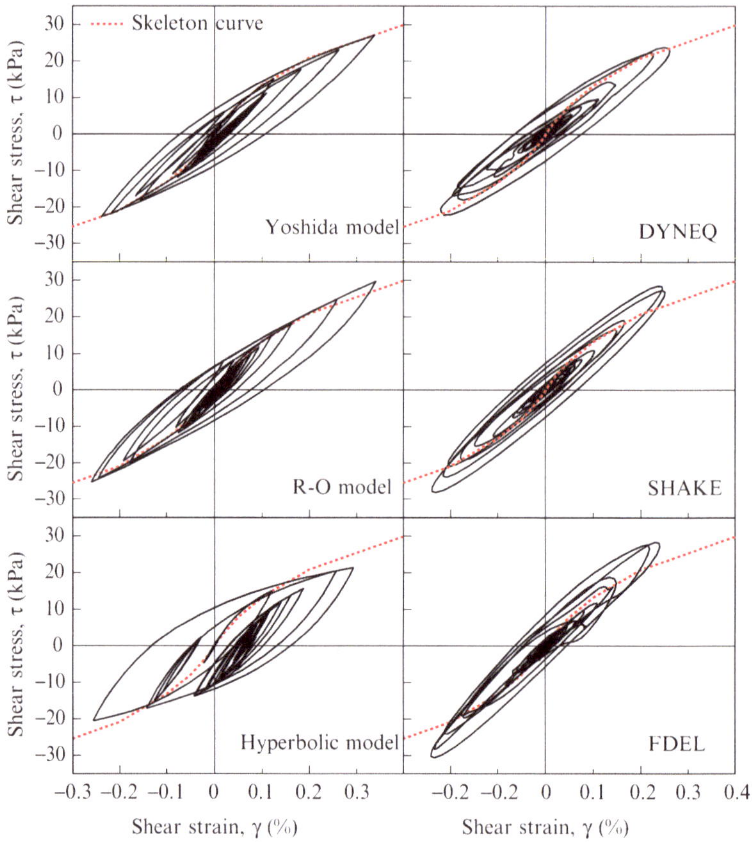

Fig. 14.12 Comparison of stress–strain curves (GL-5–7 m)

Comparisons of the acceleration at GL-28 m are shown in Fig. 14.13a. Agreement by DYNEQ and FDEL is good. There are some differences from the observed record after 4 s, which is supposed to occur as the acceleration at the ground surface saturated around this time. On the other hand, a large pulse wave with maximum 8.3 m/s^2 appears in the acceleration by SHAKE. Then time scale is enlarged around this time and is shown in Fig. 14.13b. There are many high-frequency waves around this time, which suggest that the second disadvantage of SHAKE, the underestimation of the amplification at high frequency, is the reason of this behavior.

The maximum strain 0.4 % is a little outside the applicable range of the equivalent linear method, but simulation is good in general except near the local maximum place. In this sense, this strain range is also an applicable range of the equivalent linear method. Especially DYNEQ shows good result.

The applicability of the deconvolution analysis by SHAKE is discussed in Sect. 11.4.5 and is concluded that it is less than 0.1 %. Actually, this case study with maximum strain 0.4 % is clearly outside the applicable range.

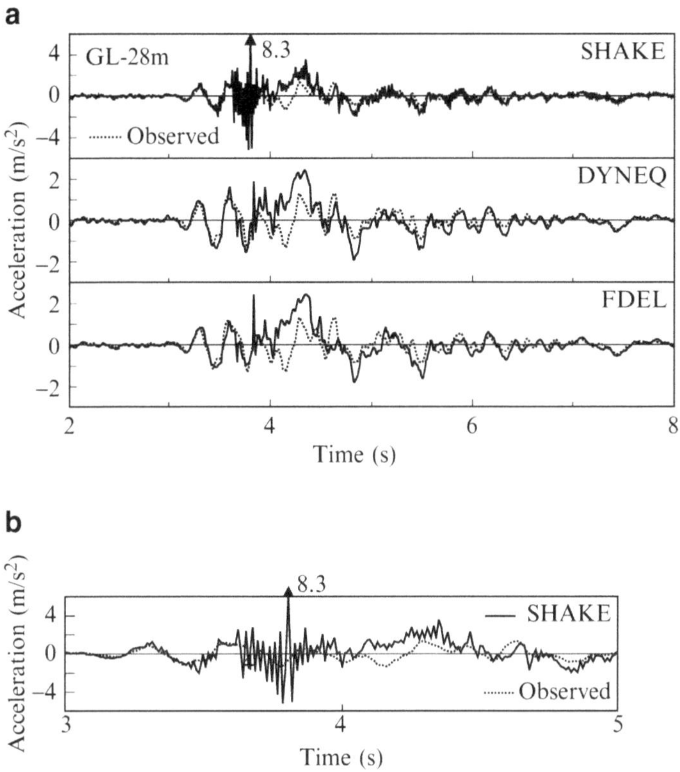

Fig. 14.13 Acceleration waveforms at GL-28 m (**a**) Comparisons (**b**) Enlarge time scale

14.3 Response at Large Strains

Vertical array record at the Port Island during the 1995 Kobe earthquake is introduced in Sect. 1.2, and soil profiles and observed record are already shown in Fig. 1.12. One of the big topics at the Port Island was liquefaction; the ground was covered by the boiled water and sand resulting in about 50 cm settlement of the ground surface. As discussed in Sect. 1.2, however, significant nonlinear behavior occurred in the soft Holocene clay layer called Ma13 beneath the fill, and it affects the response at the ground surface as well as the liquefied layer (Yoshida 1995). The analysis is made by focusing on the nonlinear behavior in this layer (Yoshida et al. 2002) and by using the GL-32.4 m record as input motion, and the behaviors between GL-32.4 m and 16.4 m are interested in. Three equivalent linear methods are used in this case study. Since liquefaction was of great interest at this site, there were many liquefaction analyses and they seem successful in expressing the liquefaction behavior, but nonlinear behavior in the Holocene clay layer was not interested in many of these researches.

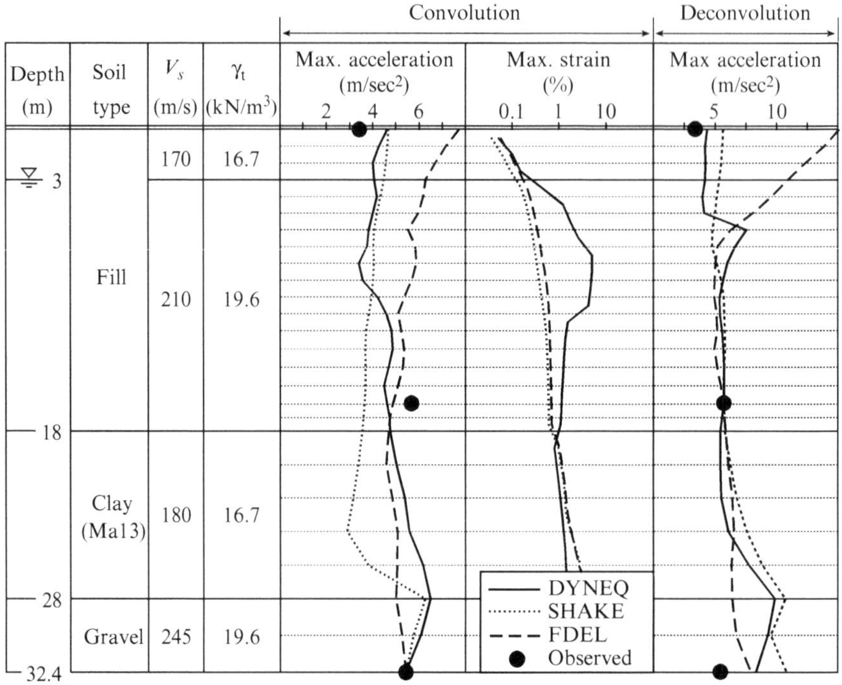

Fig. 14.14 Maximum response

The maximum response is shown in Fig. 14.14; there are significant differences between the analysis and the observed record. It is natural because the fill by the decomposed granite liquefied significantly all thickness. The seismograph at GL-16.4 m also lies within the liquefied layer, but the effect of the liquefaction may not be significant partly because this depth is near the boundary of the Holocene clay layer and partly because it may not liquefy because of the large overburden stress (Yoshida and Ito 1999). Seismographs are not installed within the Holocene clay layer. However, the depth of the seismograph, GL-16.4 m, is close to the clay layer, effect of liquefaction is small. Anyway, behavior of clay layer at large strain is focused on and agreement with the observed record is not discussed. The difference between methods for large strains is focused in this section, not the agreement with the observed record.

The acceleration time histories at the ground surface and those at GL-16.4 m are shown in Fig. 14.15. Analysis by SHAKE does not include high-frequency component, which comes from the underestimation of the amplification at large strain (second disadvantage of SHAKE). Waves with about 5 Hz significantly appear in the FDEL analysis, which is quite a different feature compared with the observed record. Similar high-frequency wave is also seen in DYNEQ analysis, but the magnitude is much smaller. DYNEQ succeeds pulse wave around 6.5 and 8 s well, but other methods cannot explain these pulses. Examples of the stress–strain curves

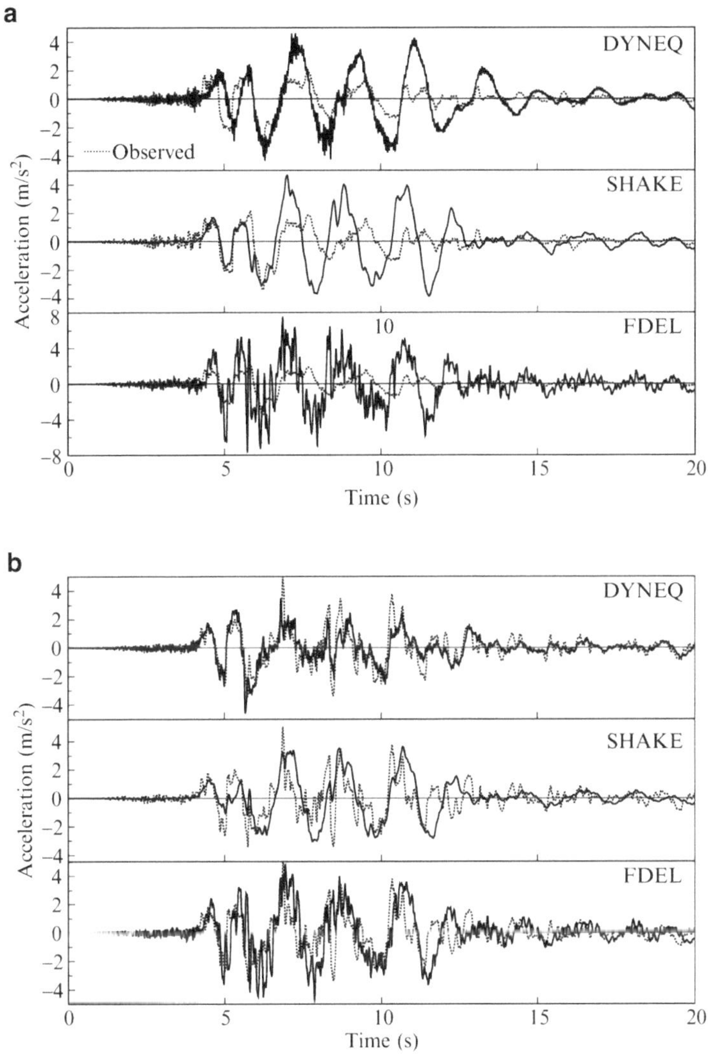

Fig. 14.15 Comparison of acceleration time histories (**a**) Ground surface (**b**) GL-16.4 m

of the fill (liquefied) and the clay layers are shown in Fig. 14.16. It is an elliptic shape in the equivalent linear method. DYNEQ shows a little high-frequency component, but the shape is similar to SHAKE. On the other hand, the disturbance from SHAKE is very large, which is a similar behavior observed in the acceleration time history.

Frequency dependencies are shown in Fig. 14.17. It is noted that stiffness decreases a little around 5 Hz in FDEL, which is strongly related to the large amplitude response around 5 Hz. Fourier amplitude of strain is small around this frequency, which is the reason why FDEL showed large amplification. As both

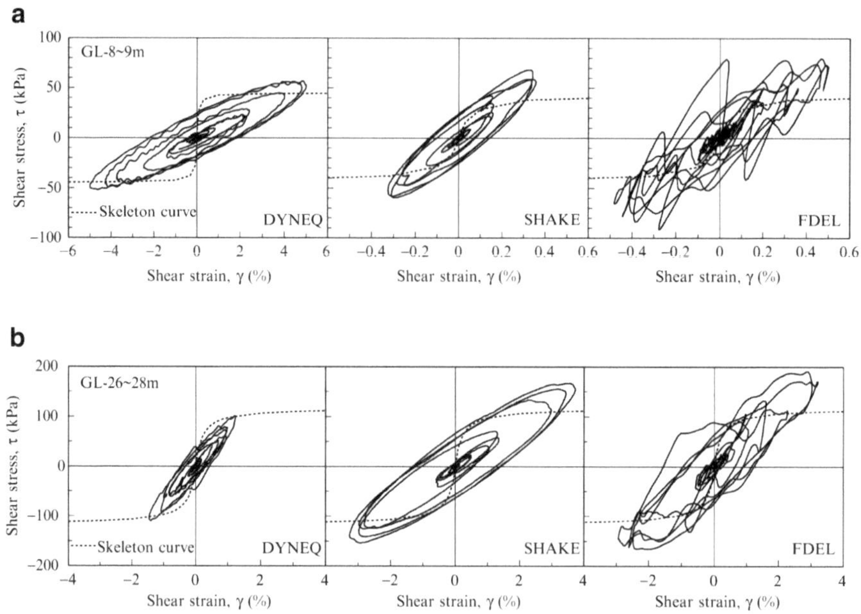

Fig. 14.16 Comparison of stress–strain curve (**a**) Liquefied layer (**b**) Holocene clay layer

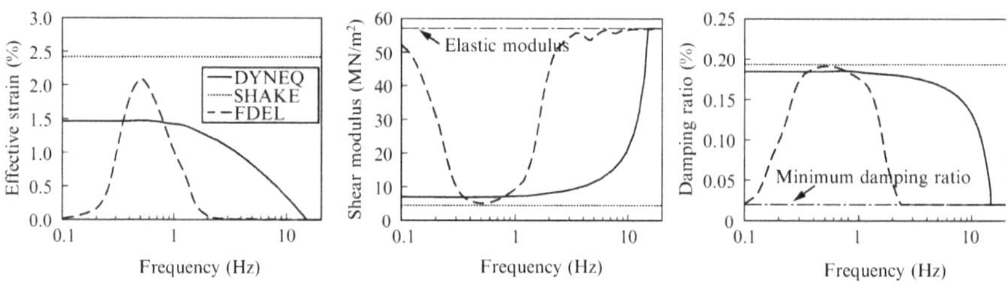

Fig. 14.17 Comparison of frequency-dependent characteristics

DYNEQ and SHAKE use larger effective strain, amplification does not become large in these methods. This indicates that the elastic or nearly elastic region is to be larger when the nonlinear behavior becomes prone.

14.4 Problems in Setting Elastic Modulus

Vertical array record was obtained at the Central Research Institute, Kansai Eclectic Power Company, in the Amagasaki City, Japan, during the 1995 Kobe earthquake. Method to set the elastic modulus is examined through the analysis at this site.

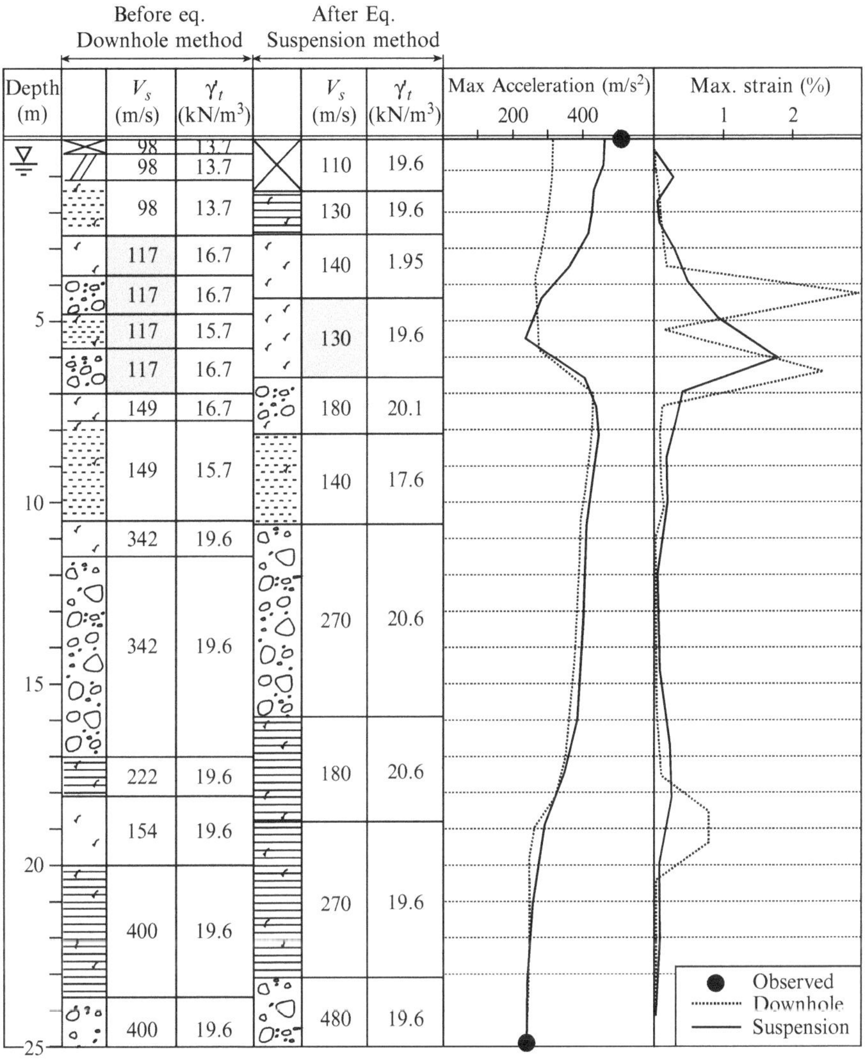

Fig. 14.18 Soil profile and maximum response

Borehole survey including the PS logging was carried out before and after the earthquake at this site (Soeda et al. 1996; Promotion committee on database of strong motion array observation 1998), which is shown in Fig. 14.18 with maximum response by means of SHAKE. Acceleration time histories are compared in Fig. 14.19. Generally, waveforms are similar with the observed record, but the maximum acceleration at the ground surface is significantly underestimated. The reason of this underestimation is discussed in this section.

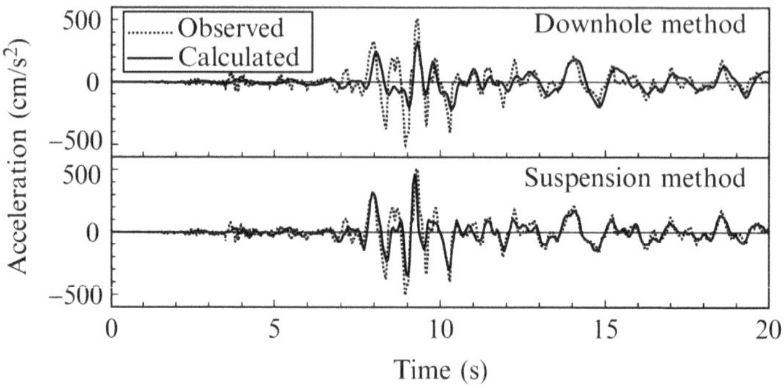

Fig. 14.19 Acceleration time history at ground surface

Fig. 14.20 Schematic figure showing nonlinear behavior of soil (**a**) Normalized by elastic modulus (**b**) Actual behavior

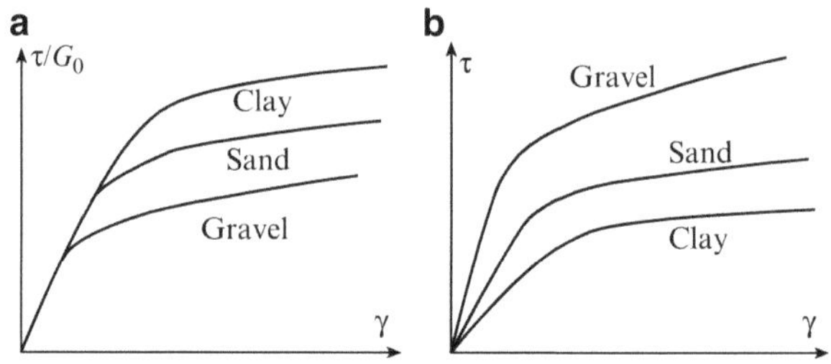

Firstly, the analysis by using the soil profile obtained by borehole data before the earthquake is investigated. The maximum shear strain reaches about 2 % at GL-4 and GL-7 m. Correspondingly, the maximum acceleration becomes small at these layers, which indicates that the earthquake motion reaches upper bound (see Sect. 1.3). By the way, the soil profile indicates that the soil at these layers is gravel, but strong nonlinear behavior in the gravel layer is not natural.

The S-wave velocity is constant to be 117 m/s between GL-3.8–GL-7.0 m, in which silt, fine sand, and gravel are included. The stress–strain curves used in this analysis are schematically shown in Fig. 14.20a. The ordinate is a shear stress normalized by the elastic shear modulus. The nonlinear behavior begins earlier in the gravel and then the sand and the clay. Since the shear moduli are the same between GL-3.8 and GL-7.0 m, the gravel shows smaller shear strength than the sand and the silt, which is the reason why strain at gravel layer becomes large.

By the way, it is very difficult to believe that the shear moduli of three different materials between GL-3.8 and GL-7.0 m are identical. Generally speaking, the elastic modulus of gravel is larger than that of silt. Therefore, actual stress–strain curves are supposed to be as that shown in Fig. 14.20b. If so, shear strength of the gravel is large although the yield strain is small. In summary, constant S-wave velocity caused this problem.

Actually, the result of the suspension method shown in Fig. 14.18, which was carried out after the earthquake, shows that the S-wave velocity of the gravel is

180 m/s and is larger than other layers. The maximum acceleration at the ground surface is still smaller than that of the observed one although a new soil profile is used. Overestimation of the PGA is frequently explained as one of the disadvantages of SHAKE, but this result conflicts it. Modeling of the ground seems not to be relevant in this problem.

The strain at GL-6 m is very large in the strain distribution in Fig. 14.18; this layer is a key layer to control the upper bound acceleration. The S-wave velocity of this layer is 130 m/s, and those above and below this layer are 140 m/s and 180 m/s, respectively; this layer is sandwiched by the layers with large S-wave velocity. As explained in Sect. 1.2, more waves reflect when the wave enters the layer with larger S-wave velocity. Then, kinematic energy concentrates in this layer as the waves reflect more at the upper and lower layers, which result in the large strain in this layer. This is the reason why the strain in this layer is very large.

If this kind of sandwich structure continues in the horizontal direction, the phenomenon explained above will actually occur. However, it is very difficult to believe it. Actually, borehole logs before and after the earthquake are fairly different to each other. This issue seems to be a disadvantage of the suspension method. The borehole log by the suspension method may be more accurate than that by the downhole method. However, there is no guarantee that the same soil profile spread in the horizontal direction.

As can be seen from this example, soil profiles obtained by the accurate method for measuring the elastic modulus are not relevant for the seismic ground response analysis. Therefore, if phenomena explained in this section are obtained, it is better to confirm that it actually spreads in the horizontal direction.

By the way, two methods in the preceding are not successful. Analysis in this site is reported to be very difficult (Soeda et al. 1996), and it is pointed out that the soil data as well as the laboratory test data are necessary in detail density. It is also reported that the prediction of the maximum acceleration is difficult even if the detailed constitutive models are used based on the detailed test data (Soeda et al. 2001). This layer may not be modeled to a horizontally layered deposit as can be seen in Fig. 14.18.

14.5 Effect of Layer Thickness and Choice of Property

Earthquake observation was made at the Shin-Ohta electric power substation in the Tokyo Bay reclaimed land during 1982 and 1989. In addition, detailed soil survey was also carried out as explained in Sect. 6.5. The analysis of this site during the 1987 Chibaken toho-oki earthquake is introduced (Masuda et al. 2001).

Five soil models are made reflecting the detailed soil surveys, which are shown in Table 14.2. Model ① is developed based on the downhole method and the thickness of the layer is 1 m. On the other hand, model ② is developed based on the suspension method and the thickness of the layer is 1 m, too; the S-wave velocity is evaluated layer by layer. A more detailed S-wave structure is evaluated

Table 14.2 Models used in the analysis

		Model ①		Model ②		Model ③		Model ④		Model ⑤	
z (m)	γ'_t (t/m^3)	N	V_s (m/s)	N	V_s (m/s)	N	V_s (m/s)	N	V_s (m/s)	N	V_s (m/s)
2	1.7	2	100	2	Every 1 m	10	Every 20 cm	10	Every 20 cm	2	Every 1 m
10	1.7	8	100	8		40		40		8	
14	1.9	4	190	4		20		20		4	
16	1.75	2	320	2		4		4		2	
19	1.75	3	320	3		6		6		3	
Base	1.8		410		410		410		410		410
Comment		Downhole method		Suspension method		Inversion analysis		Fine mesh		Modify GL-7–GL-8 m	

z depth, γ'_t density, N number of layers

by the inversion analysis from the average V_s between the pickups by the suspension method and used in model ③; V_s is evaluated every 20 cm. This is the most detailed model. Model ④ uses the same layer thickness with model ③, but V_s distribution is the same with model ②. Finally, model ⑤ is the same model with model ②, except that V_s between GL-7 and GL-8 m is replaced to the average of those at the upper and the lower layer. As this layer is sandwiched by the layers with larger V_s, this layer may become a key layer for the upper bound acceleration. The S-wave velocities near the ground surface where V_s is not measured are set 100 m/s. The nonlinear characteristics are the same with all the models.

The analysis is made by SHAKE. Maximum response under the EW input is shown in Fig. 14.21. It is noted that a log axis is used for the strain and that the soil type and V_s distribution are not coincide.

Model ① is the most popular model. It shows somewhat different behavior than the other models; the maximum strain is smaller and the maximum acceleration is larger than those by other models. The maximum strain is largest at GL-9 to GL-10 m deep because V_s is constant to be 100 m/s down to GL-10 m (see 10.3 for the reason).

The accelerations at the ground surface are significantly different between model ① (downhole method) and model ② (suspension method). Strain at GL-4–GL-5 m and GL-7-GL-8 m is very large in model ②, which becomes a key layer to control the maximum acceleration at the ground surface. The maximum acceleration

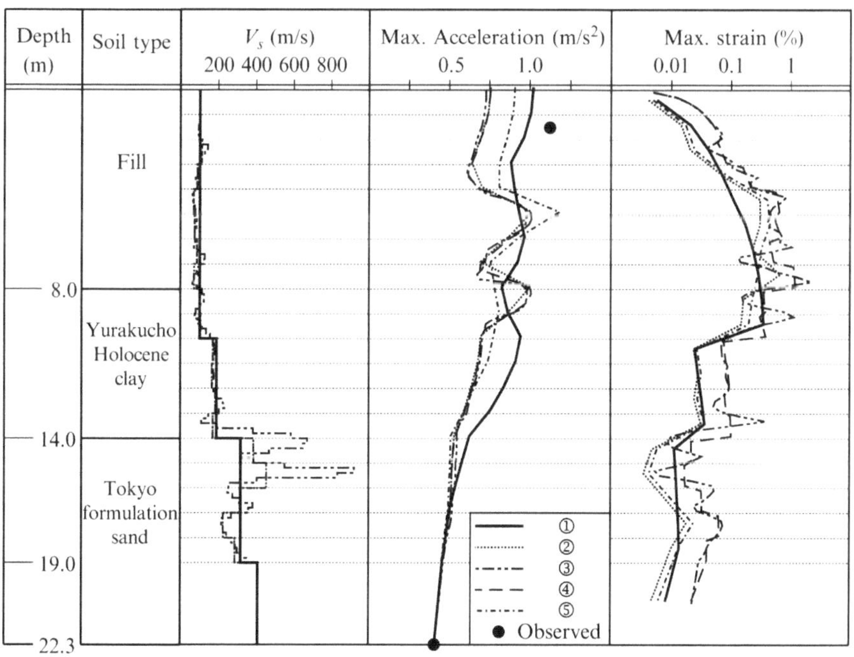

Fig. 14.21 Comparison of maximum responses

frequently becomes large just beneath the key layer, probably because the upper boundary becomes close to a free boundary in which case the composite wave becomes double of the incident wave (see Sect. 1.2.2).

As mentioned, there may be two reasons why strain in these layers becomes large. The one is that the nonlinear behavior is easy to occur as the V_s is small. The other reason is that kinematic energy is accumulated as layers with smaller V_s are sandwiched by the layers with larger V_s (see Sect. 15.6). Therefore, it is important to investigate that the soft soil layer spreads in the horizontal direction when there exists soft layer such as above. Since this site is a fill ground, it is difficult to consider that this kind of soft soil layers spreads in the horizontal directions. Model ② is not a relevant modeling in this sense.

The maximum strain is much larger in model ② in the layers between GL-7 m and GL-8 m compared with that in model ⑤, although models ② and ⑤ are identical except the elastic moduli between GL-7 and GL-8 m. The effect of this difference appears to be about 25 % difference of the maximum acceleration at the ground surface. This example indicates importance to set the elastic modulus.

Finally, there is little difference of the maximum accelerations among the models ②, ③, and ④, which suggest that the thickness of layers of 1 m is sufficiently thin in the equivalent linear method.

Comparison of the observed and the calculated accelerations is shown Fig. 14.22 for model ① where the maximum acceleration is largest. Waveforms agree well in general in the EW direction. However, there are some phase differences. Since the phase by the analysis is late, V_s is supposed to be underestimated. On the other hand, good agreement is obtained in the NS direction for both the amplitude and the phase. This difference between the EW and NS directions may be interesting because it is the same site. Since the acceleration is smaller in the NS direction input, the shear strain is also small. It indicates that average S-wave velocity is well estimated. On the other hand, nonlinear behavior becomes prone under the EW motion as input motion becomes large. The nonlinear behavior at GL-9–GL-10 m

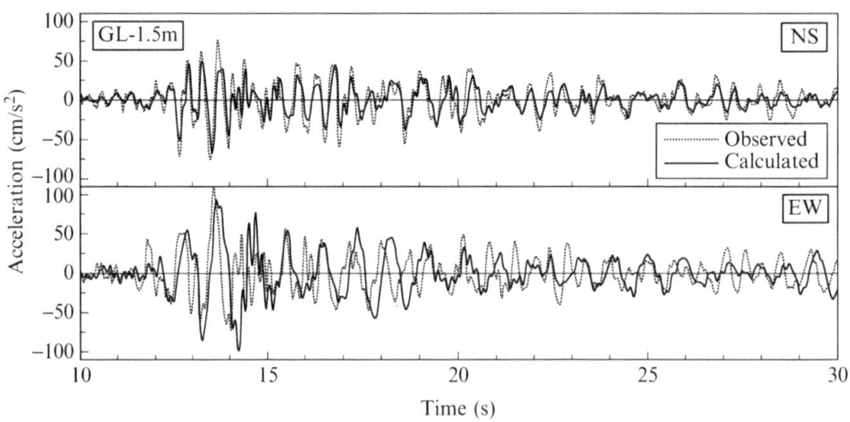

Fig. 14.22 Comparison between observed and calculated accelerations at ground surface

is the most predominant as V_s is set constant down to GL-10 m (see 10.3 for the reason). These observations suggest that a well agreed result is obtained if V_s near GL-10 m is larger keeping the average V_s constant.

As can be seen from the example in this section, setting of the elastic constant affects the nonlinear behavior and the shear strength in the conventional method, in which nonlinear characteristics are evaluated by a product of the in situ shear modulus G_0 and the G/G_0–γ relationships obtained in the laboratory. Therefore special care is required to evaluate the elastic modulus, especially in the weak layer.

References

Chang CY, Mok CM, Power MS, Tang YK, Tang HT, Stepp JC (1991) Development of shear modulus reduction curves based on Lotung downhole ground motion data. In: Proceedings of the 2nd international conference on recent advances in geotechnical earthquake engineering and soil dynamics, St. Louis, pp 111–118

Ishihara K (1982) Evaluation of soil properties for use in earthquake response analysis. In: International symposium on numerical models in geomechanics, Zurich, pp 237–259

Ishihara K, Shimizu K, Yamada Y (1981) Pore water pressures measured in sands deposits during an earthquake. Soil Found 21(4):85–100

Iwasaki T, Tokita K, Yoshida S (1979) Dynamic deformation characteristics of Holocene clay, PWRI report, No. 1504, PWRI (in Japanese)

Iwasaki T, Tatsuoka F, Takagi Y (1980) Experimental study on dynamic deformation characteristics of soils (II), Report of PWRI, vol 153-2 (in Japanese)

Katayama T, Yamazaki F (1990) Array observation and database of earthquake motion, Seminar text of course 154, Institute of Industrial Science the University of Tokyo, pp 1–93 (in Japanese)

Katayama T, Yamazaki F, Nagata S, Sato N (1990) Earthquake observation by a three-dimensional seismometer array and its strong motion database. J Struct Mech Earthq Eng Proc JSCE 422/I-14:361–369 (in Japanese)

Kiban-Kyoshin Network (KiK-net) http://www.kik.bosai.go.jp/kik/ [2010]

Lu Y, Yamazaki F, Katayama T (1989) Earthquake observation at Chiba Test Site, part 5 Amplification characteristics of free field. In: Proceedings of the 20th JSCE earthquake engineering symposium-1989, JSCE, Tokyo, pp 93–96 (in Japanese)

Masuda T, Yasuda S, Yoshida N, Sato M (2001) Field investigations and laboratory soil tests on heterogeneous nature of alluvial deposits. Soils Found 41(4):1–16

Promotion committee on database of strong motion array observation (1998) Strong motion array observation, no. 3, Association for Earthquake Prevention, Tokyo

Satoh T, Horike M, Takeuchi Y, Uetake T, Suzuki H (1997) Nonlinear behavior of scoria soil sediments evaluated from borehole record in eastern Shizuoka prefecture Japan. Earthq Eng Struct Dyn 26:781–795

Soeda E, Kato Y, Matsuda T, Takezawa S, Maekawa F (1996) Example of earthquake response analysis of vertical array record. In: Proceedings of the 51st annual conference of the Japan Society of Civil Engineering, Nagoya, vol III, JSCE, Tokyo, pp 356–357 (in Japanese)

Soeda E, Tamai H, Tanaka M, Takezawa S, Maekawa F. Amano S (2001) An consideration on applicability of nonlinear earthquake response by vertical array observation. In: Proceedings of the 26th JSCE earthquake engineering symposium-2001, JSCE, Tokyo, pp 485–488 (in Japanese)

Suzuki H, Tokimatsu K, Arai H, Midorikawa S (2009) Nonlinear behavior of cohesive soils estimated from strong motion records at K-net Anamizu and JMA Wajima in the 2007 Noto-hanto earthquake. J Struct Constr Eng AIJ 74(645):2003–2010 (in Japanese)

Tazoh T, Sato M, Hatakeyama A (1987) Nonlinear earthquake response analysis based on strong motion seismograph record. In: Proceedings of the 19th JSCE earthquake engineering symposium, JSCE, Tokyo, pp 101–104 (in Japanese)
Yoshida N (1995) Earthquake response analysis at Port Island during the 1995 Hyogoken-nanbu earthquake. Tsuchi-to-Kiso 43(10):49–54 (in Japanese)
Yoshida N, Ito K (1999) Liquefaction of improved ground at Port Island, Japan, during the 1995 Hyogoken-nanbu earthquake. In: Proceedings of the 2nd international conference on earthquake geotechnical engineering, Lisboa, Portugal, pp 531–536
Yoshida N, Tazoh T, Suzuki H (1995) Comparisons of nonlinear dynamic response analysis method. In: Proceedings of the 23rd JSCE earthquake engineering symposium – 1995, Tokyo, pp 49–52 (in Japanese)
Yoshida N, Kobayashi S, Suetomi I, Miura K (2002) Equivalent linear method considering frequency dependent characteristics of stiffness and damping. Soil Dyn Earthq Eng, Elsevier 22(3):205–222

Chapter 15
Effect of Various Factors from Case Studies

Effect of various parameters on the result of the seismic ground response analysis is discussed through case studies.

15.1 Scattering of Nonlinear Property

Morimoto and Suzuki (1983) collected about 50 cyclic shear deformation characteristics test results. Among them data under 98 kPa confining pressure are gathered and compiled to average and 95 % confidence interval of the test data as in Fig. 15.1. Then they carried seismic response analyses for typical grounds, among which analysis on the soft ground (Natural period >0.6 s (Japan Read 1985)) is introduced. Earthquake record at the Kaihoku Bridge (see Sect. 3.2) is scaled so that the maximum acceleration becomes 150 cm/s^2 and is used as the input motion. The maximum acceleration distributions as well as the soil profile and the SPT N-value are shown in Fig. 15.2. Three analyses are made by using the largest stiffness (smallest damping), the smallest stiffness (largest damping), and the average shown in Fig. 15.1, respectively.

Two extreme case studies show a half or double of the acceleration compared with the average property case. The maximum strains in these analyses are about 1 %, and the correspondence between the difference of stiffness and the difference of maximum acceleration seems good; the acceleration is larger as the stiffness increases. This case study indicates that evaluation of stiffness at large strain is important.

They also reported that the difference of the response accelerations between three analyses is small at the stiff ground and it becomes large as the ground becomes softer; the difference is largest in the case in Fig. 15.2. This can be explained that difference of the stiffness is small in the stiff ground as the maximum strain is smaller compared with the soft ground.

N. Yoshida, *Seismic Ground Response Analysis*, Geotechnical, Geological and Earthquake Engineering 36, DOI 10.1007/978-94-017-9460-2_15

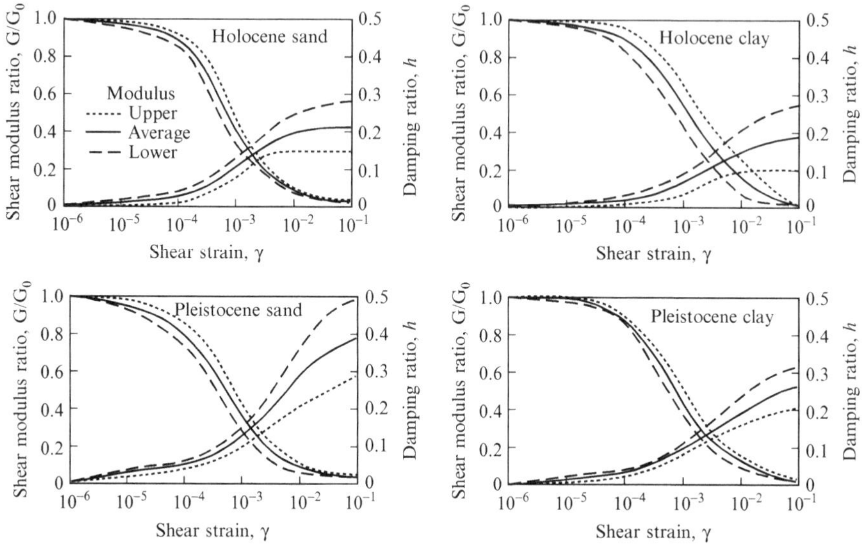

Fig. 15.1 Scattering of cyclic shear deformation characteristics

Fig. 15.2 Result of analysis on soft ground

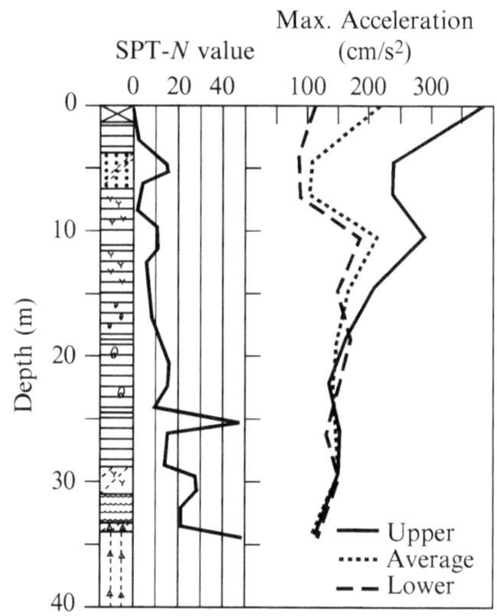

15.2 Scattering of Wave Velocity

Towhata and Ronteix (1988) made a case study by changing the S-wave velocity V_s at small strain in order to see the effect of the empirical equations to evaluate the V_s from the SPT N-value as evaluated V_s scatters between 1/2 and 2 of the actual value. Here V_s is determined by the Monte Carlo method. Other quantities such as layer

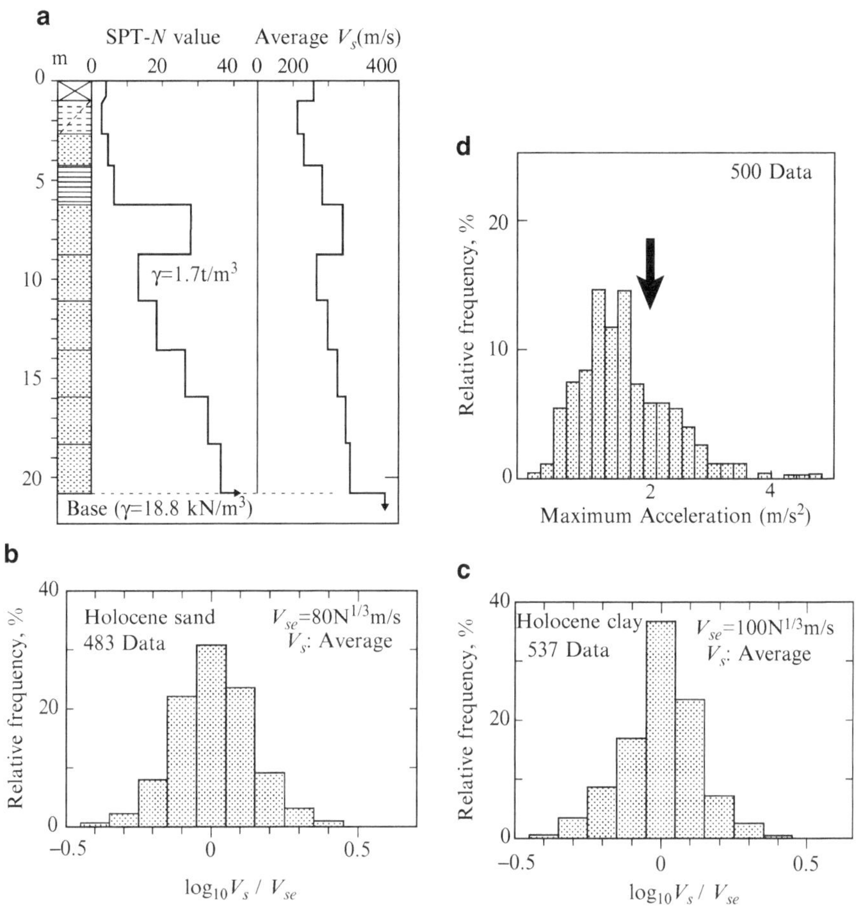

Fig. 15.3 Example of numerical analyses (**a**) Soil profile (**b**) Distribution of PGA (**c**) Relative frequency vs. Vs/Vse for Holocene sand (**d**) Relative frequency vs. Vs/Vse for Holocene sclay

division, SPT N-value, unit weight, and cyclic shear deformation characteristics are kept constant. Three earthquake motions recorded in the Chiba test factory, University of Tokyo (see Sect. 14.1) including the earthquake record during the 1987 Chibaken toho-oki earthquake are used in the analysis. Four grounds used in the past researches are used, among which a typical example is shown in Fig. 15.3.

The soil profiles are shown in Fig. 15.3a; SPT N-value and V_s become smaller in the upward in general, but there is a layer whose SPT N-value and V_s are large in the mid-depth. Distributions of the V_s for the Holocene clay and the Holocene sand are shown in Fig. 15.3b, c, respectively. The abscissa $\log V_s/V_{se}$ scatters within ± 0.5; V_s scatters between 1/3 and 3 of the average V_s, which is a little larger scatter observed in the empirical equations.

The maximum accelerations at the ground surface are shown in Fig. 15.3d against 500 calculations. The arrow in the figure is the result using the average V_s and is about 200 cm/s^2. On the other hand, the average of 500 case studies is about 150 cm/s^2, which is much smaller than the one by the average V_s case.

One of the reasons of this smaller average value is a definition of stress–strain relationships. As described in Sect. 1.3, there is an upper bound acceleration that is controlled by the shear strength of the soft layer as shown in Eq. (1.10), which was

$$\alpha_{\max} = \frac{G\tau_f}{\sigma_v} \tag{15.1}$$

On the other hand, the stress is calculated by the product of G/G_0 and V_s. Since G/G_0–γ relationships are kept unchanged, shear stress becomes smaller as the V_s becomes smaller. All the layers have possibility of the key layer to control the upper bound acceleration because the V_s and the inertia force become smaller in the upper word direction in general.

This case study indicates importance to evaluate the V_s accurately as the V_s affects the shear strength, hence the maximum acceleration at the ground surface.

15.3 Past Blind Tests

Accuracy of the method of the seismic response analyses can be evaluated from the blind test by which data for the seismic ground response analysis is shown without showing the rest. This method, however, is not a perfect method to evaluate the accuracy as there are several reasons that show well agreed result. It can be obtained by, for example, accidentally or good engineering judgment. Therefore, it just guarantees that the method has capability to explain the result. However, many engineers have the tendency to believe that the method or the computer program is good regardless of the reason (Yoshida 2008). Past blind tests or comparison analyses indicate that result scatters even if the same computer program is used.

Figure 15.4 shows summary of the blind test at the Ashigara site (IASPEI/IAEE joint working Group on ESG 1992). Figure 15.4a shows the ground conditions. Here, the earthquake record at the KR1 site where rock outcrops is given and responses at other sites such as KS2 and KD2 are required to output. It is noted that scale is different in the vertical and the horizontal directions as shown in Fig. 15.4a and there is no multidimensional behavior (IASPEI/IAEE joint working Group on ESG 1992). In other words, this is a textbookish example. The distribution of the maximum accelerations is shown in Fig. 15.4b. It is seen that the standard deviation is very wide and the range within 1/4 of the data is also wide. It is also seen that there are significant differences between the observed and the calculated accelerations. They indicate that accuracy of the analysis is not so good in general.

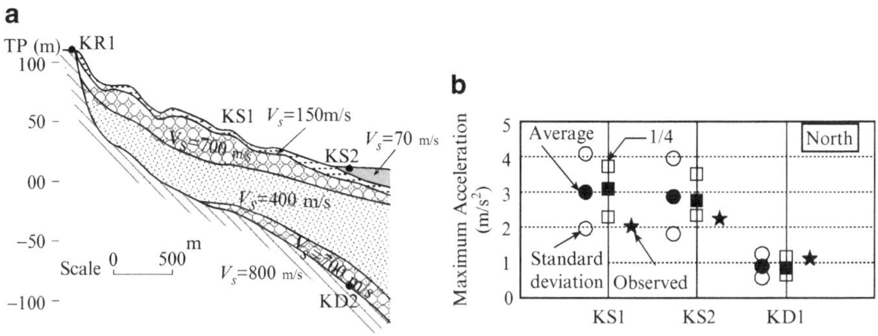

Fig. 15.4 Blind test at Ashigara site. (**a**) Ground condition. (**b**) Scattering of maximum accelerations

There are not a few ones that are supposed to use SHAKE, but results are known to scatter. As discussed later in Sect. 15.8, one of the reasons of this scattering is the definition of the base where the incident wave is supposed to be common at both KR1 and KR2 sites (Kaneko 1993). As typically seen in this example, accuracy of the seismic ground response analysis depends not only on the potential of the computer program but also on the engineering judgment.

Beside this example, there are not a few blind tests and simultaneous analyses (result is known a priori). There were significant scatterings in all these comparisons. As an example, the blind test by VELACS, in which various centrifuge test results are required to simulate, showed that there is no computer program that shows good result for all types of tests and there is also no test that all the computer programs showed good result (Arulanandan and Scott 1993). As such, seismic ground response analysis is a difficult problem even for the professional researchers.

Then the problem occurs on how to make a seismic ground response analysis for the ordinary engineers. It cannot be expected that good result is obtained by using a particular computer program. The author strongly suggests cumulating experience of the seismic response analyses. Vertical array records, shake table tests, and centrifugal tests are good examples to cumulate the experience. Characteristics of the computer programs can be understood by making many simulations. The experience of the accuracy is also cumulated by solving the known problems.

In the next chapter, several case studies are introduced which will help the engineer where to be focused and how to recognize the results.

15.4 Pulse Waves as Result of Numerical Integration

A pulse wave may appear in the nonlinear seismic ground response analysis. An example is shown in Fig. 15.5 where soil profiles and maximum responses of the nonlinear analysis at the Tokyo area are shown (Association of the Development

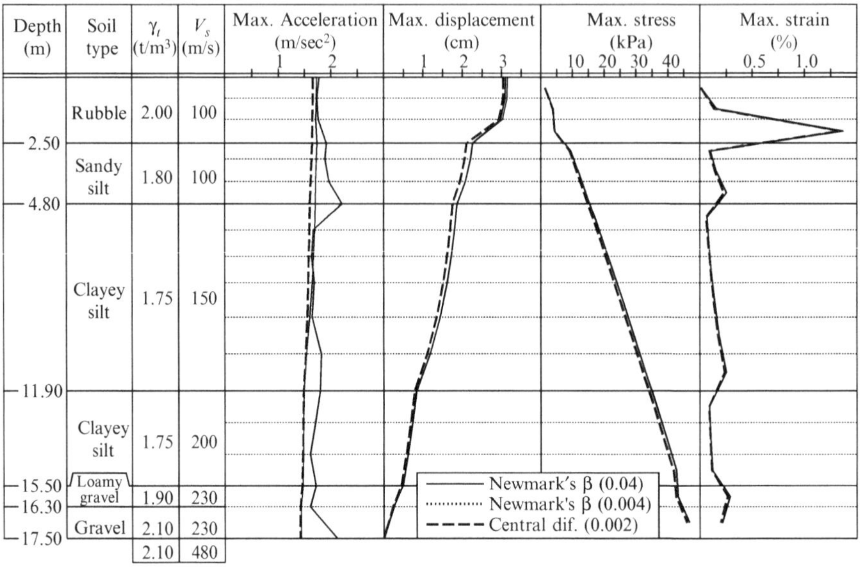

Fig. 15.5 Soil profile and maximum responses

of Architectural Research 1998). The earthquake motion is a synthesized wave that is made considering the fault in this area, and the original time increment Δt of this wave is 0.04 s. Three nonlinear analyses are made. Two analyses use Newmark's β method with $\Delta t = 0.04$ s (original) and Δt=0.004 s, and the rest uses central difference method with $\Delta t = 0.002$ s. Cases by Newmark's β method with $\Delta t = 0.004$ and the central difference method agree well, but maximum acceleration by Newmark's β method with $\Delta t = 0.04$ s shows much larger accelerations than that by other analyses. One can recognize that this case is extraordinary by comparing the results of the other analyses, but it may be difficult to believe it if only the result of this case is shown.

The acceleration time histories obtained by Newmark's β method are compared at two depths (GL-4.8 and 17.5 m) where accelerations are much larger as in Fig. 15.6a in order to find the reason why this extraordinary result appears. Generally two accelerations agree well except a local maximum places. The analysis with $\Delta t = 0.04$ shows pulse wave that attenuates quickly. This pulses are the reason why the case with $\Delta t = 0.04$ shows larger maximum accelerations. For example, the maximum acceleration was 2.21 and 1.42 m/s^2 when $\Delta t = 0.04$ and $\Delta t = 0.004$, respectively.

The stress–strain curves at layers just above and below the large acceleration are shown in Fig. 15.7 for the case of $\Delta t = 0.04$. The maximum acceleration (pulse) occurs at 32.92 s at GL-4.8 m. State points at this time are shown by an arrow in Fig. 15.7a; strain reversal or unloading occurs at this time. The stress–strain behavior between the time ① and ② in Fig. 15.6 where the acceleration changes

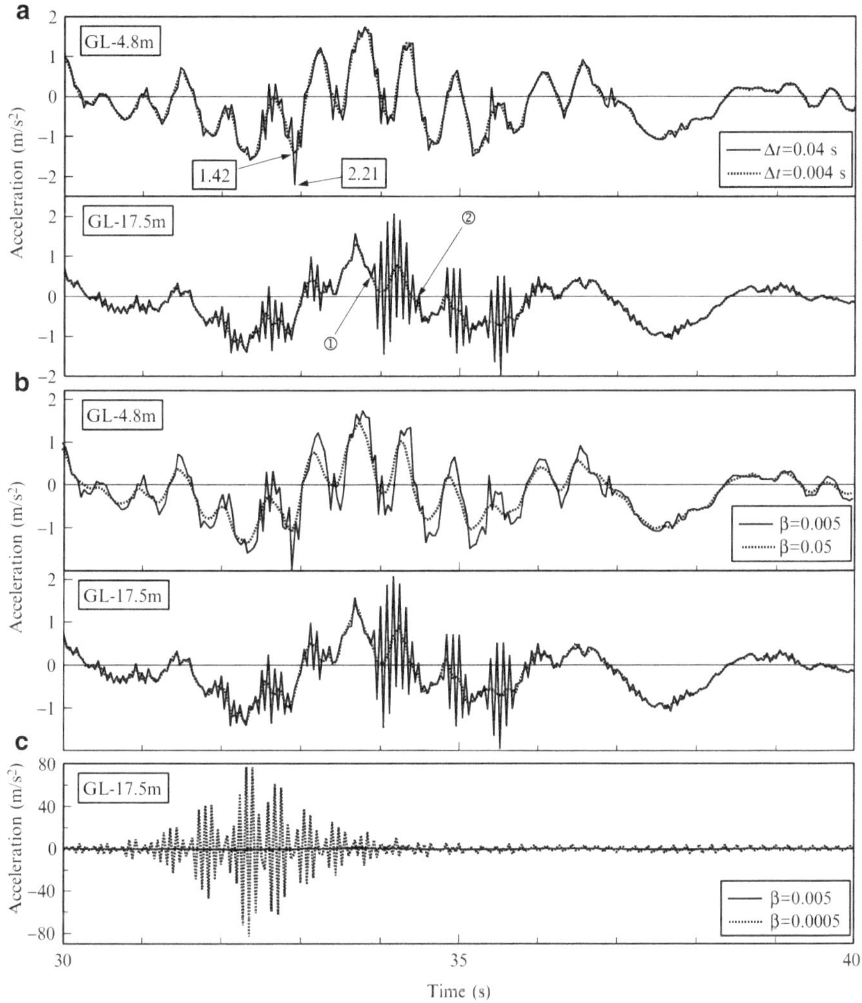

Fig. 15.6 Acceleration time history during 30–40 s (**a**) Effect of time increment (**b**) Effect of stiffness proportional damping (**c**) Case with very small damping ratio

significantly is shown by a solid dashed line in Fig. 15.7a, but extraordinary behavior is not seen in Fig. 15.7. It suggests that unloading relates the pulse wave.

The mechanism of the pulse wave is schematically shown in Fig. 15.8. Let us consider that the behavior until point A is already calculated and the next time increment from point A is going to be calculated. At the beginning, the program uses the tangent modulus for the numerical integration, which is the line n. Then a strain increment $\Delta\gamma$ is obtained by solving the equation of motion. At this stage, since the program used the line n, the resulting point is point B. In other words, stress at point B is in equilibrium with the external load (inertia force). On the other hand, since

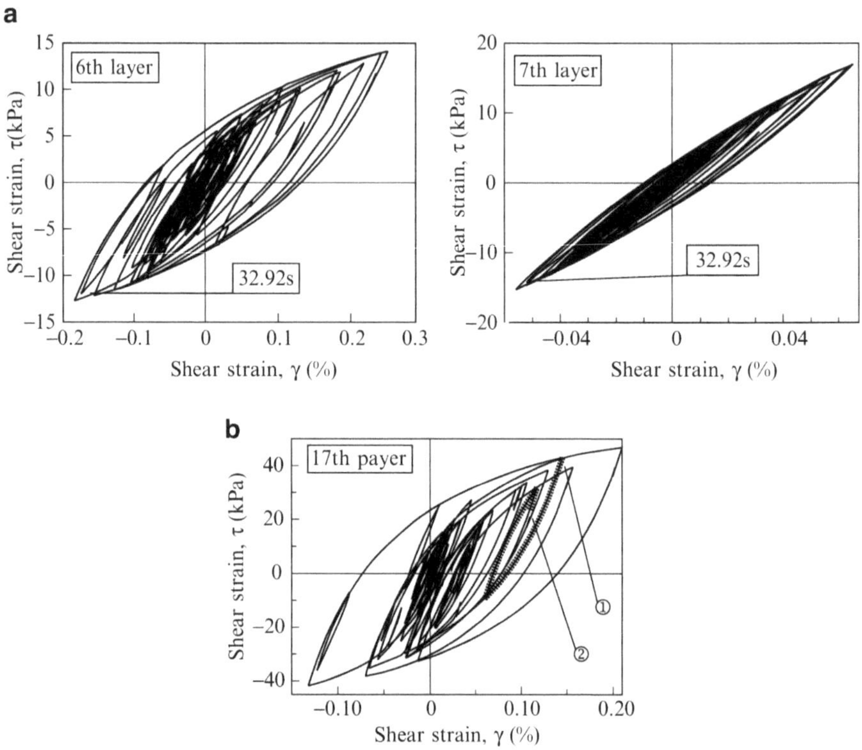

Fig. 15.7 Stress–strain curve when $\Delta t = 0.04$ s. (**a**) Layers just above and below GL-4.8 m. (**b**) Layer above GL-17.5 m

Fig. 15.8 Schematic figure showing unbalanced force at unload

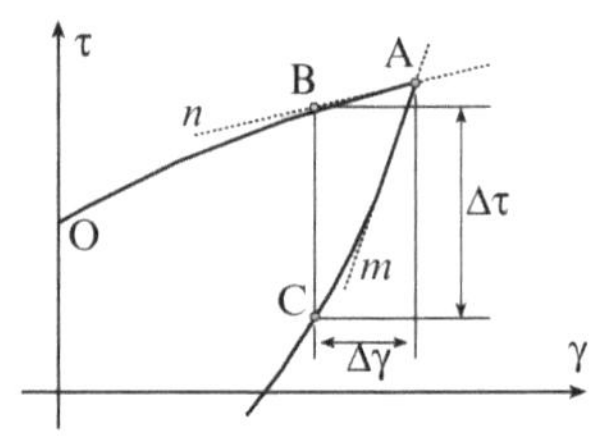

strain reversal occurred, the actual tangent modulus is line m and the resulting state point is point C. Therefore, external load and internal stress are different by $\Delta\tau$. This stress difference is called an unbalanced force, and it should be zero in the theory.

There are several methods on how to treat the unbalanced force; one may choose a method depending on the computer program. In the static analysis, iterative procedure is frequently employed in each incremental analysis to decrease the unbalanced force. In the dynamic analysis, however, simpler methods are frequently employed. The simplest method is to neglect the unbalanced force. Since the cyclic loads are applied in the seismic ground response analysis, unbalanced force has a tendency to be canceled at the loading in the opposite direction. Therefore, this method is frequently sufficient in the practical use. The next simpler method is to carry the unbalanced stress into the next time increment; the unbalanced force is added to the inertia force in the calculation of the next increment. The unbalance may appear at each stage, but the total amount is kept small in this method. The most exact method is to employ the iterative procedure same as the static analysis in order to make the unbalanced force as small as possible.

The analyses in this subsection employ the second method; unbalanced force is carried into the next increment. When unloading occurs, large external load is applied because of the large unbalanced force after the point A, and it creates a pulse wave. Exactly speaking, this pulse wave can occur in the ordinary case when there is unbalanced stress, because the external load increment is applied as a step function in the step-by-step time integration method. However, forces are absorbed in the damping term at first. So, a pulse wave does not appear if the load increment is sufficiently small, but it appears when the load increment becomes large or stiffness becomes small. In this sense, the pulse wave can appear in the ordinary state when the acceleration increment becomes large, especially in the liquefaction analysis where stiffness becomes nearly zero. It is, however, noted that this pulse wave does not affect other quantities such as displacement, stress, and strains as can be seen in Fig. 15.5. In addition, pulse wave attenuates quickly such as a damped free vibration.

This kind of pulse waves occurs at the base layer easier. Lumped mass is a half at the base compared with other layers because it is a half of the above layer at the base, whereas it is sum of mass of the upper and lower layers. Therefore, change of acceleration is about double of the other layer. In addition, unbalanced force just above and below the layer may cancel out at the nodes above the base. Pulse wave can appear even when iterative procedure is used if the number of iteration is not sufficient.

Pulse wave can be erased by using a smaller time increment. Actually, pulse wave is not seen in the analysis with $\Delta t = 0.004$ s, because the magnitude of the unbalanced force becomes small. The time increment $\Delta t = 0.04$ s is a fairly large value, which is the reason why pulse wave appears frequently. The time increment is usually 0.005 s or 0.01 s in the practical analysis, in which case possibility of the pulse wave becomes much smaller.

As mentioned at the beginning of this subsection, it is difficult to find the appearance of the pulse wave by looking at the maximum acceleration distribution such as Fig. 15.5. It can be found by drawing the acceleration time histories, but drawing the acceleration time history at all nodes requires significant work. Alternately, one can solve this problem by adding an analysis in which the time increment is reduced a half or less. If the results of two analyses are similar, one can confirm that pulse wave does not occur.

Analysis using a half time increment, for example, requires computing time two times longer. Generally speaking, one-dimensional analysis is very fast, several seconds at maximum, in which case double or several times longer computing time is accepted in the engineering practice. However, it may not be accepted for the multidimensional analysis because computing time is usually much longer than that of one-dimensional analysis. This problem may be solved by adjusting the damping term.

The stiffness proportional damping, as a part of the Rayleigh damping, is used in the analyses in this subsection ($\beta = 0.005$). Since average V_s of this ground is 155.7 m/s, predominant period T is about

$$T = \frac{V_s}{4H} = \frac{2\pi}{\omega_i} = 2.24\ \text{s} \tag{15.2}$$

Then damping ratio is about 0.7 % at the first mode. The result of the analysis in which damping ratio is 10 times larger than it ($\beta = 0.05$) is compared in Fig. 15.6b. Change of the waveform may be acceptable, and pulse wave is successfully suppressed. On the other hand, damping ratio is reduced 1/10 ($\beta = 0.0005$) in Fig. 15.6c, in which significant pulses appear and actual solution (solid line) cannot be seen well. This example shows that there is a certain range of the damping ratio under which result of the seismic ground response analysis is not affected much and can suppress the pulse waves.

15.5 Equivalent Linear vs. Truly Nonlinear

As explained in the preceding, equivalent linear method is not an equivalent method but an approximation method. Therefore, there are discussions on the applicability of the analysis. SHAKE and FDEL are explained to overestimate maximum acceleration and maximum shear stress even when the maximum strain is only 0.08 % in Sect. 14.1. Discussion whether this overestimation is acceptable or not is not made here because it depends on how the result is used.

Difference of the damping term is hardly discussed in the past researches to compare the equivalent linear method and the nonlinear method. The Rayleigh damping is usually or frequently used in the nonlinear analysis, whereas the damping term of this kind is not considered in the equivalent linear analysis. The stiffness proportional damping is the most important damping in the Rayleigh damping, because it suppresses pulse wave at high frequency as the damping ratio increases with frequency.

In the case study here, stiffness proportional or constant damping is introduced, and both Rayleigh damping (stiffness proportional damping) and constant damping are considered by introducing the modal damping (see Sect. 12.2.2 in detail), by which comparison between equivalent linear and nonlinear analyses can be made under the same conditions.

The improved equivalent linear method DYNEQ and truly nonlinear method by using DYNES3D (Japan Read 1985) are used for comparison. In addition, the analysis by SHAKE is also made as a representative conventional method. The damping ratio is set to 2 % at the first mode; the value of β is 0.0026 corresponding to the natural period 0.418 s. Both the constant and the frequency proportional damping are considered. In the following figures, the result by the equivalent linear analysis is called "frequency" and that by the nonlinear analysis is called "time," whereas the result by the frequency proportional damping (or stiffness proportional damping) is called "proportional" and the constant damping is called "constant."

The ground and the earthquake motion used in the previous section are used in this analysis, too. The time increments are set sufficiently small so that pulse does not appear. It is determined considering the constant damping case because the pulse wave is the easiest to occur as the damping at high frequency is much smaller than that of the Rayleigh damping. The time increment is set to $\Delta t = 0.001$ s.

Soil profiles and the maximum response are shown in Fig. 15.9. Results of the elastic analysis are shown in the middle of the figure. The maximum acceleration is almost the same at GL-4.8 m, but this is supposed to be accidental. The maximum acceleration distribution is classified into two groups: the cases with the constant damping (solid line and dashed line) show larger acceleration and strain than the cases with the frequency proportional damping (dotted line and chained line). Waveforms are not shown in this book, but they are almost identical except that high-frequency waves are seen in the cases with constant damping. These high-frequency waves result in the difference between the maximum accelerations.

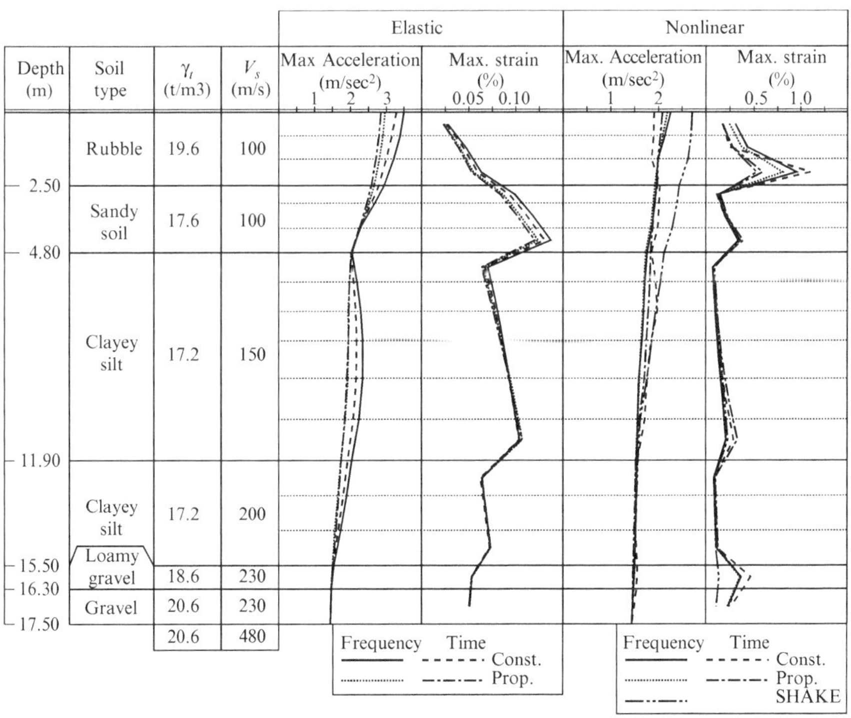

Fig. 15.9 Soil profiles and maximum response

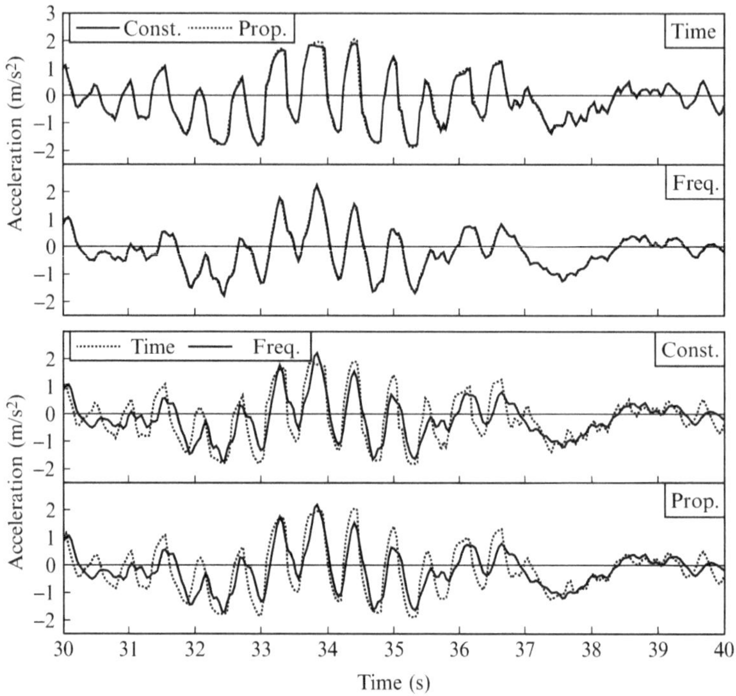

Fig. 15.10 Comparison of accelerations at the ground surface

The result by SHAKE is added in the left side of Fig. 15.9. SHAKE shows larger acceleration than other analyses, which mechanism is explained frequently in this book. The maximum accelerations by the other analyses are almost the same, which may be upper bound acceleration as a result of the nonlinear behavior between GL-1.6 and 2.5 m as the maximum strain in this layer is large.

Figure 15.10 shows comparison of the acceleration waveform at the ground surface near the time at the maximum acceleration so that the differences can be seen easily.

Comparisons between the constant damping and the proportional damping are shown for the upper two figures in the nonlinear and equivalent linear analyses, respectively. Acceleration waveforms are almost identical in the nonlinear analysis except about 34 s where accelerations become maximum. If waveforms are investigated in detail, however, it is found that waveform by the proportional damping analysis is smoother than that by the constant damping analysis. The same tendency can be seen in the equivalent linear analyses, but differences are very small.

Next, differences between the numerical integration schemes are compared in the lower two figures. Slope of the acceleration quickly becomes small near the peak in the nonlinear analysis which appears as a plateau of the acceleration. As can be seen in Fig. 15.11, it occurs because shear stress reaches close to shear strength,

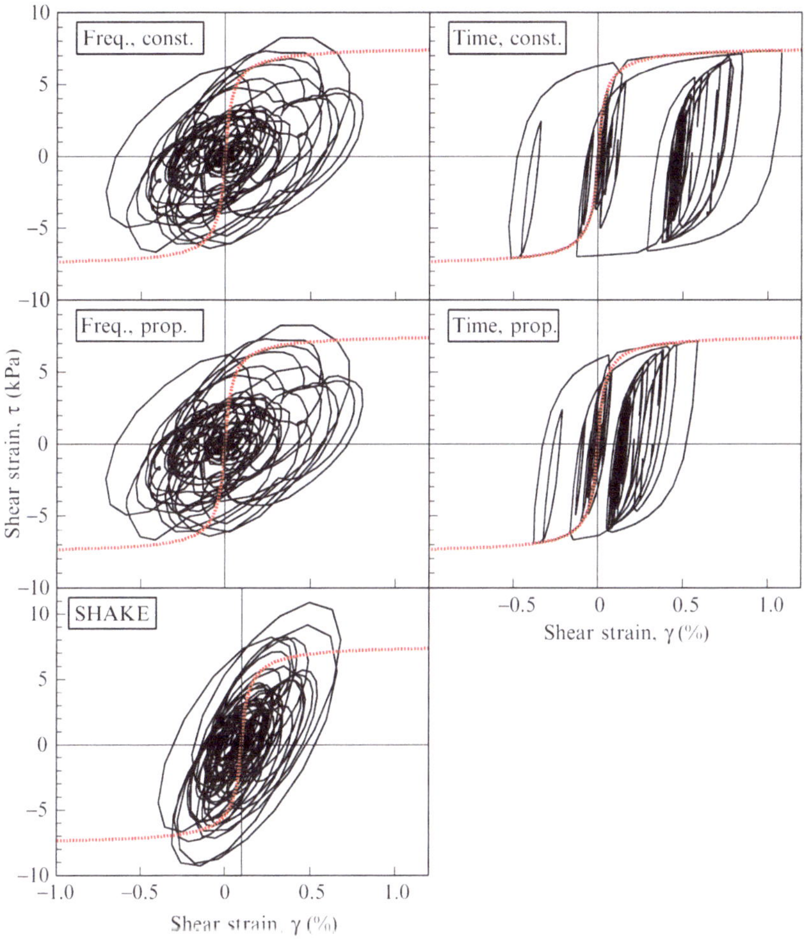

Fig. 15.11 Comparison of stress–strain curves

and the acceleration reaches the upper bound acceleration. On the other hand, these phenomena are not seen in the equivalent linear analysis. As stress–strain curve is an elliptic function in the equivalent linear analysis, plateau region does not appear in the shear stress as well as the acceleration.

Stress–strain curves at the third layer (GL-1.6–2.5 m) where strain is the largest are compared in Fig. 15.11. It is noted that pure stress is used in the nonlinear analysis, whereas the real part of the stress is used in the equivalent linear analysis. Therefore, constant or proportional damping terms are included in addition to the hysteretic damping in the stresses.

Since hysteresis curve is elliptic in the equivalent linear analysis, as explained several times in the preceding, there are some differences between the specified skeleton curve and stress–strain curve. However, it is clearly seen that stress–strain curve by DYNEQ is much improved than that by SHAKE.

The strain by the proportional damping case is smaller than that of the constant damping cases in both the equivalent linear analysis and the nonlinear analysis. Especially, differences are large in the nonlinear analysis. This can be explained by considering the load carrying capacity of three terms, i.e., the inertia, damping, and the restoring force terms that compose the equation of motion. In the nonlinear analysis, as can be seen in Fig. 15.11, stress–strain curve shows plateau when strain becomes large. It means that stiffness term does not carry load. Then the inertia force increment is carried by the damping term. Therefore, the velocity increment is smaller in the proportional damping than the constant damping as the damping is larger in the proportional damping, which results in smaller displacement increment. In other words, the proportional damping suppresses the strain or displacement at large strains.

Finally, the spectra are compared in Fig. 15.12. The amplification and the response spectra are quite similar to each other in the frequency region less than the predominant frequency (2.4 Hz). This can be recognized by the fact that the acceleration waves are quite similar as shown in Fig. 15.10. Some differences appear at high-frequency region, but this difference may be acceptable in many practical problems, although limitations of SHAKE (overestimation of maximum acceleration and underestimation of amplification at high frequency) can also be seen clearly.

Considering these observations, the nonlinear analysis and the improved equivalent linear methods have the same accuracy. It is noted that this does not indicate that applicability of the equivalent linear method is up to 1 %; it is necessary to probe that the nonlinear analysis is applicable up to 1 % strain. Applicability of the nonlinear analysis depends on the accuracy of the stress–strain model and the damping term.

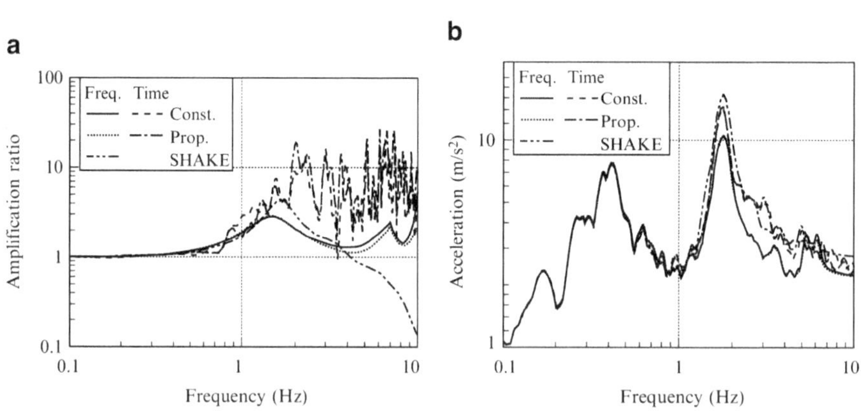

Fig. 15.12 Comparison of spectrum. (**a**) Amplification ratio. (**b**) Response spectrum

15.6 Determination of Damping for Deep Bedrock Problem

A case study in which base is located at very deep depths is investigated in this section. Both the nonlinear analysis and SHAKE are used in the analysis. Soil profiles and the maximum responses are shown in Fig. 15.13. The depth of the base is more than 300 m, which is the difference from the other analyses in this book.

Models in this analysis are developed by the following idea. At first, elastic behavior is assumed at depth greater than 39.4 m because V_s is larger than 300 m/s, and the Q value (see Sect. 12.3) is set to 100. On the other hand, the Q value is set to 35 at the layers above GL-39.4 m, and the nonlinear behavior is also considered. The Q values 100 and 35 correspond to the damping ratio 0.5 % and 1.43 %, respectively. In the nonlinear analysis, the Rayleigh damping (stiffness proportional damping) is employed so that the damping ratio becomes the abovementioned values at the predominant frequency. Here, as the damping ratio is different in depth, 0.5 % damping is considered by the global stiffness matrix and the rest 0.93 % in the surface layer is considered by the element stiffness matrix.

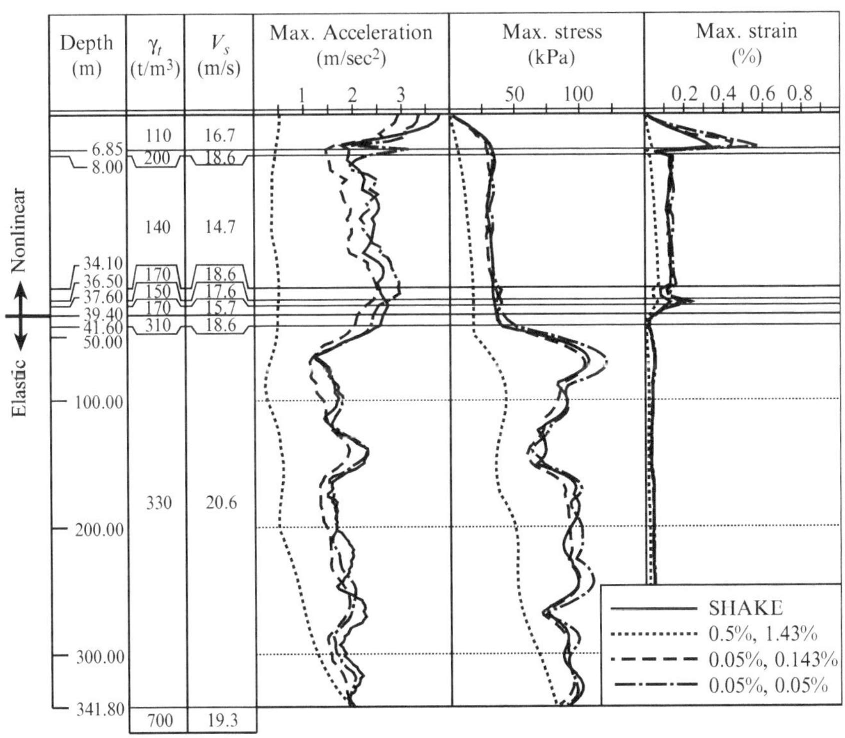

Fig. 15.13 Soil profiles and maximum response

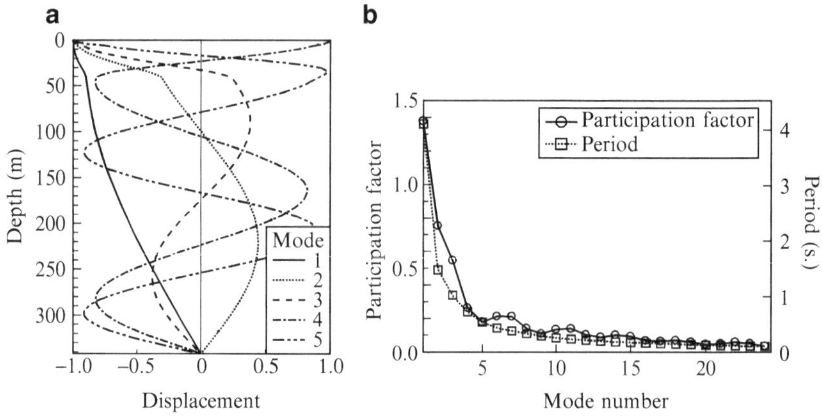

Fig. 15.14 Eigenvalue analysis. (**a**) Eigenvector. (**b**) Natural period and participation factor

Significant difference is observed between SHAKE and the result shown by dotted line (0.5 %, 1.43 %); acceleration by the nonlinear analysis is much smaller than that by SHAKE.

An eigenvalue analysis is made in order to find the reason, which is shown in Fig. 15.14. Natural period is about 4 s (natural frequency is about 0.25 Hz) and the wavelength at this mode is about 1,500 m. The participation factor is largest at the first mode, which indicates that this mode is predominant.

By the way, engineers are not interested in this kind of long period wave, but much higher frequency ranges are interested in. For example, the JMA instrumental seismic intensity is evaluated in the frequency range about 0.5–8 Hz. In conclusion, as the coefficient of stiffness proportional damping is evaluated based on the small frequency region, the damping is overestimated in the frequency range important in the engineering practice. Actually, as can be seen in Fig. 15.13, result of the nonlinear analysis moves close to that of SHAKE as damping ratio becomes smaller.

This example indicates that the damping ratio is not discussed with the mode but is evaluated based on the frequency range that the engineers are interested in. In many cases, the method based on the eigenvalue works well partly because the natural frequency of the ground is in the interested frequency range and partly because the damping ratio does not affect the result of the seismic ground response analysis in the ordinary cases as pointed out in Sect. 15.4.

15.7 Role of Hysteretic Damping Term

Both the hyperbolic model and the Ramberg–Osgood model are frequently used in stress–strain models in the nonlinear analysis of the ground. When Masing's rule is employed in the hyperbolic model, the maximum damping ratio becomes $2/\pi$, which is much larger than that of the ordinary soil; damping ratio is generally

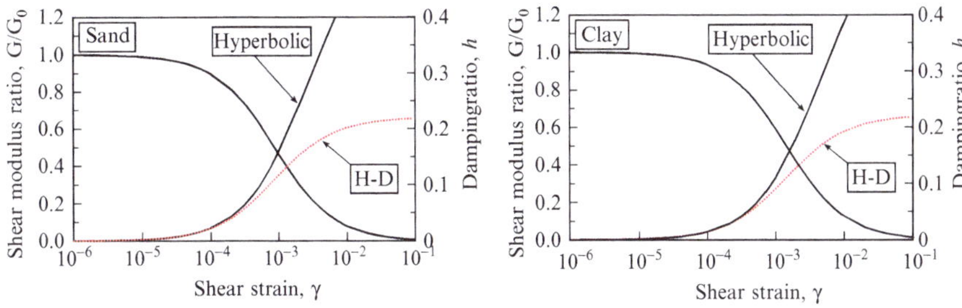

Fig. 15.15 Hyperbolic model and H–D model

overestimated at large strains. In addition, when comparisons by the different stress–strain models or different methods are made, maximum acceleration by the hyperbolic model frequently becomes smallest. Then, sometimes it is explained that the response by the hyperbolic model is suppressed as it overestimate the damping ratio. This explanation, however, is not correct, which is shown in this section (Yoshida et al. 2003).

Another factor is also investigated in this section, which is the stiffness at unload. Almost all the constitutive models assumes that the stiffness at unload is the same with the elastic modulus. Stiffnesses at unload are read from the cyclic shear deformation test result and are shown in Fig. 8.15, which shows that it decreases as strain. It can be expressed by modifying the hyperbolic model a little:

$$\frac{G_0}{G_{max}} = \frac{1 - G_{min}/G_{max}}{1 + \gamma/\gamma_r} + \frac{G_{min}}{G_{max}} \tag{15.3}$$

The hysteresis curves cannot be defined different from the skeleton curve in many constitutive models because they use Masing's rule. It is, however, possible by Yoshida's method (Sect. 8.2.5). Then three stress–strain models are examined. All models use the hyperbolic model for the skeleton curve, but definition of the hysteresis curve is different. They are named as hyperbolic, H–D, and H–D/E and the damping ratio is evaluated as follows:

1. Hyperbolic: Masing's rule
2. H–D: Hardin–Drnevich model
3. H–D/E: Hardin–Drnevich model with control of stiffness at unload based on Eq. (15.3)

The cyclic deformation characteristics of the clay and the sand are shown in Fig. 15.15. As the H–D and the H–D/E are of the same cyclic deformation characteristics, only H–D is shown in the figure. The stress–strain curves are shown in Fig. 15.16, in which difference of the damping characteristics can be seen as the difference of the area of the hysteresis loop. It is also noted that the hysteresis loops of the H–D and the H–D/E are similar to each other. The hysteresis loop is essentially a spindle shape. Therefore, there is a few freedom on the shape when

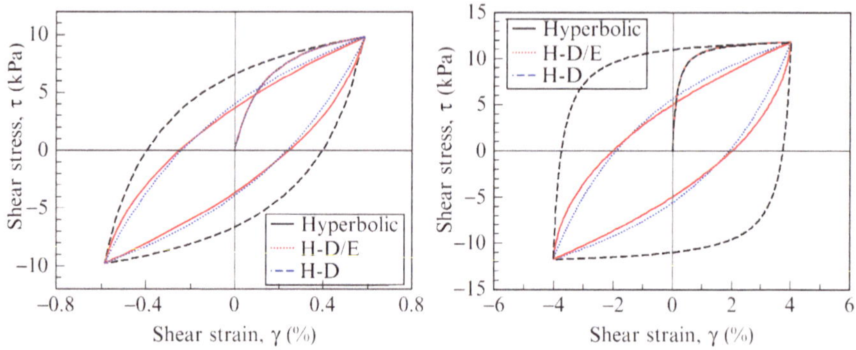

Fig. 15.16 Example of stress–strain curves

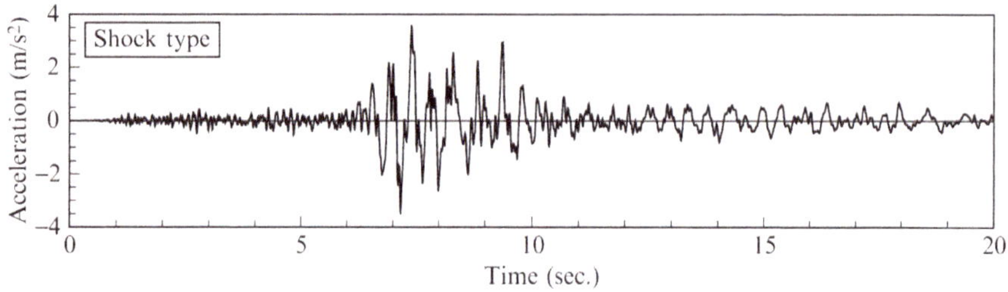

Fig. 15.17 Input earthquake motion

areas are the same, which is the reason why H–D and H–D/E show similar shape. Moreover, results of analysis are also almost similar, as the hysteresis curves are similar.

Acceleration used in this analysis is shown in Fig. 15.17. Both the shock type and the vibration type waves are used in the original paper, but only the result of the shock type earthquake is shown here because both waves lead similar conclusion.

The maximum responses are compared in Fig. 15.18. The maximum strain exceeds 5 %, which is a similar situation with the right-hand side of Fig. 15.16; the stress–strain curves are quite different as the damping ratios are significantly different. However, the maximum accelerations at the ground surface are almost the same in three cases. In other words, difference of the damping ratios cannot be seen in the maximum acceleration, which is against a common sense described above.

Exactly speaking, the maximum acceleration at the ground surface is the upper bound acceleration. Then it is natural that the maximum accelerations are similar as the three models use the same shear strength. As explained in Sect. 8.2.3, the hyperbolic model has a tendency to show the nonlinear behavior rather quickly and shear stress tends to shear strength slowly or gradually. Therefore, the apparent upper bound acceleration is somewhat smaller than the upper bound acceleration evaluated from the shear strength. This is the reason why the hyperbolic model

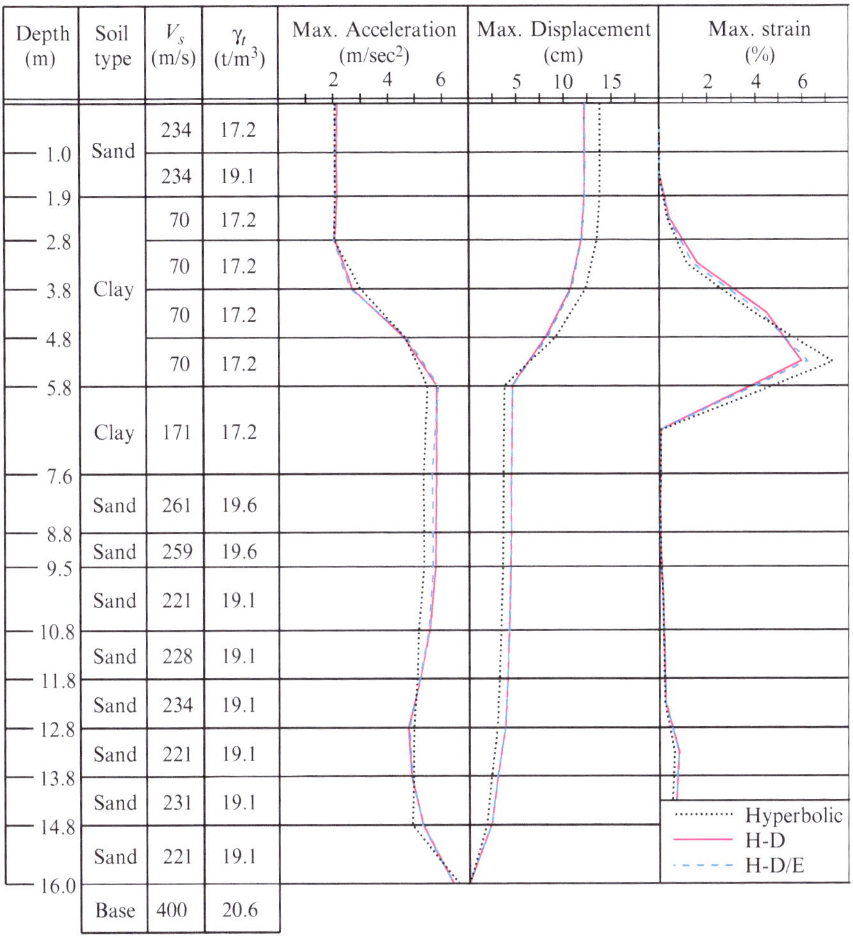

Fig. 15.18 Soil profile and maximum response

shows smaller acceleration than the other models/methods. It is also pointed out that the equivalent linear method overestimates the maximum acceleration and there is no upper bound acceleration in the Ramberg–Osgood model, which also works so that the hyperbolic model shows smaller acceleration.

The damping ratio (hysteretic damping) affects the behavior after unload. On the other hand, the maximum response usually occurs when the state point lies on the skeleton curve. This also suggests that the damping ratio does not affect the maximum acceleration at large strains.

Although the damping characteristics do not affect the maximum accelerations, they affect other behaviors. There are two significant differences in the acceleration time history at the ground surface between the hyperbolic and the H–D models shown in Fig. 15.19. There are many large amplitude waves in the result of the

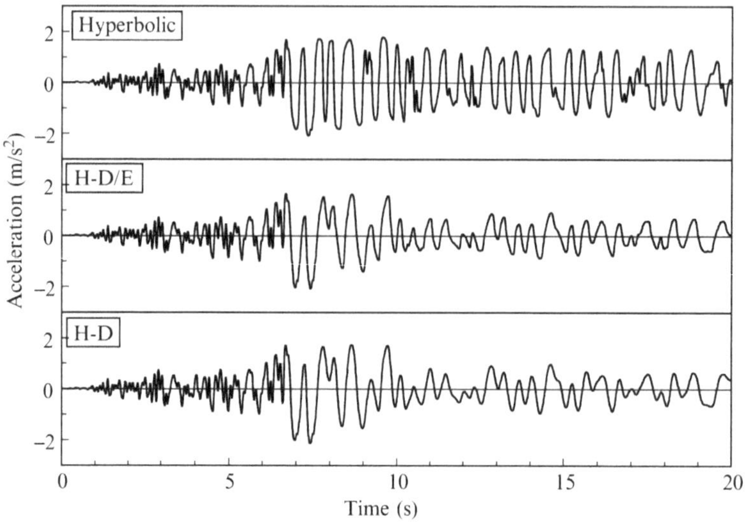

Fig. 15.19 Acceleration time history at ground surface

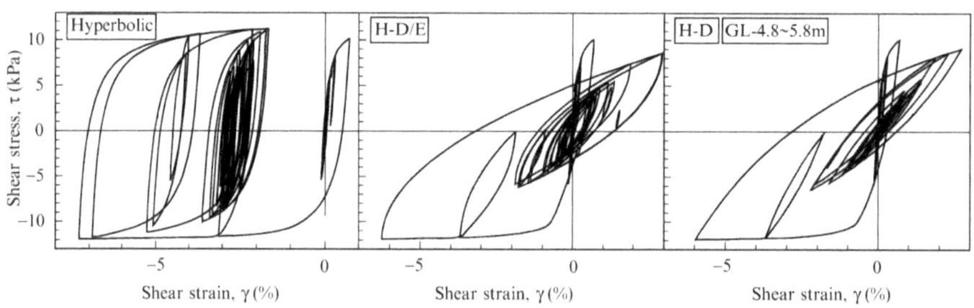

Fig. 15.20 Stress–strain curves

hyperbolic model, whereas the waves with long period are prone in the result of the H–D model. This difference can be easily recognized by looking at Fig. 15.20. The modulus at unload is large and it does not become small quickly in the hyperbolic model as the area of the hysteresis loop is large. Therefore, large change occurs in the stress and the stiffness under the small strain changes. On the other hand, the stiffness of the hysteresis loop quickly becomes small after unload in the H–D model. Therefore, the apparent stiffness is small and the stress does not reach the shear strength at both sides.

Since the earthquake motion is a cyclic load, direction of loading changes after some time. In addition, some time is necessary for strain to be produced. The above phenomenon can be explained from these two mechanisms. Since the stiffness is large in the hyperbolic model, large change of the stress occurs within the short time interval which causes large change of the acceleration. On the other hand, since the

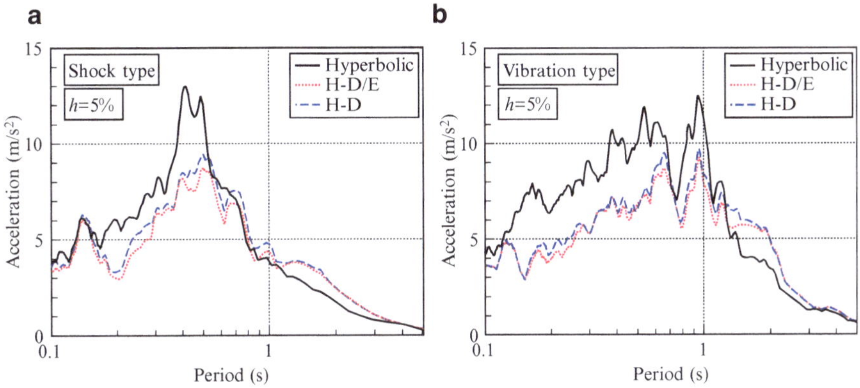

Fig. 15.21 Response spectra. (**a**) Shock type input. (**b**) Vibration type input

stiffness is small, the stress cannot become large before the subsequent unload. In other words, high-frequency component is included and the stress amplitude is large in the hyperbolic model because the stiffness at unload is large.

Difference of this mechanism in the hysteresis loop does not appear in the maximum acceleration, but it appears clearly in the response analysis. Surprisingly, the hyperbolic model that has the largest damping ratio shows larger response acceleration in the important frequency range as shown in Fig. 15.21a. This large response acceleration becomes larger against the vibration type earthquake motion shown in Fig. 15.21b.

As explained in the preceding, the upper bound acceleration and the effect of the hysteretic damping must be discussed separately. The damping ratio as a velocity proportional term and that as a hysteretic damping are treated separately. They can be treated the same only in the very simple case such as one-degree-of-freedom system subjected to the sinusoidal wave, but it is not true on the response of a more complicated systems.

15.8 Location of Engineering Seismic Base Layer

As explained in Chap. 1, it is ideal to solve the behavior from the fault to the ground surface, but it is almost impossible. Therefore, it is practical to separate the whole region into three regions and to analyze separately. The engineering seismic base layer is necessary for this reason. Problems to use the engineering seismic base layer are discussed in this section.

A site in the Hachinohe City, Japan, is investigated. Soil profiles at this site up to the seismic bedrock were evaluated by Midorikawa and Kobayashi (1978) and are shown in Table 15.1, where h denotes the depth from the ground surface, V_s denotes an S-wave velocity, ρ denotes mass density, and Q denotes a quality factor or the Q value.

Table 15.1 Soil profiles

Layer no.	Depth (m)	V_s (m/s)	ρ (t/m^3)	Q	Comment
1	2.0	107	1.80	14	
2	4.0	176	1.80	13	
3	6.5	201	1.90	12	
4	9.0	193	1.90	12	
5	15.5	239	1.70	12	
6	22.0	234	1.70	9	
7	32.0	248	1.80	7	
8	40.0	309	1.80	7	
9	50.0	378	1.80	7	EB2
10	180.0	379	1.70	100	
11	360.0	690	2.00	100	EB1
12	380.0	1,100	2.10	100	
13		2,800	2.50	200	SB

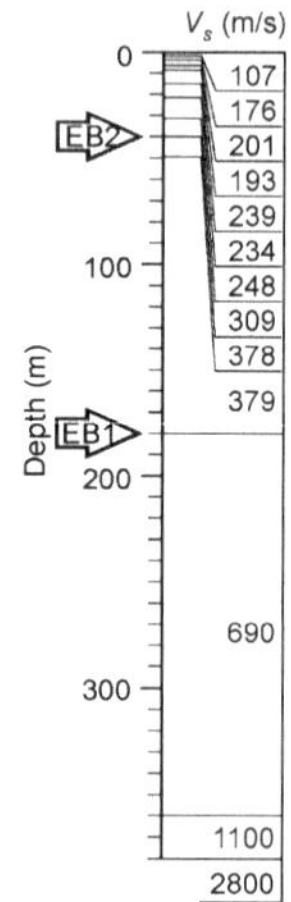

Modified from Midorikawa and Kobayashi (1978)
V_s S-wave velocity, ρ mass density

15.8.1 Problem to Separate at Engineering Seismic Base Layer

The engineering seismic base layer is set at two depths. The one is the layer below GL-180 m; the S-wave velocity is 690 m/s, which corresponds to the definition in the design specification of the nuclear power plant (Electric Technical Guideline Committee 1987). The other is the layer below GL-50 m, whose S-wave velocity is 378 m/s and corresponds to the definition of the engineering seismic base layer in many design specifications. These two engineering seismic base layers are distinguished by the name EB1 and EB2 as shown in Table 15.1. The S-wave velocity structure is also shown at the right of Table 15.1 in the natural scale so that locations can be easily recognized.

Two one-dimensional models are used in analyzing the ground in Table 15.1. The one is a continuous analysis in which the ground from the seismic bedrock to the ground surface is solved simultaneously and will be referred as continuous analysis. The other is a separate analysis in which the whole ground is analyzed by dividing into two parts, i.e., from the seismic bedrock to the engineering seismic base layer and the subsurface layers (above the engineering seismic base layer). The analysis is carried out in two steps in the latter case. The outcrop motion at the engineering seismic base layer is first computed, and a half of the outcrop motion is applied at the bottom of the surface layer as the incident wave. This analysis is called a separate analysis, and the lower part and the upper part of the model will be referred as the

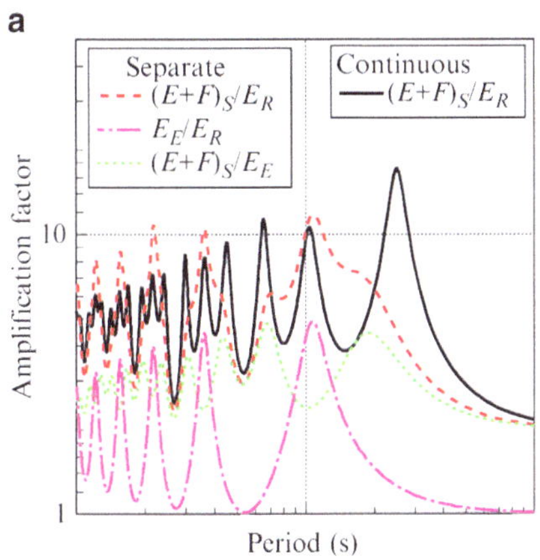

Fig. 15.22 Comparison of amplification characteristics. (**a**) EB1 model. (**b**) EB2 model

lower and the upper grounds, respectively. The upper ground is the same meaning with the subsurface layers.

A multi-reflection theory is employed in the following analyses in order to distinguish the incident and the reflected waves. The improved equivalent linear method DYNEQ (Yoshida et al. 2002) is used in the nonlinear analysis.

Figure 15.22 shows the amplification factor from the seismic bedrock to the ground surface. Here, E and F denote the incident and the reflected waves, respectively, and subscripts S, E, and G indicate the seismic bedrock, the engineering seismic base layer, and the ground surface, respectively. For example, E_E/E_S is a ratio of Fourier amplitudes at the engineering base layer to those at the seismic bedrock. The amplification in the separate analysis is computed by multiplying the amplification factor in the lower ground (E_E/E_S) and that in the upper ground $(E+F)_G/E_E$). If $(E+F)_G/E_S$ by both continuous and separate analyses coincides to each other, the separate analysis is justified.

The separate analysis by the EB1 model, shown in Fig. 15.22a, does not have the peak amplification at around 2.5 and 0.65 s that the continuous analysis has. Disagreement is also seen in short period, too. The reason why the predominant period of 2.5 s does not appear seems clear; there is no peak amplification in the separate analysis in this period. Since the amplification ratio at the ground surface is a product of two amplifications in the separate analyses, it is impossible to produce a large amplification without peak response in one or both amplification characteristics. The same discussion can be made for other peaks. For example, amplification around 1 s is produced as there is peak amplification at about 1 s in the lower ground, and amplification ratio of the surface layer is greater than two. On the other hand, amplification at about 0.65 s is almost minimal in both the upper and lower grounds and it is the reason why the peak of amplification does not appear in the separate analysis.

In the case of the EB2 model, amplification is nearly reproduced at the periods of 2.5 and 1 s although periods are a little shorter. However, there is no peak around 0.65 s, and there are significant differences in peak amplification at periods shorter than 0.65 s. As can be seen in Table 15.1, since thickness of the lower ground is very large compared with the surface layer in the EB2 model, the peaks of the amplification at 2.5 and 1 s are predominantly controlled by the lower ground. This explains the reason why these peaks are reproduced in the separate analysis and why the peak period is a little shorter than that of the continuous analysis. Actually, amplification ratio in the subsurface layers is nearly identical (about 2.0), and it is the reason why agreement between the continuous and the separate analyses is better than that of the EB1 model. However, amplification of the ground surface is larger than that of the continuous analysis at periods where the subsurface layers show maximal amplification. It indicates that the separate analysis shows error at the predominant periods of the subsurface layers.

15.8.1.1 Investigation Through Eigenvalue Problem

Typical eigenvalues and eigenvectors are summarized in Fig. 15.23. The circled number is a mode number. Figure 15.23a is a result of the continuous analysis; eigenmode up to fifth order is shown in the figure. On the other hand, Fig. 15.23b shows results of the separate analyses. Left side figure shows the EB1 model and right side figure shows the EB2 model. Eigenvectors are shown in each model. Eigenmodes are shown up to the mode whose shortest predominant period is less than that of the fifth mode of the continuous analysis. Periods shown in the figure are the same with the ones derived from the periods with peak amplification in Fig. 15.22.

The overall ground from the seismic base layer to the ground surface vibrates without reflection point in the first mode. Since there is no such vibration mode in the two separate analyses, it is natural that the separate analysis cannot produce this mode. On the other hand, in the case of the EB2 model, since the upper ground is shallow, its contribution of the overall behavior is small. In addition, since the period at the first mode of the subsurface layers is 0.665 s, which is shorter than the first mode period of the continuous analysis, 2.5 s, it hardly affects the behavior in longer periods. This is the reason why, in Fig. 15.22b, amplification characteristics in the EB2 model are similar to the ones in the continuous analysis around the first mode period of the whole ground, although periods at maximal amplification are a little shorter. Actually, the first mode shape on the continuous analysis is quite similar to the ones in the lower ground.

The second mode period is 1.04 s in the continuous analysis. On the other hand, the first mode period of the lower ground in the EB1 model is 1.09 s. They are close to each other and this agreement is the reason why the period at maximal amplification about 1 s appears in both continuous and separate analyses. In this sense, this agreement can be said to be accidental.

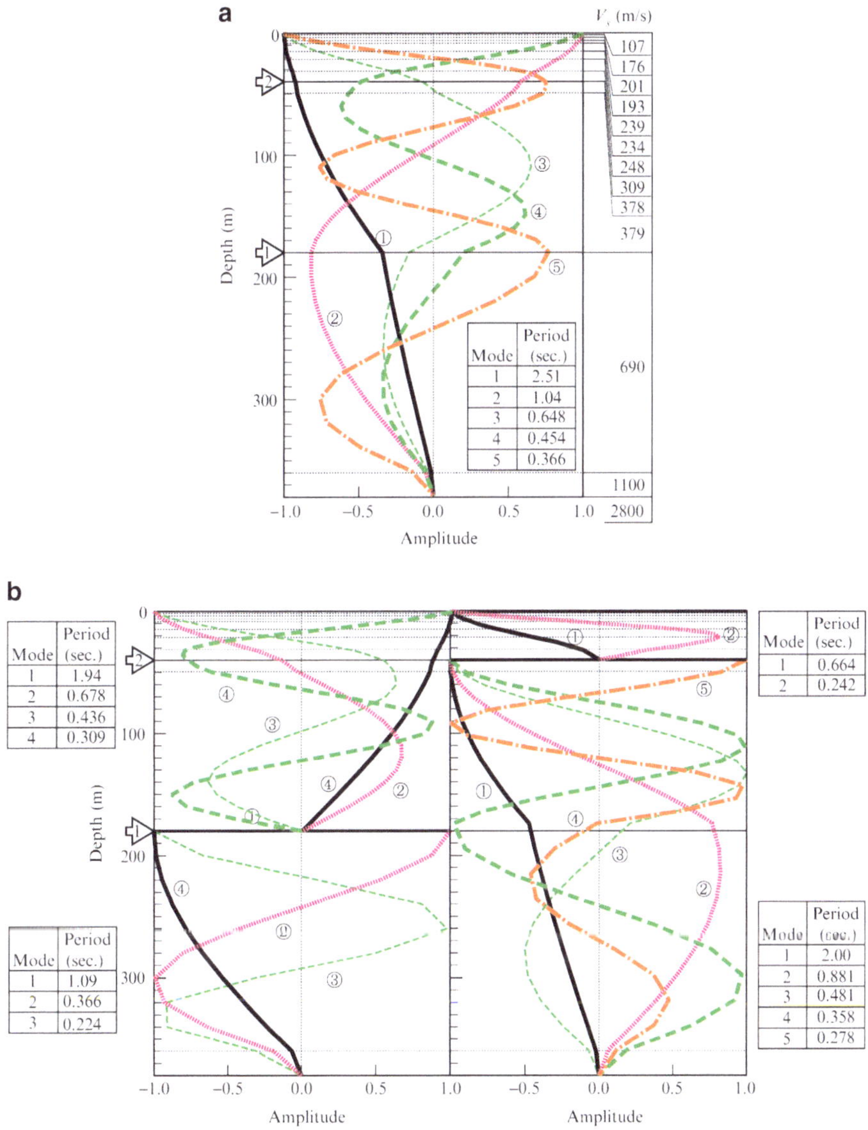

Fig. 15.23 Eigenvectors and predominant periods. (**a**) Continuous analysis. (**b**) Separate analysis (EB1 at *left* and EB2 at *left*)

As can be seen in this discussion, agreement and disagreement of the natural periods between the upper and lower grounds determine the ground vibration characteristics of the overall ground by the separate analysis, and predominant periods are determined individually without an interaction to each other. Therefore, there is no guarantee that the predominant period of the continuous analysis agrees with the ones by the separate analysis.

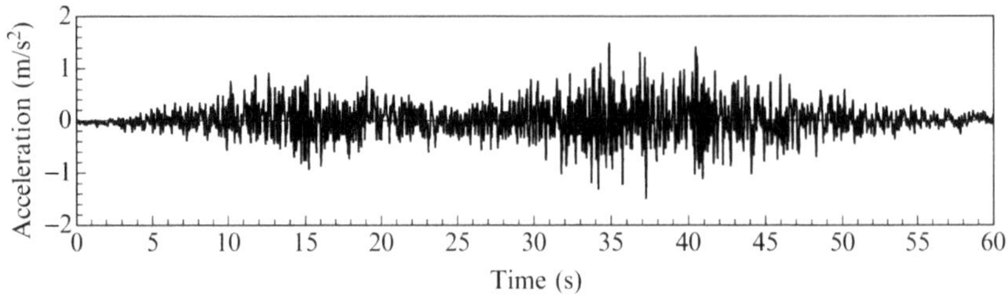

Fig. 15.24 Outcrop waveform at seismic bedrock

15.8.1.2 Wave Propagation: Incident and Reflected Waves

Separate analysis is justified if the interaction between the lower ground and the subsurface layers does not occur. This will occur in two cases. The one is that all the energy of the incident wave is absorbed in the subsurface layers such as by the nonlinear effect, and the other is that all reflected wave does not pass the engineering seismic base layer as all the wave reflects at the engineering seismic base layer. The former case will be discussed later. In the latter case, the incident waves at the engineering seismic base layer in the continuous analysis and the separate analysis are identical, and downward wave does not travel from the subsurface layers into the engineering seismic base layer.

The earthquake motion at Furofushi site (rock site) obtained during the 1983 Nihonkai-chubu earthquake, shown in Fig. 15.24, is used in the investigation. The waveforms at the engineering seismic base layer are shown in Fig. 15.25 for the EB1 model. Here, response between 35 and 45 s is enlarged in order to see the differences easily.

The reflected wave is compared with the incident wave in Fig. 15.25a. Magnitude of the reflected wave is smaller than that of the incident wave in general, but it is sufficiently large compared with zero or they are the same order. In other words, all wave traveled downward is not completely reflected at the engineering seismic base layer.

The incident and the reflected waves by the separate analysis are compared with those by the continuous analysis in Fig. 15.25b, c. It is clear that differences of the incident wave caused difference of the seismic motion at the ground surface by two analyses.

15.8.1.3 Nonlinear Analysis

Finally, the effect of the separate analysis is investigated by the nonlinear analysis; the separate analysis may be justified if all the energy is absorbed by the hysteretic nonlinear behavior. It is taken into account in the layers up to GL-50 m. The ground below GL-50 m is treated as elastic media with Q value of 10. Average of the dynamic deformation characteristics shown in Sect. 7.2.10 is used as dynamic deformation characteristics of soil. Accelerations at the ground surface by the

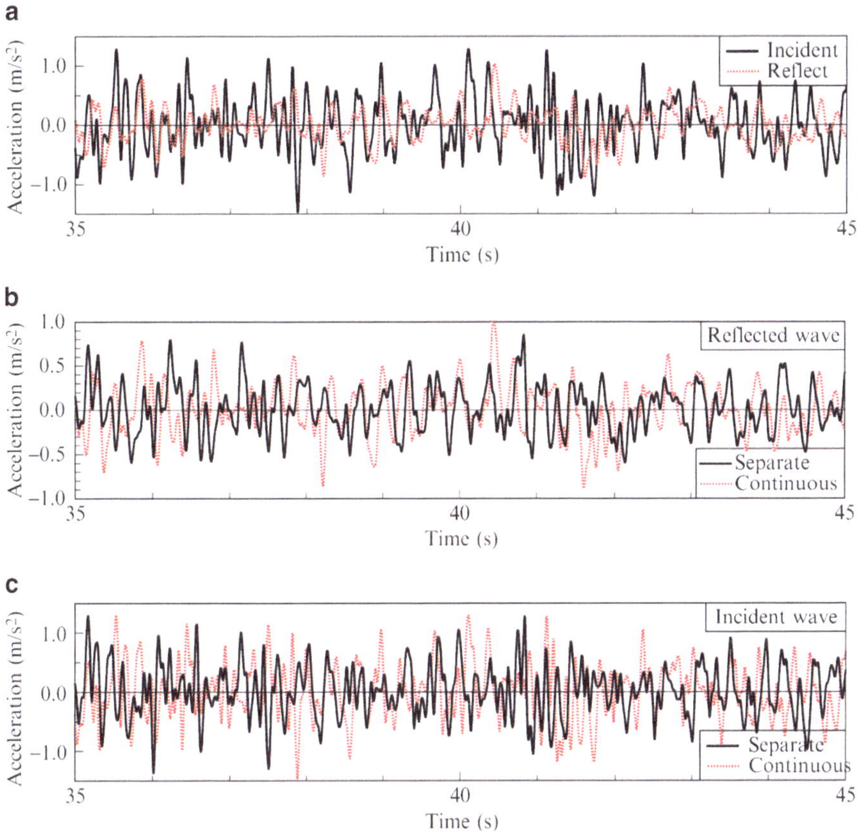

Fig. 15.25 Ground shaking at engineering seismic base layer. (**a**) Upward and downward waves (continuous analysis). (**b**) Reflected waveforms in separate and continuous analyses. (**c**) Incident waveforms in separate and continuous analyses

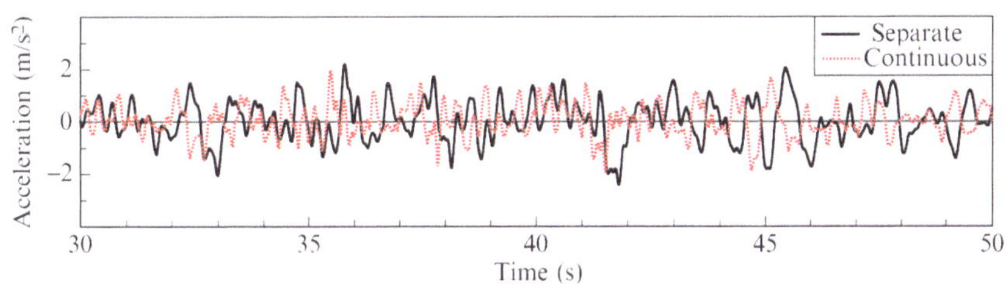

Fig. 15.26 Waveforms at the ground surface by nonlinear analysis

continuous and the separate analyses are compared in Figs. 15.26 and 15.27. They cannot be said to be identical or similar. The amplification characteristics and the acceleration response spectrum with 5 % damping ratio are compared in Fig. 15.28. Difference of the amplification characteristics has the same tendency

Fig. 15.27 Maximum acceleration and strain by nonlinear analysis

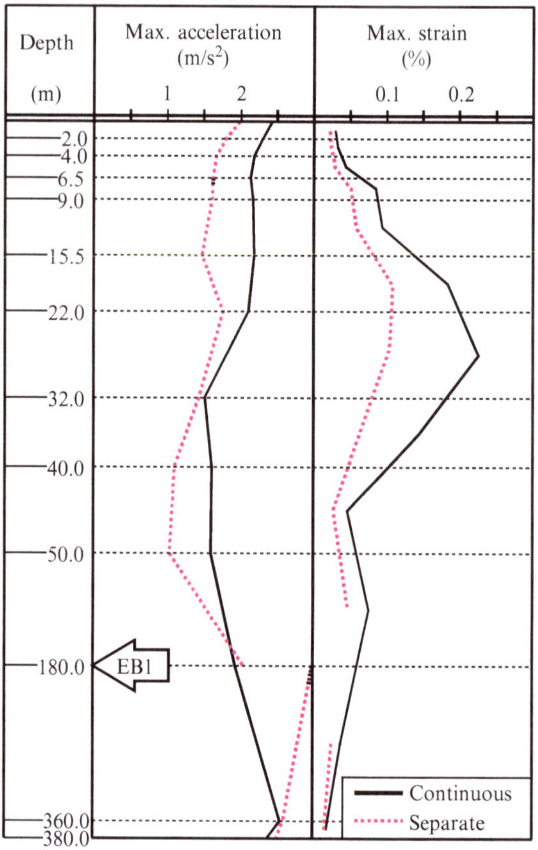

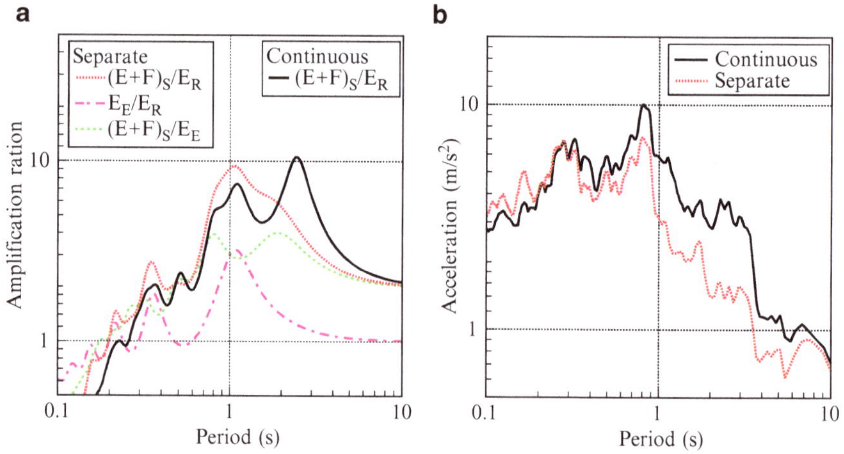

Fig. 15.28 Comparison of response in nonlinear analysis (EB1 model). (**a**) Amplification factor. (**b**) Acceleration response spectrum

with Fig. 6.21a; peaks of the amplification at period 2.5 and 0.65 s are not reproduced in the separate analysis. The acceleration response is underestimated at periods longer than about 0.3 s and that is overestimated at shorter periods. Differences are especially large at a period longer than about 1 s.

15.8.2 Setting Design Earthquake Motion

Separation by means of the seismic base layer is shown not to work well in the preceding. Then how does one specify the design earthquake motion?

15.8.2.1 Various Methods for Design Earthquake Motion

Figure 15.29 shows various possibility of the location of the design earthquake motion schematically. The dotted line indicates the analyzed region. In addition, ◎ is the location of the original earthquake motion (seismic bedrock), and ○ is the location of the incident wave in the analysis. The incident wave is given as the design earthquake motion when E is circled, and it is obtained as a result of the analysis in the whole region from the seismic bedrock to the ground surface. For example, the model shown by "Continuous" in the left side is the case of the preceding one as continuous analysis. The incident wave at the engineering seismic base layer is calculated by this model. On the other hand, the model "Separate" in the right side is a separate analysis in the preceding, in which incident wave at the engineering seismic base layer is calculated from the analysis with outcropped base layer and resultant incident wave is used as the incident motion of the subsurface layers.

In addition to the previous two cases, other two cases are added.

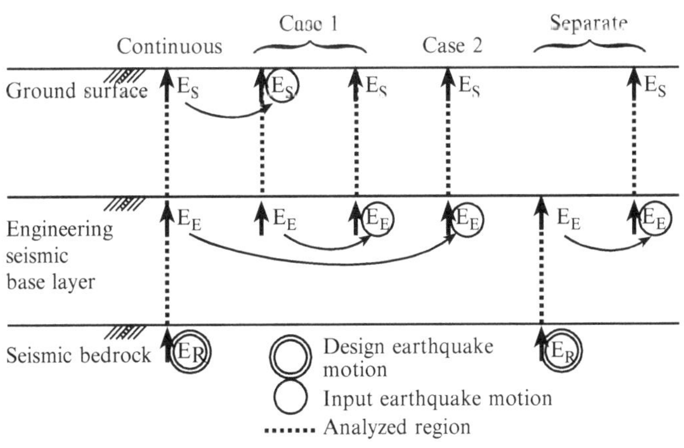

Fig. 15.29 Location of design seismic motion

The response at the ground surface obtained in the continuous analysis is used in order to evaluate the incident wave at the engineering seismic base layer by using the subsurface layer model, and obtained incident wave is applied at the surface model in Case 1. This method works perfect in the elastic system. It is used in the engineering practice in the following cases, for example. A small earthquake record obtained at the ground surface of the site is deconvoluted to obtain the incident wave at the engineering seismic base layer, and the incident wave is scaled and used in the analysis of the subsurface layers in order to see the nonlinear behavior of the subsurface layers. It is also used in the synthesized ground motion developed by the wave synthesis method from the small earthquake record. The synthesized motion is the earthquake motion at the ground surface under the large ground shaking, but it must be once deconvoluted for the nonlinear analysis because the original small record does not contain the effect of the nonlinear behavior.

The incident wave at the engineering seismic layer in the continuous analysis is used as the incident wave to the surface model in Case 2.

The same soil profiles shown in Table 15.1 are used and the engineering seismic base layer is set as EB1.

Amplification characteristics are shown in Fig. 15.30a for the elastic case. Three results, Continuous, Case 1, and Case 2, coincide very much and cannot be distinguished in the figure. As described, in the linear analysis, Case 1 agrees with the original wave perfectly from the theory. Agreement of Case 2 indicates that the effect of the reflected wave is not significant. This issue will be discussed again soon after.

Next, nonlinear behavior of the subsurface layers is investigated by using the original wave as the incident wave to the base layer, which corresponds to nearly double of the original input motion. The equivalent linear method SHAKE (Schnabel et al. 1972) is used in the analysis, and the average of the empirical equation shown in Sect. 7.2.10 is used for the nonlinear property. The layer below the engineering seismic layer is set elastic.

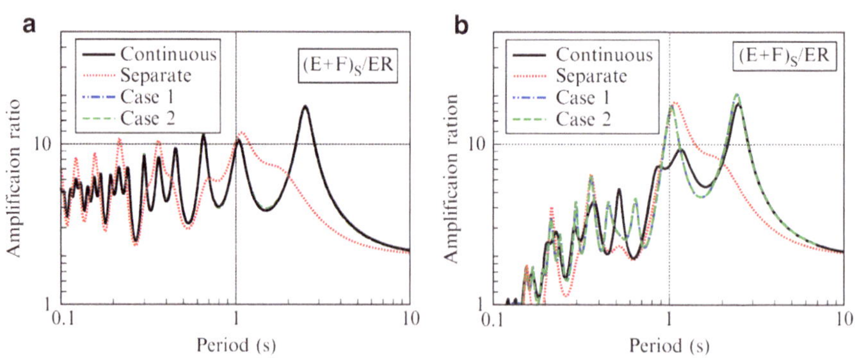

Fig. 15.30 Comparison of amplification ratio (elastic analysis). (**a**) Elastic analysis. (**b**) Nonlinear analysis

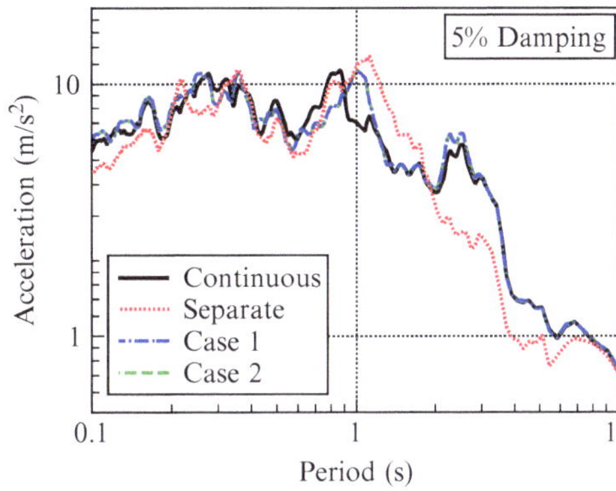

Fig. 15.31 Comparison of response spectra (nonlinear analysis)

Amplification characteristics are shown in Fig. 15.30b. Although the first natural period does not change, the second natural period increases and the amplitude becomes small in the continuous analysis. Amplification at shorter period becomes much smaller than that of the elastic analysis, which comes from the limitation of SHAKE (Sect. 11.4.2).

The separate analysis cannot explain the peak at 2.5 s same as the elastic analysis. The new two methods agree to each other, and their agreement is much better than that by the separate method although some peaks cannot be reproduced. Good agreement can also be seen in the response spectrum in Fig. 15.31.

15.8.2.2 Effect of Layers Beneath Engineering Seismic Layer

It is shown in the preceding that the behavior beneath the engineering seismic base layer affects the behavior in the surface layer. Then how does the deep structure affect the behavior of the subsurface layers when the design earthquake motion is defined at the engineering seismic base layer?

The soil profile shown in Table 15.1 is again used in the case study. The engineering base layer is set as EB2 or the ninth layer with $V_s = 379$ m/s considering that the V_s is about 300–400 m/s in many design specifications. Then the thickness and the V_s are chosen as parameters as is shown in Table 15.2. At first, V_s is set to 200 m/s or 700 m/s so that it is smaller or larger than the original V_s, respectively. In addition, the thickness of this layer is also varied as 5, 50, and 100 m. Then the 11th layer is set as the seismic bedrock with $V_s = 3$ km/s. Other analytical conditions are the same as that in the preceding analyses. In addition to the cases shown in Table 15.2, the continuous model in the previous section is made, which is called a standard case. In all the analyses the wave in Fig. 15.24 is applied as the outcrop motion at the seismic base layer; a half of them is applied as the incident wave at the engineering seismic base layer.

The maximum responses are shown in Figs. 15.32 and 15.33 for the case with smaller V_s (=200 m/s) and larger V_s (=700 m/s), respectively. Here, it is noted that

Table 15.2 Parameters at 10th layer

Case no.	1	2	3	4	5	6
V_s (m/s)	200			700		
Thickness (m)	5	50	100	5	50	100

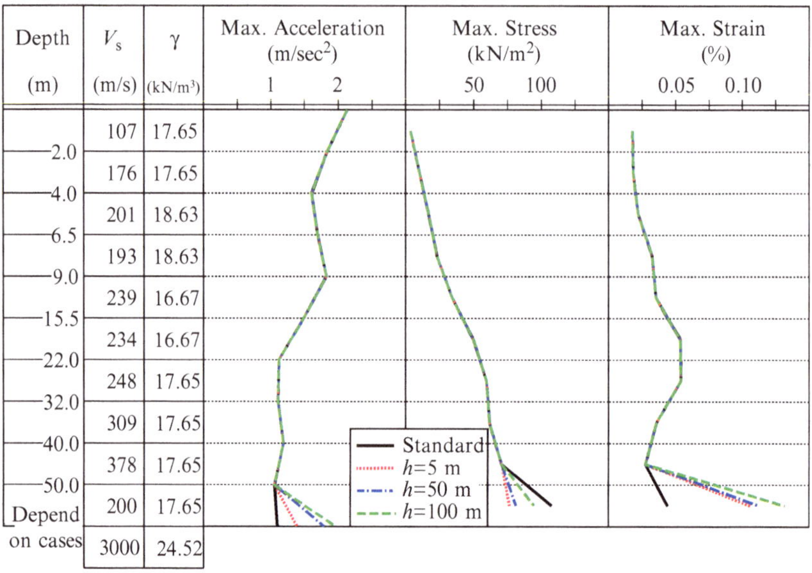

Fig. 15.32 Maximum response ($V_s = 200$ m/s)

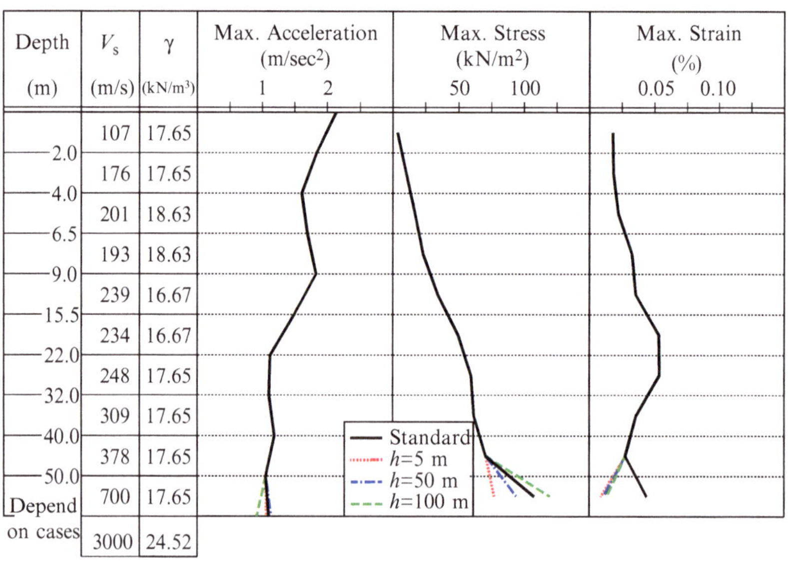

Fig. 15.33 Maximum response ($V_s = 700$ m/s)

scale of the depth is modified below the engineering seismic base layer so that all the data can be seen at the same depth.

Behaviors above the engineering seismic base layer are almost identical in all cases and one cannot distinguish them in the figure. Acceleration time histories are also the same in all cases.

The maximum acceleration in the standard case is smallest below the engineering seismic base layer, and it increases as the thickness of the engineering seismic base layer increases. In the same manner, the maximum stress decreases and the maximum strain increases as the thickness of the engineering seismic base layer increases. Since the earthquake wave does not amplify in this layer as V_S is constant, this difference is just an apparent difference associated by the difference of the thickness.

On the other hand, the maximum acceleration is the largest in the standard case, but the difference seems smaller than the case with the smaller V_S.

15.8.3 Remarks

It is natural that the behavior from the seismic bedrock cannot be reproduced by the separate analysis because, as discussed in Sect. 3.1, the seismic engineering base layer is defined so that the predominant period is a little longer than the structure that is interested in. In this example, the predominant period of the whole ground from the seismic bedrock to the ground is about 2.5 s, whereas that of the subsurface layers above the engineering seismic base layer is about 2 s in the EB1 model and 0.6 s (Fig. 15.22). Therefore, in the conventional concept, the use of the engineering seismic base layer is justified if the predominant period of the structure is smaller than 2.5 s (EB1 model) or 0.6 s (EB2 model).

As can be seen in this example, however, the frequency response characteristics cannot be reproduced by the separate method. It indicates that the wave radiated from the subsurface layers to the seismic bedrock cannot be ignored. In this sense, the use of the engineering seismic base layer is not justified.

This limitation can be improved by conducting the analysis considering the seismic base layer at one place in the process to develop the design earthquake motion or by using the design wave that is developed including the effect of the seismic bedrock. In this sense, it is important to recognize the existence of the engineering seismic base layer.

It is also shown that the behavior of the subsurface layers is not affected by the velocity structure below the engineering seismic base layer. In this sense, the use of the engineering seismic base layer has meaning in the engineering practice. In other words, it may be rational if the design earthquake motion at the engineering seismic base layer is decided considering the deep structure below the engineering seismic base layer. It is also pointed out that it is sufficient as boundary condition because the behavior of the surface layer is independent from the behavior below the base.

References

Arulanandan K, Scott RF (eds) (1993) Proceedings on verification of numerical procedures for the analysis of soil liquefaction problems. Balkema, Davis

Association of the Development of Architectural Research (1998) Report of design standard against various load based on performance based design, Tokyo (in Japanese)

Electric Technical Guideline Committee (1987) Technical guidelines for aseismic design of nuclear power plants, Japan Electric Association, Tokyo

IASPEI/IAEE joint working group (1992) Proceedings of the international symposium on the effect of surface geology on seismic motion, Odawara, IASPEI/IAEE

Japan Read Association (1985) Specifications for highway bridges and commentary, Part V, Earthquake resistant design

Kaneko F (1993) Sensitivity analysis of one-dimensional analysis, research of the effects of surface geology on seismic motion. Report to the Ministry of Education, Takeuchi Y (ed), pp 78–90 (in Japanese)

Midorikawa S, Kobayashi H (1978) Spectral characteristics of incident wave from seismic bedrock due to earthquake. Transactions of AIJ, No. 273, pp 43–54 (in Japanese)

Morimoto I, Suzuki S (1983) Effect of fluctuation of soil dynamic properties on results of one dimensional response analysis. In: Proceedings of the 18th annual meeting of JSSMFE, pp 635–636 (in Japanese)

Schnabel PB, Lysmer J, Seed HB (1972) SHAKE a computer program for earthquake response analysis of horizontally layered sites, Report no. EERC72-12. University of California, Berkeley

Towhata I, Ronteix S (1988) Probabilistic estimation of shear wave velocity of soils in terms of SPT-N and its application to seismic response analysis. In: Proceedings of the 23rd Japan national conference of soil mechanics and foundation engineering, pp 825–826 (in Japanese)

Yoshida N (2008) Accuracy of numerical analysis, performance based design of ground and structure considering liquefaction, Kanto Branch of JGS, pp 91–103 (in Japanese)

Yoshida N, Kobayashi S, Suetomi I, Miura K (2002) Equivalent linear method considering frequency dependent characteristics of stiffness and damping. Soil Dyn Earthq Eng 22(3):205–222

Yoshida N, Sawada S, Takeshima Y, Mikami T, Sawada S (2003) Effect of hysteretic damping characteristics on earthquake response of ground. JSCE J Earthq Eng Symp, vol. 28, Paper No. 158 (in Japanese)

Index

N. Yoshida, *Seismic Ground Response Analysis*, Geotechnical, Geological and Earthquake Engineering 36, DOI 10.1007/978-94-017-9460-2